Information and Communication Technologies (ICT)
for Development in Africa

European University Studies
Europäische Hochschulschriften
Publications Universitaires Européennes

Series XXXI
Political Science

Reihe XXXI Série XXXI
Politikwissenschaft
Sciences politiques

Vol./Bd. 533

PETER LANG
Frankfurt am Main · Berlin · Bern · Bruxelles · New York · Oxford · Wien

Olaf Nielinger

Information and Communication Technologies (ICT) for Development in Africa

An Assessment of ICT Strategies and ICT Utilisation in Tanzania

PETER LANG
Europäischer Verlag der Wissenschaften

Bibliographic Information published by Die Deutsche Bibliothek
Die Deutsche Bibliothek lists this publication in the Deutsche Nationalbibliografie; detailed bibliographic data is available in the internet at <http://dnb.ddb.de>.

Zugl.: Hamburg, Univ., Diss., 2006

Gratefully acknowledging financial support by
Fritz Thyssen Stiftung, Cologne.

D 18
ISSN 0721-3654
ISBN 3-631-55069-3
US-ISBN 0-8204-9877-7

www.peterlang.de

Acknowledgements

This study was truly a companion during the last three years, and, that it reached some form of completion is due to the support and help of many people and institutions. I am particularly indebted to my academic supervisor, Cord Jakobeit, who gave the research project an institutional home at the Institute of African Affairs in Hamburg, and whose constant support greatly helped to write this study. I also would like to thank Hans J. Kleinsteuber for his readiness to act as a second examiner.

My sincere gratitude goes to the Fritz Thyssen Foundation for financing the research project and the publication of this book.

Furthermore, I would like to thank all my interviewees. Their readiness and their commitment to assist were essential in conducting the study. Particular assistance was provided by David Sawe, Simbo Ntiro, Ralph Karhammar, Nils Jensen and Riku Akisainen. Special thanks are also due to Francis Matambalya and Issmail Nnafie who generously shared parts of their research findings, and, thereby, enriched the fundament of this research project. I am also indebted to Vasco Fournadjiev, who acted as my local advisor during the field visits to Tanzania and became a valuable source of constant support.

I also would like to express my gratitude to Peter Häußler and Valeria Salles for their great hospitality. They gave me a home in Tanzania and made things lots of easier.

Constructive comments on the manuscript were made by Rolf Hofmeier, Stefan Brüne and Armin Osmanovic. Their valuable contributions greatly improved the quality of the study. I highly appreciated their support. I also want to acknowledge Zoe Matserat and Judy Smith, who carefully read the draft, or specific chapters, and shaped up language and style of the study.The mistakes that remain, however, are entirely mine.

Finally, Daniela, Jakob and Lino had their footprint in the project. Their keen sense of patience and encouragement has been vital for the finalisation of the project.

Olaf Nielinger

Hamburg

April 2006

Table of Content

Abbreviations

AA	African Analysis
AAAS	American Association for the Advancement of Science
AAU	Association of African Universities
ABT	Agreement on Basic Telecommunication
ADB	African Development Bank
AISI	African Information Society Initiative
APC	Association for Progressive Communication
ARB	African Research Bulletin
ARPU	Average Revenue Per User
AVU	African Virtual University
BA	Balancing Act's New Update
BEST	Business Environment Strengthening in Tanzania
BIS	Business Informations Services
BOT	Bank of Tanzania
CABECA	Capacity Building for Electronic Communication in Africa
CATS	Computer and Telecom Systems Limited
CCR	Call Completion Rate
CDMA	Code Division Multiple Access [Standard for 3rd generation mobile communication]
CIDA	Canadian International Development Agency
CN	Cellular News
COAT	Communication Operators Association of Tanzania
COMESA	Common Market for Eastern and Southern Africa
COSTECH	Commission for Science and Technology
CROMABU	Crop Marketing Bureau
CSD	Civil Service Department
CSRP	Civil Service Reform Programme
DANIDA	Danish International Development Agency
DFID	Department for International Development
DILES	Distance Learning Education Services

DIT	Dar es Salaam Institute of Technology
DOI	Development Opportunity Initiative
DotForce	Digital Opportunity Taskforce
EAC	East African Community
EAP&TC	East African Posts and Telecommunication Corporation
EAPTA	East African Posts and Telecommunications Administration
EASP	Economic and Social Action Programme
EASSY	East African Submarine System
ECOWAS	Economic Community Of West African States
EIU	Economic Intelligence Unit
ERP	Economic Recovery Programme
ERP	Economic Recovery Programme
ESAMI	Eastern and Southern African Management Institute
ESRF	Economic and Social Research Foundation
EU	European Union
GATS	General Agreement of Trade in Services
GATT	General Agreement on Tariffs and Trade
GCF	Global Catalyst Foundation
GDLN	Global Development Learning Network
GDP	Gross Domestic Product
GNI	Gross National Income
GPRS	General Packet Radio Service [Standard for Mobile Communication]
GSM	Global System for Mobile Communication
GTZ	Gesellschaft für Technische Zusammenarbeit
GVF	Global VSAT Forum Economic and Social Action Programme
HIPC	Heavily Indebted Poor Countries
HIVOS	Humanistic Institute for Developing Countries
ICDL	International Computer Driving Licence
ICL	International Computers Limited

ICT	Information and Communication Technologies
IDA	International Development Association
IDM	Institute of Development Management
IDRC	International Development Research Centre
IFC	International Finance Corporation
IFM	Institute of Finance Management
IFMS	Integrated Financial Management System
IICD	Institute of Communication and Development
IIT	Institute of Information Technology
ILO	International Labour Organization
IMF	International Monetary Fund
IMIT	Institute of Management and Information Technology
IMP	Information Master Plan
ION	Indian Ocean Newsletter
IP	Internet Protocol
IPP	Information Policy Plan
ISP	Internet Service Provider
IT	Information Technology
ITU	International Telecommunication Union
JICA	Japan International Cooperation Agency
KFAED	Kuwait Fund for Arab Economic Development
KfW	Kreditanstalt für Wiederaufbau
KIP	Kasulu Internet Project
LAN	Local Area Network
LDC	Least Developed Countries
MCT	Multipurpose Community Telecentre
MDG	Millenium Development Goal
MIC	Millicom International Cellular
MIT	Ministry of Trade
MoCT	Ministry of Communication and Transport
MoDaNS	Ministry of Defence and National Services

MoEC	Ministry of Education and Culture
MoF	Ministry of Finance
MoH	Ministry of Health
MSI	Mobile Systems International (Celtel)
MT	Marchés Tropicaux
MTN	Mobile Telephone Networks
MTUHA	Mfumo wa Taarifa za Uendeshaji wa Huduma za Afya
MUCHS	Muhimbili University College of Health Sciences
NBS	National Bureau of Statistics
NEPAD	New Partnership for Africa's Development
NGO	Non-governmental Organisation
NICTCO	National ICT Council / Commission
NIMR	National Institute for Medical Research
NSRC	National Science Resources Center
NTP	National Telecommunication Policy
OSS	Open Source Software
OUT	Open University of Tanzania
PoPs	Points of Presence [Internet access points]
PRGF	Poverty Reduction and Growth Facility
PRS	Poverty Reduction Strategy
PRSP	Poverty Reduction Strategy Papers
PSRC	Presidential Parastatal Sector Reform Commission
PSRP	Public Service Reform Programme
PSTN	Public Switched Telephone Networks
PTO	Public Telecommunications Operators
PWC	PriceWaterhousCoopers
RINAF	Regional Informatics Network for Africa
RMIT	Royal Melbourne Institute of Technology
SADC	South African Development Cooperation
SIDA	Swedish International Development Agency
SME	Small and Medium Enterprise

SMS	Short Message Serivices
SNO	Second National Operator
TACS	Total Access Communications System
TANESCO	Tanzania Electric Supply Company
TAZARA	Tanzania Zambia Railways Authority
TBC	Tanzania Broadcasting Commission
TCC	Tanzania Communications Commission
TCCIA	Tanzania Chamber of Commerce, Industry and Agriculture
TCRA	Tanzania Communication Regulatory Authority
TDLC	Tanzania Development Learning Centre
TEHIP	Tanzania Essential Health Intervention Project
TEIL	Technology Enhanced Independent Learning Environment
TIC	Tanzania Investment Centre
TISPA	Tanzania Internet Service Providers Association
TIXP	Tanzanian Internet Exchange Point
TPC	Tanzania Posts Corporation
TPTC	Tanzania Posts and Telecommunications Corporation
TRA	Tanzania Revenue Authority
TRASA	Telecommunication Regulators Association of Southern Africa
TRC	Tanzania Railways Corporation
TRP	Telecommunication Restructuring Programme
TTC	[Kasulu] Teacher Training College
TTCL	Tanzania Telecommunications Company Ltd
UCC	Uganda Communications Commission
UCC	University Computing Centre
UCLAS	University College of Lands, Architecture and Survey
UDSM	University of Dar es Salaam
UEMOA	Union Économique et Monétaire Ouest Africaine
UN	United Nations
UNCITRAL	United Nations Commission on International Trade Law

UNCSTD	United Nations Commission on Science and Technology for Development
UNCTAD	United Nations Conference on Trade and Development
UNDP	United Nations Development Programme
UNECA	United Nations Economic Commission for Africa
UNESCO	United Nations Educational, Scientific and Cultural Organisations
UNHCR	United Nations High Commissioner for Refugees
URT	United Republic of Tanzania
VAT	Valued Added Tax
VETA	Vocational Education and Training Authority
VoIP	Voice over the Internet Protocol
VPN	Virtual Private Networks
WEF	World Economic Forum
WLAN	Wireless Local Area Networks
WLL	Wireless Local Loop
WMAN	Wireless Metropolitan Networks
WSIS	World Summit on the Information Society
WTDC	World Telecommunication Development Conference
WTO	World Trade Organisation
WWW	World Wide Web

List of Figures

List of Tables

List of Boxes

1. Introduction

The 1990s witnessed major changes in the scope and reach of information and communication technologies (ICT). The accelerated diffusion of ICT, embedded in an overall process of increasing globalisation, had a profound impact on societies, as utility and functionality of the new technologies have increased significantly in the last decade. ICT became a powerful driving force of social and economic transformation (c.f. Bell 1999, Castells 2000, Steinmüller 1993, Toffler 1980). At first, contributors, attempting to analyse the potential impact of ICT, focused on the developed world, but subsequently, attention was also shifted to the developing world and opened a new field for researchers and development practitioners to address issues of ICT and their role in the development process. Particularly in the early phase, when the debate commenced in the 1990s,[1] critics have emphasised the opportunities for developing countries to harness ICT for development, later, this view was challenged by commentators, who pointed at the emergence of a 'digital divide' between information 'haves' and 'have-nots', actually exacerbating social and economic exclusion (Bridges.org 2001). However, both views were different sides of the same coin, saying that there is room to benefit from increased ICT utilisation, but at the same time, there are risks and substantial challenges for developing countries. This study "Information and Communication Technologies (ICT) for Development in Africa. An Assessment of ICT Strategies and ICT Utilisation in Tanzania" would like to contribute to the ongoing discussion. It explores the prospects and the likely impact of ICT in Africa, with particular reference to the country case Tanzania. It deals with strategies attempting to utilise new information and communication technologies for broader development goals, and with the role of ICT in a number of areas of social or economic concern, such as governance, private sector development or education.

Considering the enormous challenge harnessing the potentials of ICT, from the very beginning, the goal to make ICT work for development has faced considerable scepticism. However, deploying ICT is not a choice of 'either or', at least some minimal provisions for basic connectivity will be essential for developing countries to participate in the global exchange of information and ideas. Beyond an in-depth analysis of ICT potentials, some good arguments, explaining the need to develop a proactive ICT development strategy, have been introduced right from the start:

---

[1] Some of the early texts that have addressed the impact of ICT in developing countries were UNCSTD (1995), Roffe et al. (1995), Bezanson / Sagasti (1995) and Talero / Gaudette (1996).

- There is sufficient evidence that developing countries can take advantage of modern communication technologies and that there is a need to translate individual potentials in a coherent approach to ICT,
- There is a need to provide, at least, some basic ICT connectivity to secure junctions to an emerging global information society in order to counter tendencies of ongoing marginalisation, particularly in Africa. The availability of ICT, at least in some critical areas, will be a prerequisite to participate in global working procedures, that will become increasingly important in a globalised world.
- It is shortsighted to demand from the developed world to adjust to the challenges of information technologies, and at the same time, claim these are only luxury goods for the developing world,
- Finally, although the costs of using ICT are high, the costs of not doing so are likely to be much higher (c.f. UNCSTD 1997, Mansell/When 1998: 7).

The critical implication of these stipulations is that it is not so much the question of whether ICT can have a positive impact on development, but how the technologies can foster advances in development and what policy interventions are appropriate to respond to the risks of marginalisation and social exclusion. Thus, policies and strategies, adjusting the technological potentials to the needs of broader development goals, came to the forefront of the debate. Evidently, ICT will not change the development path by themselves, this will need processes of deliberate political decision-making aiming at maximising the benefits and minimising the risks of technology deployment. Benefits have been associated with new opportunities in economic development, in delivering effective and transparent governance, or improved service delivery in health or education. Risks were related to the perils of economic and social exclusion, as well as trade-offs with other urgent goals on the development agenda (c.f. Analysys 2000, Kenny et al. 2000, DOI 2001, Grace et al. 2001).

In concordance to the current approach to development, individual components of a possible strategy have materialised: a private-sector led development strategy, particularly highlighting the role of liberalisation, privatisation and competition. To this end, governments need to assume strategic leadership and create enabling environments for private sector growth and provide a level playing field for all market participants by effective, fair and transparent regulation (World Bank 1998/1999, Melody 2001, Bipe 2000). Thus, creating efficient markets is regarded as an essential ingredient to harness ICT for development. However, this should not be taken as a pure market approach faithfully based on liberal textbook thinking. Firstly because the agenda, to a large extent, responded to defunct, state-owned telecommunication monopolies, that were increasingly unable to supply the most basic demand of telecommunication services, secondly, the strategy put particular emphasis on regulation, as the perhaps strongest individual tool to guide markets in an intended direction and addressing issues of market failures, and finally, the suggested strategy very well recognises the tendency of (unregulated) markets in failing to provide telecommunication services for the poor, especially in rural and remote areas that are, from an operator's point of view, largely considered to be unprofitable. Thus, strategies take into consideration the existence of segments not covered by the market, and generally make special provisions that enable operators to service poor and remote areas with appropriate and viable telecom solutions. Meeting minimum needs and demand for private and public access to ICT – or components thereof – for the disadvantaged segments of the population has been the key feature of what has been labelled universal access strategies (c.f. Navas-Sabater et al. 2002).

Derived from that background, this study is based on the assumption that ICT can have a positive impact on development. However, ICT do not offer automatic solutions to all development problems, but will need considerable political, economic and social decision-making and decision-making capacities. This study accepted the strategic approach outlined above and takes this as the central hypothesis that can be read as follows:

Extending market efficiency by good governance, effective regulation and sufficient implementation and enforcement capacities is a vital prerequisite to creating a basis for increased ICT diffusion and effective technology utilisation. To take full advantage of ICT, however, the focus on private sector-led development models must be extended by framing goals for universal access, strategic interventions to build the necessary human capacities and efforts to build demand-driven content ensuring the appropriateness and relevance of available ICT solutions. Only if these conditions are met, there is a good likelihood that ICT will contribute to achieving broader development goals.

However, general strategic outlines provide only a rough orientation. Prospects of success will depend on the given environment of an individual country case. The focus of this analysis is on Sub-Saharan Africa, a continent that due to weak infrastructure, a low skill base and weak states[2] will face multiple problems in

[2] The concept of weak states was introduced by Myrdal (1968), arguing that inefficient states, characterised by, for example, corruption and nepotism, were the main cause for the poor development record of developing countries; for a discussion on weak states in Africa see also Bratton / Walle (1994) and World Bank (1989).

making effective use of the technologies. This does not only relate to challenges of strategy development, policy planning and implementation, countries also have to cope with a new freedom of communication – or put in place (electronic) means to prevent it – and have to give away their control and possession of telecommunication operators, and thus might loose considerable power and a stable source of revenue. On the other hand, countries might significantly benefit from economic, political and social returns from viable ICT strategies, enhancing capabilities for efficient decision-making and integrating Africa in the global exchange flows of information, knowledge and ideas. How a country copes with such challenges will be shown with special reference to the country case Tanzania. Due to the following reasons, the East African state is particularly suitable for an in-depth assessment:

- Tanzania is characterised by a considerable level of stability and thus provides a secure and constant framework for the analysis,
- Since the mid 1980s Tanzania has undergone profound reforms targeting a market-based development approach and democratic governance. Reforms can be directly linked to recommended ICT promoting policies based on market principles and good governance,
- Tanzania was among the first African countries that introduced pro-active ICT development policies, resulting in a considerable ICT performance record since the mid 1990s,
- Finally, Tanzania is one of the large territorial states in Africa,[3] in which, in addition, the majority of the population is rural. Thus, more than other African states, Tanzania faces challenges of upcountry ICT deployment in order to bring ICT in reach of rural areas.

However, the focus on Tanzania is not solely meant to be a case study. Putting the country case into an African perspective also allows a comparative outlook, in which not only the strengths and weaknesses of the country case can be detected, but which also helps to identify basic trends in African ICT developments. In the light of the given country case, the study assesses the viability of the outlined strategy and analyses the prospects of ICT utilisation in individual sectors in order to identify those applications, in which some political, economic and social return of investment can be expected.

Research methodologies comprised a critical review of existing literature and the examination of primary source, such as legal texts, policy documents or evaluation reports, collected during two field visits to Tanzania, the first from

[3] For a map of Africa, see appendix 2, a detailed map of Tanzania is given in appendix 3.

April to August 2002, the second between April and June 2003. Secondly, the research findings are based on empirical research conducted during the field visits. The policy analysis is mainly founded on expert interviews conducted with 73 decision-makers from the public and the private sector, as well as with civil society representatives.[4] Investigations of ICT utilisation in different sectors made use of quantitative assessments[5] and complementing qualitative approaches including expert interviews and the analyses of primary sources.[6]

With this research design, the study is among the first to attempt a comprehensive analysis of ICT in Tanzania, linking the levels of policy-making, market developments and utilisation assessments. Although there is a increasing body of literature dealing with ICT in developing countries, there is an apparent lack of empirical country studies necessary to draw conclusions about the viability and sustainability of ICT and to identify those factors in the political process that support or infringe increasing ICT diffusion. Notwithstanding the relevance of existing reports on ICT in Tanzania, the most valuable are Miller (2003), Intelecon (2002), Fahamu (2002), GTZ / ZEF (2002), Nnafie (2002), Miller Esselaar and Associates (2001) plus a number of others, these focus mainly on individual issues and are rather limited in scope.[7] Thus, this study seeks to fill the gap in empirical research at the country level and it attempts to introduce a more comprehensive approach to ICT combining policy and regulatory issues with an analysis of responding market developments and assessments of ICT utilisation in given sectors such as governance, health or education.

The chapters of the study are organised in the following way: Chapter 2 introduces the theoretical framework of the ICT for development debate. It gives the background of the debate, defines the terms and provides some basic

[4] Sources based on expert interviews are labelled with an 'Int.' and the date of the interview. A list of interviews can be found in appendix 4.

[5] Surveys were conducted with regard to an IT vendor, NGO, and rural ICT utilisation survey. Details on the research design are briefly introduced in the respective chapters. After pre-tests in the field, the surveys on IT vendors and NGOs were submitted online. In addition, raw data on SME's ICT utilisation and public Internet cafes was provided by Issmail Nnafie and Francis Matambalya, who generously made their data available. The data was re-edited and re-processed to meet the objectives of this study.

[6] All figures in the study without a clear reference to the source, are the author's own compilations.

[7] Furthermore, the mentioned reports were mainly written for consultancy or evaluation purposes, so that they did not always provide the sources and references of their findings. Here, considerable effort has been made to increase the reliability of sources by checking them thoroughly and by focusing on primary sources wherever possible.

assumptions framing the idea of information and knowledge societies. Furthermore, it links the agenda of ICT promotion to the overall approach to development, characterised by good governance and private sector-led development models, in order to explain the current strategic approach to ICT. Chapter 3 shifts the focus to Africa and the African perception of the agenda. It explains the origins of the African reform context and outlines the role of different organisations and initiatives that shaped the development of ICT on the continent. The main emphasis, however, is on changes in African telecommunications and the analyses of the ICT core segments regulation, fixed line telephony, mobile cellular and the Internet to identify basic tendencies in the African telecommunication landscape. Chapter 4 introduces the country case Tanzania and elaborates on those factors identified as being key enablers for increased ICT diffusion: policy, regulation, markets and education. For a better orientation, the history of ICT in Tanzania is divided into three phases – recognition (until 1993), orientation (1994-1999), and departure (since 2000). The chronological blocks, in which policy, regulation, market and education are assessed for the given period, help to identify the main developments in each of the analysed areas and also show the increasing sectoral differentiation as a basis for new functionality and adjustment capabilities in the emerging telecommunication market. Chapter 5 continues the country case, but changes the perspective away from building frameworks and looks at patterns of ICT utilisation in the country, focusing on computers and Internet usage, as well as on electronic networking. Areas covered are ICT in governance, health, and education, the role of ICT utilisation for NGOs and SMEs, and, finally, the prospects of rural ICT utilisation are elaborated on. The chapter is closed by some concluding remarks on ICT utilisation in Tanzania. Chapter 6 returns to the general picture. The conclusion relates the hypotheses of the study to the findings of the research to allow some remarks on the viability of the Tanzanian ICT strategy and some general thoughts on the prospects of ICT in Africa.

2. Theoretical Framework of the "ICT for Development Debate"

2.1. Definitions and Context

Particularly since the late 1980s and early 1990s, advances in information technology reinforced economic and social changes and information and communication technologies were regarded as a powerful driving force for globalisation. Global changes were sometimes labelled as revolutionary and called for major adjustments in all countries to harness the new technologies for development. Earlier in the developed world, later in the developing world. For example, in 1994, the so-called Bangemann Report, submitted as white paper to the European Commission, and named after the chairman of a high-level group on the information society stated:

> "Throughout the world, information and communication technologies are generating a new industrial revolution already as significant and far-reaching as those of the past.
>
> It is a revolution based on information, itself the expression of human knowledge. Technological progress now enables us to process, store, retrieve and communicate information in whatever form it may take, unconstrained by distance, time and volume.
>
> This revolution adds huge new capacities to human intelligence and constitutes a resource which changes the way we work together and the way we live together" (Bangemann Report 1994).

Six years later, the Okinawa Charter on the Global Information Society adopted by the G-8 Heads of State during the G-8 summit in Okinawa, extended the notion of a coming information age to the developing world. The charter said:

> "Information and Communications Technology (IT) is one of the most potent forces in shaping the twenty-first century. Its revolutionary impact affects the way people live, learn and work and the way government interacts with civil society. (...) The essence of the IT driven economic and social transformation is its power to help individuals and societies to use knowledge and ideas. Our vision of an information society is one that better enables people to fulfil their potential and realise their aspirations. To this end we must ensure that IT serves the mutually supportive goals of creating sustainable economic growth, enhancing the public welfare, and fostering social cohesion, and work to fully realise its potential to strengthen democracy, increase transparency and accountability in governance, promote human rights, enhance cultural diversity, and to foster international peace and stability. Meeting these goals

will and addressing emerging challenges will require effective national and international strategies." (G-8 Okinawa Charter 2000)

There is an apparent similarity between the two documents,[8] both in the language and in the direction. A visionary, pathetic tone combined with characterisation of ICT as the major driving force of the twenty-first century, at the same time accelerator and facilitator of globalisation and global development. Though many of the promises made, still await their realisation, it must be emphasised that both documents were political documents and politically motivated; in both cases meant to bring a debate from a rather technocratic, isolated basis to a wider audience. The documents should raise awareness and this was made by emphasising the visionary element of an information society. But apart from the promises, what makes ICT so powerful? To answer this question, firstly some clarification on terms needs to be done.

Firstly, there is the mere technological aspect, defining information technologies as electronic means of capturing, collecting, processing, storing, transmitting and presenting information. Thus, information technologies basically comprise telephones, computers and networks (c.f. World Bank 2002a: 3, Heeks 1999). In addition, information and communication technologies (ICT) has developed as a generic term that includes features beyond the technology itself, also paying attention to the environment in which the technology operates. This more inclusive definition was particularly elaborated by Richard Heeks (1999). He explicitly added two further components: processes of purposeful activity and people to undertake those processes. Both, processes and people, forming an information system in which users are able to purposely process, retrieve or disseminate information. This also includes a context of a giving environment of institutions, such as organisations, groups, markets and of other influencing factors like political, economic, socio-cultural, technical and legal (Heeks 1999: 3). Putting all these together, as visualised in figure 2.1, a systematic view of ICT can take shape.

[8] Other major policy documents originating from the early debate were also characterised by a visionary fascination of new communication tools, among them, Denmark's Info Society 2000 (1994), the United States' Global Information Infrastructure reports (1993) or Singapore's IT 2000 (c.f. Thomas et al. 1999).

Figure 2.1: A Systemic View of ICT

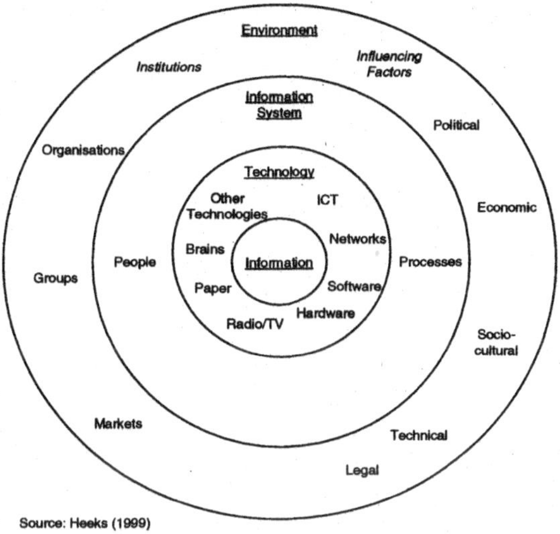

Source: Heeks (1999)

This model explains the scope of ICT touching upon different types of human interaction based on electronic means, but also includes supportive information systems of people and processes, and environmental factors as benefiting or disturbing factors. These broad impact factors have increased the pervasiveness of ICT, making it potentially applicable in various dimensions of human communication.[9] The pervasive moment of ICT was, in addition, accelerated by technological dynamism, to which two factors have contributed: the increasing convergence of telecommunications and computing rapidly advanced technological developments, and this fusion launched ever new opportunities for networking.[10] Together with new developments in hard and software, that

[9] The feature of pervasiveness, of course, makes the technology application neutral, that is it might be utilised for development and the well-being of people, but also might be utilised to increase the effectiveness of warfare and crime or strengthen the capacities of states to keep a watch on its citizens.

[10] The most extraordinary qualitative change in the new technological infrastructure came with the Internet and the invention of the World Wide Web (WWW). The Internet has grown at a rate unprecedented in the history of communications. No other communication advance has gained widespread acceptance so quickly (c.f. UNDP 1999: 58). The Internet was developed in the late 1960s by the US Department of Defence as a centreless web of computer networks, called ARPANET (Advanced Research Projects Agency Networks). Soon it was used to link science and university communities and successively, user-friendly innovations like the WWW as a graphic user interface that replaced computer language with easy to use point and click devices (mouse), made the Internet more widely accessible. For a

continually expanded the power of computing, converging technologies laid the basis for ICT utilisation in new variety and complexity. At the same time, the costs of computing and transmitting information have decreased dramatically, offering increasing capacities at relatively declining costs. Figure 2.2 shows these advancements.

Figure 2.2.: Advancements and cost development of global ICT

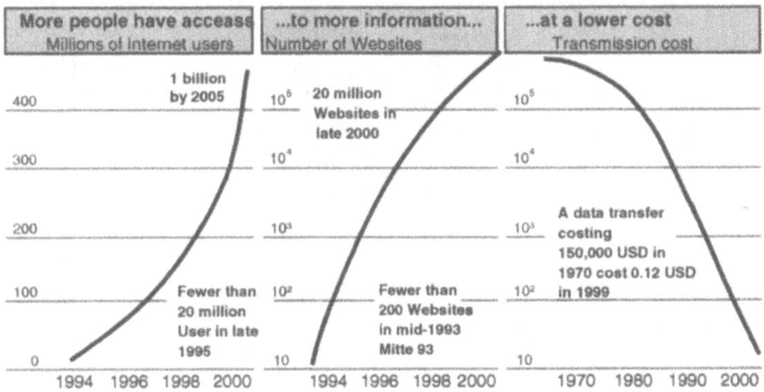

Source UNDP 2001: 32

Accordingly, changes in communication were due to three main reasons: (i) the expansion of computing power, (ii) the convergence of computing and telecommunications, and, (iii) decreasing costs. They radically increased the value of global information processing with effects on the pattern of global exchanges of goods and services as well as in the configuration of human interaction. The new scope of global exchange demanded new explanations, and, as a result, new terms of information or knowledge societies emerged, attempting to put the developments into perspective and to outline the main features and likely consequences. In the practice of the development discourse, both terms have often been used interchangeably, but the distinction between the two touches upon some of the fundamentals of what is addressed in the ICT for development debate.

In contrast to pure information, the acquisition of knowledge based on individual experience and capacities to apply information deliberately in a given context, so that knowledge acquisition empowers its possessors with the capacity for intellectual or physical action. Thus, "knowledge is fundamentally a matter of cognitive capability" (David / Forray 2003: 4). In his pioneering book

history of the Internet see Kahn (1999), Leiner et al. (2003) and the introducing texts at http://www.isoc.org/internet/history/.

on the post-industrial society Daniel Bell defines knowledge as "a set of organised statements of facts or ideas, presenting a reasoned judgement or an experimental result, which is transmitted to others through some communication medium in some systematic form" (Bell 1999: 175). Information, on the other hand, appears only as more or less structured or formatted data that remain passive until used by those with the knowledge needed to interpret and process them. Thus, information can be reduced to data that have been organised and communicated.[11] Although the declining costs and growing capacities to reproduce or disseminate information enabled a new quality of global exchange, the reproduction of knowledge has been less pervasive, because it is much more expensive and challenging to build cognitive capabilities that are not easy to articulate or to transfer.

There is no broadly accepted concept of a global information or knowledge society, but taking into consideration the above distinction, the following approximations can be made: knowledge society forms a broader concept that is organised around global networks of capital, management and information, whose access to technological know-how is at the roots of productivity and competitiveness (c.f. Castells 2000: 502). In this concept, knowledge is regarded as a commodity and the production of knowledge as a key determinant of success in the capitalist economy. In contrast, the notion of information society implies a weaker approach, in which social and economic gains for societies result from relatively free and unfettered communication (c.f. Powell 2004: 2). Similarly, the Digital Opportunity Initiative (DOI) has distinguished between two major approaches to ICT, which, with some restrictions, coincides with the differences between knowledge and information. Thereby, ICT can be addressed as a production sector focusing on the development or strengthening of ICT related industries, and ICT can be addressed as an enabler for social-economic development where cross-sectoral strategies aim to harness the uniqueness of ICT to accelerate a wider development process (DOI 2001: 19). The broader or narrower definitions do not represent a mere difference of opinion, rather they point towards the coexistence of different models of information or knowledge societies. There is no single model, but, instead, both knowledge-based economies and information-centred human interaction compete within the same societies. This text prefers the term information society, because it represents the less demanding concept more suitable for an analysis of ICT in Africa and, in addition, it is also the term that became generally accepted in the course of the discussion on ICT for development. However, whatever shape and direction it will take, there is also a growing recognition, that whatever kind of information

---

[11] The definition is taken from Castells (2000: 17), who borrowed it from Marc Porat's work "The Information Economy" published in 1977.

society will emerge, it will be shaped by deliberate decision-making, or as Bell put it: "Technology does not determine social change; technology provides instrumentalities and potentialities. The ways these are used involve social choices" (Bell 1999: xxxviii).

Whereas there is no doubt that technological changes shaped the conditions of social action around the world, and writers like Bell, Castells and others[12] pioneered the work to explain the fundamental patterns of these changes, it must be emphasised that their main focus was on the developed world. Bell, for example drew upon US American experiences of the early 1970s and identified five dimensions that explained the coming of a post-industrial society:

1. "Economic sector: the change from a goods producing to a service economy;
2. Occupational distribution: the pre-eminence of the professional and technical class;
3. Axial principle: the centrality of theoretical knowledge as a source of innovation and of policy formulation for society;
4. Future orientation: the control of technology and technological assessment;
5. Decision-making: the creation of a new "intellectual technology." (Bell 1999: 14).

Thus, the changeover to a knowledge society based on changes (i) in the economy that shifts from manufacturing to services, (ii) in technology that centres around science-based industries and (iii) in sociological terms that is the rise of new technical elites (c.f. Bell 1999: 487).

The most recent and most ambitious effort to redraw a map of modern society is Manuel Castells three volume "The Information Age: Economy, Society and Culture." Castell's starting point is the information technology revolution based on a converging set of technologies in micro-electronics, computing and telecommunication, their exponential growth and their pervasiveness penetrating all domains of human activity (c.f. Castells 2000: 29). For Castells, the character of the current technological revolution is not the "centrality of knowledge and information, but the application of such knowledge and information to knowledge generation and information processing/communication devices, in a cumulative feedback loop between innovation and the use of innovation" (Castells 2000: 31). The innovative feedback loop was facilitated by another

[12] Bell singled out a number of pathbreaking books on the issue, see Bell (1999: xxi-xxiv).

major technological shift, from decentralised, stand-alone devices to pervasive computing and interconnected information processing capabilities able to build powerful networks in all spheres of human activity. In that sense, knowledge application and information processing became a productivity factor.

> "Networks are appropriate instruments for a capitalist economy based on innovation, globalization, and decentralized concentration; for work, workers, and firms based on flexibility and adaptability; for a culture of endless deconstruction and reconstruction; for a polity geared toward the instant processing of new values and public moods; and for new social organisation aiming at the supersession of space and the annihilation of time (Castells 2000: 502).

Accordingly, the social structure of the information age is made up of networks of production, power and experience – this is why Castells prefers the term network society – and though not all dimensions and institutions follow the logic of the network society, all societies are penetrated, with different intensity, by the pervasive logic of dynamically expanding network societies (c.f. Castells 2000a: 381).

Both authors, Castells and Bell, have contributed to a better understanding of today's global information society. But, at the same time, their prerequisites to enter the information age call for high standard technological diffusion, a significant availability of human capacities and a sophisticated level of societal organisation – features that can be attributed to the industrialised world, but hardly to developing countries that often lack the infrastructure and ability to use and adapt ICT to generate wealth, power and knowledge. This is why both authors have been rather pessimistic about Africa. Bell concludes that "Africa is in serious trouble" dealing with the large socio-economic consequences of emerging information societies – and to make the transition to post-industrial sectors (Bell 1999: xlix). Similar observations were made by Castells, who stated that particularly Africa – with only South Africa offering some hope for the continent – has become increasingly marginalised in the global economy by each leap forward in technological change. "The disinformation of Africa at the dawn of the information age may be the most lasting wound inflicted on this continent by new patterns of dependency, aggravated by the policies of the predatory state" (Castells 2000a: 95). Both authors saw only little room for Africa (and other developing regions) in the information society.[13] However,

[13] This would at least have applied at the time, when both books were published. Castells book "End of the Millenium" was first published in 1998, the quotation from Daniel Bell was taken from his 1999 foreword to a new edition of his work "The Coming of Post-Industrial Society".

recent history proved that their forecasting, to some extent, underestimated the dynamisms of the technology in the developing world. Thus, it makes sense, to shift the perspective from a global or industrialised world point of view to a developing country perspective. Within the development context, how have issues of an emerging information society been perceived and adapted. This will be the theme of the following sections.

2.2. Making ICT Work for Development

The role of information, knowledge or technology for development has been recurrent topics in the development debate and beyond. Advances in communication have transformed societies throughout history from the early times with the invention of writing and printing techniques to telegraphy, telephone and television.[14] With the onset of first computerisation in the 1960s and 1970s, the issue of informatics and the use of computers in developing countries began to materialise rudimentarily with the first computer installations for government use. But it was not before the technological quantum leap, when computers became a mass product and borderlines between computerisation and telecommunication vanished that the new term of information and communication technologies became the focus of the debate.

In this context, computers and computer usage started to overlap with the development discourse, focusing on technologies and their potential impact on, for example, democratic decision-making and more effective governance, better health care delivery, improved education or lifelong learning. Thus, enabling people to utilise ICT effectively and efficiently, that has been the assumption, will contribute to capacity building and the empowerment of people. This, of course, touches upon the concept of development, itself, because development constitutes the raison d'etre of all ICT promoting initiatives. Though it is common understanding that each country must reach its own consensus on the meaning of development, there is, nevertheless, a traceable shift in development thinking since the onset of the development discourse in a narrower sense after the Second World War (c.f. Menzel 1995: 21). Whereas the early concepts of development essentially referred to processes of economic growth and industrialisation, today, it is much more widely recognised that the social setting

[14] One of the most influential books on the effects of different media or modes of communication on society has been Marshall McLuhan's classic 'Understanding Media', published in 1964. McLuhan analysed media as the extension of men and argued that the personal and social consequences of any medium "result from the new scale that is introduced into our affairs by each extension of ourselves, or by any new technology", thus, that it is the medium that shapes and controls the scale and form of human association" (McLuhan 1964: 9).

and people's capabilities are crucial determinants for advances in development.[15] These broader approaches particularly highlight the role of knowledge and human capital as essential to all aspects of development. Bezanson and Sagasti, for example, stated that "the capacity to acquire and generate knowledge in all its forms, including the recovery and upgrading of traditional knowledge, is perhaps the most important factor in the improvement of the human condition" (quoted in Mansell / Wehn 1998: 9).

Putting a new emphasis on capacities or capacity-building implies, however, that, at first, some prerequisites must be met, that is to provide opportunities to access, acquire and process information and to put in place the freedom to communicate, to make human capabilities a vital component of development. Nobel laureate Amartaya Sen has advanced this thought most radically, when he conceptualised development from a perspective of human freedom. Sen brought in the argument that real development consists of the removal of various types of unfreedoms – "poverty as well as tyranny, poor economic opportunities as well as systematic social deprivation, neglect of public facilities as well as intolerance or overactivity of repressive states" (Sen 1999: 3). His main argument is that human freedom is the overarching objective of development and he identifies rights and opportunities that characterise human freedom and classifies these interconnected instrumental freedoms in five categories: political freedoms, economic facilities, social opportunities, transparency guarantees, and protective security as the five central constituencies of human freedom (Sen 1999: 38-40). Applying Sen's concept, it would become essential "to develop and support a plurality of institutions, including democratic systems, legal mechanisms, market structures, educational and health provisions, media and other communication facilities and so on" (Sen 1999: 53). Sen's understanding of development, based on the expansion of freedom (to access information and knowledge) and the enhancement of people's capability to generate information, turn information into knowledge and apply knowledge for sustainable development, is also the normative foundation of 'development' in this context.

2.2.1. Putting ICT on the Development Agenda

Early contributions to the ICT for development debate can be traced back to the early 1960s (c.f. Drew/Foster 1994), the emergence of a broader academic interest only arose in the 1990s and a systematic discussion started only in the mid 1990s and studies from UNCSTD (1995), UNCTAD (Roffe et al. 1995),

[15] Of course, this is a radical shortening of the development debate, useful summaries on different concepts of development can be found, for example, in Menzel 1995, Schuurman 1993, Easterly 2001.

IDRC (Bezanson / Sagasti 1995) and the World Bank (Talero / Gaudette 1996) brought first substantial input to the in-depth analysis of information and communication technologies and their role in and for development. Thus, it is a very young subject. Furthermore, early discussions on ICT had been mainly politically motivated rather than academically or seeking a theoretical basis for an information society of the poor. Particularly UN organisations and the World Bank pushed the discussions and argued very pragmatically. Their main concern had been that developing countries must take appropriate measures to minimise the risks of social and economic exclusion. In order to minimise risks and to identify best practice approaches, the United Nations Commission on Science and Technology for Development (UNCSTD) was set up in 1995 to "develop a better understanding of the social and economic impacts of ICT on all sectors of the economy, and on the social, cultural, and political lives of the citizens of developing countries" (UNCSTD 1995). In the light of empirical evidence on ICT utilisation in developing countries, UNCSTD presented positive examples – such as improved competitiveness or more efficient service delivery by governments – and negative examples – most people in the least developed countries have been excluded from the global information society – the commission approved some basic recommendations. These particularly call for the establishment of national ICT strategies to build capabilities to make effective use of the technologies and thereby priority needs to be given to policies, regulation and education (UNCSTD 1995).

In 1998 the World Bank dedicated the World Development report to the issue of knowledge and development proposing to look at the "problems of development in a new way – from the perspective of knowledge" (World Bank 1998/1999: 1). Accordingly, knowledge, and the acquisition of knowledge were at the core of the report. The report highlighted that decreasing costs of communications made the world's knowledge more accessible and, thus, increased developing country opportunities to acquire and adapt knowledge already available in the developed world and to create and disseminate new knowledge conducive to improving people's lives. To facilitate the acquisition, absorption and communication of knowledge, the World Bank recommended a market oriented development approach, particularly emphasising the role of privatisation, competition, and regulation, thus implementing good policies for an efficient business environment to stretch the limits of market reach. For goals in telecommunications beyond market reach, the Bank proposed the establishment of universal access funds and reiterated the then new idea of subsidies awarded by lowest demand auctions (c.f. World Bank 1998/1999: 69). Apart from market-based policy interventions, however, basic education and provisions for lifelong learning were promoted as the second, though equally important, pillar of the strategy. "Basic education is the foundation of a healthy, skilled, and agile labour force. Lifelong learning education beyond the basics enables countries to continually assess, adapt, and apply knowledge" (World Bank 1998/1999: 8). Therewith, the World Bank introduced the basic elements of a today widely accepted ICT for development approach. Although many of the ideas had been formulated before, partly inside the World Bank in the preparatory work for the World Development Report, partly outside the Bank, the World Development Report 1998/1999 provided a comprehensive background and roadmap for the debate on ICT and remained one of the most valuable sourcebooks on ICT for development until today.

In 1999 the World Bank report was accorded by a number of influential voices from the UN, such as UNDP (1999), UNESCO (1999), and the ITU (1999) who covered ICT for development in major publications. The 1999 Human Development Report called for a globalisation with a human face and highlighted the vital role of ICT as a driving force for globalisation. UNDP framed ICT as a precondition to participate in an emerging network society based on the fusion of computing and communication (c.f. UNDP 1999: 57). UNESCO's 1999 World Communication and Information Report focused on the socio-cultural impacts of ICT, particularly accentuating ICT opportunities in education and for preservation of the world's cultural heritage – supplemented by regional chapters that addressed the different challenges in the individual world regions (c.f. UNESCO 1999). Finally, the ITU's report "Challenges to the Network. Internet for Development" explored the likely impact of Internet development in a number of areas of social and economic concern, such as commerce, health and education (ITU 1999).

Though all reports also addressed challenges such as the existing gap in the global availability of ICT or the threat of digital exclusion, their overall credo was positive by emphasising the new opportunities of ICT. It would be wrong to interpret this prevailing optimism as a fade-out of critical attitudes or ignore the poor conditions of ICT infrastructure and human resources in most developing countries. Instead, these reports were meant to prepare the way for a broader ICT debate and conveyed a political message – you need to harness ICT for development. Thus, the main task at this time was agenda setting, and all those institutions, spearheading the debate,[16] stressed the relevance of ICT and, in support of their arguments, brought forward the many positive examples of successful ICT utilisation in the developing world that, to some extent, falsely conveyed the impression of most developing countries just ready to jumpstart in

[16] Apart from the mentioned organisation, UNCTAD and ILO also introduced major publications focusing on the organisation's respective work profile, UNCTAD (2000) on commerce and ILO (2001) on working conditions in the information society.

the 21st century. New opportunities were particularly identified in the following areas:

- *Improved economic efficiency and competitiveness* by better access to product and price information, global reach of production, distribution and services, reduced transaction costs and new sources of revenue by improved market reach and the creation of new business models (ITU 1999: 61, Analysys 2000).[17]

  Particular attention was given to small and medium enterprises (SMEs), because they were considered to be a main source of productivity and employment in many developing countries (c.f. Kennedy / Hobohm 1999: 7). Their gains from ICT were associated with access to new markets, but above all, in cluster-building and inter-firm networking to increase their "collective efficiency" (Schmitz 1995, see also Levitsky 1996, Hallberg 2000).

- *More efficient governance and public administration* by open, transparent, accountable and democratic governance facilitated by ICT. Specific measures usually distinguished between internally-oriented ICT applications to support the internal management, including staffing and budgeting accounts and externally-oriented applications improving the public access to government policies or provision of specific documents like laws, statistics or health information (c.f. Heeks 1999, Kenny et al. 2000, Levin 1996).

- *Improved health care delivery* due to more efficient resource allocation by using networking techniques. New opportunities relate to remote diagnoses and consultation, as well as the dissemination of medical knowledge. In addition, health logistics and interaction of health professionals benefit from affordable networking (DOI 2001: 10, Analysys 2000, ITU 1999: 73).

- *Better education systems* by taking advantage of distant learning programmes, improved teacher training, access to teaching materials and better administration of the teaching administration (c.f. UNESCO 1999: 46, ITU 1999: 90, Kenny et al. 2000).

- *Opportunities to increase social capital* by promoting active citizenship. Thereby, ICT can ease the participation in political, social or administrative affairs by information-sharing and networking. Particularly

[17] For a general overview on the correlation of ICT and economic growth, see Kenny (2000) and Zhen-Wei Qiang / Pitt /Ayers, (2003: 20), a more critical appraisal can be found in Pohjola (2002) and Kenny (2002).

NGOs could benefit from enhanced collaboration and networking and, thus, raise their voice in the political process (c.f. Talero / Gaudette 1996, Analysys 2000, DOI 2001: 15).

Notwithstanding the vital input owed to these elaboration on digital opportunities, particularly in the beginning of the debate, most ICT for development reports, were, to a large extent, motivated by the goal to set an agenda of a just commencing debate. Though they all paid attention to the risks of ICT, their basic message was one of technological optimisms. Especially in the late 1990s, a time when ICT became a broader issue and induced a business hype of the new economy in the developed world, a notion of high expectations prevailed and found its expression in the new, increasingly prominent metaphor of leapfrogging. The term referred to the chances of developing countries to leapfrog development stages and accelerate their development process by jumpstarting in the 21st century.[18] The leapfrogging argument was particularly prominent in Africa, for example the African Information Society Initiative (AISI), created in 1996 to realise a sustainable information society in Africa by 2010 (see chapter 3.1.). The AISI action plan stated that "Africa has great potentials to leapfrog development stages by adopting new strategies to build capabilities" (AISI 1996).[19] ICT were regarded as an enabler for development, and the implicit relationship to economic and social development all stem from several basic characteristics related to improved information production and sharing, and the positive effects of an improved information flow on economic, social and political development capacities (c.f. Grace et al. 2000, infoDev 2000, Analysys 2000).

[18] The model of development stages was borrowed from the thinking of modernisation theory. Most influential has been Rostow's stage theory, published in 1960, in which he argues that investment (in form of foreign aid) in strategic sectors would trickle down and eventually would take off into self-sustained growth (c.f. Rostow 1960).

[19] Interpreting leapfrogging as a shortcut to development, however, is actually based on a misunderstanding of the term. The metaphor can be traced back to an article by Soete (1985), who used the term in a sense to leapfrog technological stages in order to facilitate an appropriate diffusion of technology. In the same vein, also the World Bank which introduced the term to a greater audience referred solely to technological leapfrogging. "Developing countries [...] have seized an opportunity to leapfrog the industrial countries by going straight from underdeveloped networks to fully digitised networks, bypassing the traditional analog technology that still forms the backbone of the system in most industrial countries" (World Bank 1998/1999: 57).

2.2.2. The Digital Divide

Broadly, until 2000, the general view was mainly on the potentials of technology and produced considerable enthusiasm with which the development community rushed into ICT and ICT related programs.[20] Only as projects and initiatives did not bring the expected outcomes, there was a growing recognition that ICT promotion was more than installing computers, but part of a more complex problem that manifests itself in the lack of most needed resources – for better education, infrastructure and institutions. As a result, the pendulum of technology perception swung in the other direction, and a new metaphor took the centre stage of the debate: the digital divide.[21] Though the term was new, of course, the phenomenon was not. "Information gaps" have been noticed and mentioned since the mid-1970s and the role of information in economic development was the subject of a number of studies in the 1980s (c.f. Fillip 2001). The most recognised has been the Missing Link Report by the ITU, published in 1984, that highlighted the gaps in access to telephones and the potential impact of such lack of access for economic development (c.f. ITU 1984).[22] Thus, at first glance, the digital divide debate sized old problems, putting the uneven distribution of communication tools at the centre of debate. Of course there are many other gaps in access to many things, from food to medicines, etc. The key concern has been that ICTs have a catalysing and leveraging potential. The gap in ICTs is therefore potentially more dangerous to developing countries than gaps in access to other things (c.f. Fillip 2001). At first, the digital divide was mainly associated with global imbalances, that were illustrated by a number of vivid and persuasive statements, such as the whole of Africa has less telephone lines than either Manhattan or Tokyo (infoDev 2000: 14), in Bangladesh, a computer costs the equivalent of eight years average pay

[20] In 2000, a project database compiled by the Digital Divide Network had entries of over 20,000 ICT related projects; see the database at www.digitaldividenetwork.org.

[21] The term itself was popularised by the US Department of Commerce. In 2000, the Department published a report entitled "Falling through the Net. Toward Digital Inclusion". The report, focusing on ICT diffusion patterns in the United States, concluded that information tools, such as the personal computer and the Internet, have been increasingly critical to economic success and personal advancement and that more Americans than ever have access to telephones, computers, and the Internet. At the same time, however, the report found a significant digital divide separating American information "haves" and "have nots" (US Department of Commerce 2000: XV).

[22] In addition, the role of communication in society, comprising matters such as media ownership and governance, concepts of information as commodity or freedom of information as individual vs. collective rights was intensely debated in the Unesco led initiative for a New World Information and Communication Order (c.f. Unesco 1980). However, arguments were trapped in cold war rhetoric, so that there were only very little achievements in the end.

(Bridges.org. 2001: 12), or that one third of the world population has never made a telephone call (DotForce 2001). Successively, the digital divide also included disparities within countries, such as with regard to income, education, geographic region, gender, ethnicity or age (c.f. Bridges.org 2001, infoDev 2001, Digitaldividends 2000, OECD 2001).

Putting the statements together, the digital divide has been a reflection of existing broader socio-economic inequalities describing a patchwork of varying levels of ICT access and effective use.[23] Both terms, access and effective use, are critical, because access to basic telecommunication infrastructure is fundamental and without effective use, even the best infrastructure is of no value. But at the same time, a number of questions is raised, such as what type of access – individual or shared – or what quantity and quality of access will be sufficient to bridge the digital divide? The same applies to effective usage, where conditions for effective use must be examined. In effect, different responses to questions of access and effective use have further dismantled the agenda into different sets of individual goals and policy recommendations. The South African NGO Bridges.org, for example, advocated a broad understanding of technology access and introduced a set of issues labelled as "real access criteria" which, according to the NGO, are going beyond just physical access and make it possible for people to use technology effectively to improve their lives:

- *Physical access.* Is technology available and physically accessible?
- *Appropriate technology.* What is the appropriate technology according to local conditions, and how people need and want to put technology to use?
- *Affordability.* Is technology access affordable for people to use?
- *Capacity.* Do people understand how to use technology and its potential uses?
- *Relevant content.* Is there locally relevant content, especially in terms of language?
- *Socio-cultural factors.* Are people limited in their use of technology based on gender, race, or other socio-cultural factors?
- *Trust.* Do people have confidence in and understand the implications of the technology they use, for instance in terms of privacy, security, or cybercrime?

[23] A good summary on different views on the digital divide can be found in Gurstein (2003).

- *Legal and regulatory framework.* How do laws and regulations affect technology use and what changes are needed to create an environment that fosters its use?
- *Local economic environment.* Is there a local economy that can and will sustain technology use?
- *Macro-economic environment.* Is national economic policy conducive to widespread technology use, for example, in terms of transparency, deregulation, investment, and labour issues?
- *Political will.* Is there political will in government to do what is needed to enable the integration of technology throughout society? (Bridges.org 2001: 5).

There have been other attempts to translate the existence of a digital divide into proposals for action. Some are similarly extensive like the 7Cs summarising UNDP's approach,[24] or the recommendations by Gurstein, who highlighted the role of appropriate infrastructure and technology, relevant content, appropriate services, capacity development and an appropriate structure of finance and governance (Gurstein 2003). The DotForce concluded that the digital divide "can be characterized by insufficient infrastructure, high cost of access, inappropriate or weak policy regimes, inefficiencies in the provision of telecommunication networks and services, lack of locally created content, and uneven ability to derive economic and social benefits from information-intensive activities" (DotForce 2001: 6), and the Development Opportunity Initiative developed a strategic framework for ICT that consists of five interrelated areas: (i) infrastructure building, (ii) human capacity building, (iii) policy formulation, (iv) enterprise promotion and (v) content development (DOI 2001: 6).

Though the reports or articles differ in individual points, mainly due to a different understanding of access and effective use, they all share a set of common themes in their proposals to bridge the digital divide: that is, building infrastructure, creating conducive environments for ICT, human capacity building and development of local relevant content. Therewith, the digital divide

[24] *Connectivity* (setting up telecommunications and computer networks), *Community* (focusing on group access, not individual ownership), *Capacity* (building human skills for the knowledge society), *Content* (putting local views, news, culture and commerce on the Web), *Creativity* (adapting technology to local needs and constraints), *Collaboration* (devising Internet governance for diverse needs around the world), *Cash* (finding innovative ways to fund the knowledge society (c.f. UNDP 1999: 63).

debate, still ongoing today,[25] contributed to a deeper understanding of ICT, resulted in more strategic approaches to make ICT work for development, and induced a number of global, regional and local initiatives resulting in increasing awareness and sensitivity for ICT in developing countries (see chapter 3.1.). The elaborations of the digital divide also pointed towards the complexity of including developing countries in the information society. Beyond the concrete policy results, however, the digital divide debate also introduced a broader dimension to the theoretical framework of ICT. Whereas the earlier texts highlighted issues of opportunity, the digital divide texts focused on justice or the lack thereof in the global distribution of ICT. Ever since, both themes 'opportunity' and 'justice' have formed the brackets of today's ICT for development approaches.

2.3. Toward an ICT for Development Strategy

The discussion on ICT for development has matured considerably in recent years. Different disciplines have sized individual issues of an increasingly complex debate and addressed the multiplying matters regarding economics, politics, regulation, legal or social affairs. The overwhelming body of literature concurs with the general findings that ICT can foster development efforts and improved communication and exchange of information can create powerful political, economic or social networks as a promising basis for advances in development.[26] There is also consensus that this is no automatic process but needs deliberate action from strategy development to implementation. This view

---

[25] More recent texts argue the ICT gap between rich and poor countries is closing in relative terms. Over the last 25 years, telephone penetration has been expanding faster in low income countries, and considerably faster in middle income countries, than it has in high income countries (Fink/Kenny 2003, Orbicom 2003), whereas until 2001 most articles identified a widening gap between developing and developed countries (ILO 2001, Rodriguez/Wilson 2000, Bridges.org 2001).

[26] This does not mean that the road to ICT development is unchallenged. Some authors, for example, criticise the way, the ICT agenda is being imposed on developing countries, especially Africa, promoting "what is effectively a tax on aid budgets to pay for Northern hardware, software and associated services. In this way, the North is to be paid to smooth Africa's insertion into a globalised world order, from which the North will then profit" (Cline-Cole/Powell 2004: 6). In a similar vein, Ya'u (2004: 10) argued that ICT promotion appears to work in the interest of multinational corporations and a similar perspective was also taken up by Ishemo (2004). These authors argue mainly from a Marxist perspective and give some valuable insight regarding the problematic structure of globalisation. However, none of them, in substance, question the potential role of ICT in development, and above all, they fell short of translating their diagnosis of "virtual imperialism" into viable policy action.

goes beyond individual project designs, but aims at scaling up the potential effects of ICT in favour of broader development goals.[27]

The identification of opportunities and risks, as discussed above, has singled out the individual, but interlinked problems that hampered the deployment of ICT in developing countries in the past. On this basis, at least four areas of priority action have been detected that became the basis of most strategies to make ICT work for development:[28]

1. Creating a sound environment for ICT, emphasising the role of good policies and the establishment of a regulative framework conducive to ICT, embedded in a comprehensive national strategy for ICT,
2. Building infrastructure and, thereby, increasing access and lowering costs of ICT utilisation,
3. Building human resources through a range of targeted training and education to facilitate knowledge creation,
4. Development of relevant local content.

Though this can only introduce first general terms of reference, it, nevertheless, provides some orientation points for approaching the issue. At the same time, it is evident that the emergence of the approach was built upon some basic thinking on development, particularly with regard to the role of states and markets. This is why it is useful to look at some determinants of the current approach to development.

[27] The most ambitious goal set by the international community relates to the so-called Millenium Development Goals (MDG) adopted at the UN Millenium Summit in 2000. The eight goals and associated targets represent an agreement "to work towards a world in which sustaining development and eliminating poverty would have the highest priority. Although ICT, themselves, are no explicit development goal, they are regarded as an important tool to contribute to improving the efficiency in delivering the MDGs. "There is a growing acceptance that ICT can play an important role by providing new and more efficient methods of production, bringing previously unattainable markets within the reach of the poor, improving the delivery of government services, and facilitating management and transfer of knowledge, a key factor in reaching the MDGs" (World Bank 2003: 8). For an assessment of ICT and their contribution to the MDGs, see infoDev (2003).

[28] The list is compiled from various sources and represents the quintessence of different approaches to ICT (e.g. World Bank 2002, DotForce 2001, UNDP 2001, DOI 2001, ITU (1999), UNCTAD (2000), ILO (2001). It is neither exhaustive nor necessarily reflects all focal areas of individual approaches. Nevertheless, the four points emerged as an undisputed fundament of an ICT for development strategy.

2.3.1. The Private Sector-led Development Paradigm

To understand the current approach to ICT, it is important to see it in the light of a broader approach to development, in which the private sector has become the central focus for development due to two main reasons: the failure of public sector-led economies and the rise of globalisation. Thereby, it is not incidentally that almost all major stakeholders in the ICT for development debate rely on a set of shared core issues and a more or less joint approach. These reflect some basic assumptions of the current development paradigm, which can be labelled as "Monterrey Consensus"[29]. The new consensus strives upon many subthemes, including aid, economic issues, governance, human rights, etc. (c.f. UN 2002),[30] in this context it is sufficient to emphasise the call for good and transparent governance, sound policies and strong domestic institutions, all required to mobilise domestic resources and attract private investment to create economic opportunities (c.f. World Bank / IMF 2002). Translating the consensus in the ICT for development debate provides an outline of the agenda: implementing sound economic policies and promoting competition, putting in place good governance, investing in people, and establishing an investment climate to attract private capital. The role of the state focuses on creating an environment conducive to private sector participation, moderating, facilitating and monitoring compliance with the rules and, if necessary, sanctioning non-compliance. Thus, enabling environments allow private firms to operate efficiently and to put in place institutions and policies that promote private sector development. Enabling environments set the parameters of public policies aiming at encouraging economic and non-economic activities. According to Kennedy / Hobohm (1999: 6), these parameters include the following features:

- *Macroeconomic stability*. A key to economic growth is prudent economic management that avoids volatility and uncertainty. Policies that lead to high inflation, excessive taxation, distorted factor prices, or lack of foreign exchange discourage business formation and investment.

[29] In the development discourse, the Monterrey Consensus replaced the Washington Consensus, which was mainly based on macroeconomic considerations which were increasingly regarded as insufficient to tackle development problems. However, the consensus is no ready-to-use model, but a collection of principles that necessarily left room for interpretation (c.f. Williamson 2003, Manuel 2003). The name derived from the United Nations Conference on Finance for Development that took place in March 2002 in the Mexican town of Monterrey (UN 2002).

[30] A useful overview on Monterrey is also available on the conference website of the UN Department of Economic and Social Affairs, http://www.un.org/esa/ffd/

- *Financial System.* An efficient financial system is an important prerequisite for economic development, in mobilizing savings and channelling them into investments. For private sector firms it serves an additional purpose besides providing them credit: it is a means of enforcing discipline on firms that borrow.
- *Competitive Markets.* Even more important to economic development than private ownership is the presence of competitive markets. Competition promotes efficiency and policies that promote intense, but fair, competition will result in competitive enterprises and a strong private sector.
- *Regulation.* Regulation of the private sector is necessary to ensure competition and fair trade, but unnecessary regulation burdens the private sector and leads to fewer and less efficient enterprises, and reduced competition.
- *Physical Infrastructure.* The size, characteristics and efficiency of the private sector depend very much on the available infrastructure.
- *Political and social stability.* To operate efficiently, the private sector requires a generally conducive political and social environment. Institutions such as public administration systems, judicial systems, schools and hospitals are necessary for the stable society that provides an environment for sound economic development.
- *Legal Framework.* Business, like society in general, needs to operate under the rule of law. No strong private sector can exist in the absence of an adequate legal framework to resolve disputes, facilitate efficient transactions, and to protect property rights.
- *Policy framework.* Various policies, even those not directly related to the private sector, can affect its growth and size. Policies that provide adequate support to the private sector are essential.
- *Access to resources and support services.* Entrepreneurs and enterprise managers need access to finance, information, and various types of support services to create and operate businesses in the most efficient and competitive manner.

These factors describe basic trends in development thinking, in which the private sector is the primary source of wealth creation, economic growth and employment and, therewith, a strong private sector became essential to delivering developing countries own developmental objectives. Beyond the theoretical assumptions, the new thinking had a very practical impact, because those countries that were unable to take up these challenges became increasingly marginalised.

2.3.2. Linking ICT Promotion with Private Sector-led Development Approaches

Derived from the overall approach, basic thinking of ICT promotion presumes that good governance, rule of law and the introduction of pro-competitive policies are critical factors for a private sector-led growth strategy that creates a dynamic from which ICT promotion can commence. The approach takes advantage of assumed market potentials, flexibility and responsiveness and, at the same time, calls for regulative capacities to guide, escort and monitor the basic direction in ICT development. Thus, governments are assigned the tasks of providing strategic leadership and sectoral guidance to put in place an enabling environment to integrate ICT into the design and implementation of country and sector development strategies.

Taking the promises and pervasive impacts of ICT, countries need to embed their ideas of technology promotion into their overall development strategy to manage the benefits and risks of ICT in view of the likely outputs of increased ICT utilisation and the interaction between ICT, economic performance and poverty reduction. The call for national ICT strategies in support of a coherent approach to technology development and taking into consideration a number of key issues including a global, regional and local dimension was among the first issues raised by reports on ICT.[31] Globally, national e-strategies should be in line with the current development paradigm, where the International Monetary Fund and the World Bank have enforced basic standards of economic performance. On the technical level, approaches must be consistent with the international frameworks as provided by the World Trade Organisation (WTO) concerning the rules of global telecommunication, or the International Telecommunication Union (ITU) with regard to technical standards (see chapter 3.1.). Regionally, strategy options are influenced by regional organisations, often requiring participating countries to frame their ICT strategies in relation to regional experiences and initiatives. The national component, however, presented the most challenging tasks, because it has not only addressed key issues of building access, enhancing capabilities and skills or creating content, it also included the necessity to identify priority sectors and sequence reform steps.

National ICT strategies contained the multiple dimensions of ICT, one emphasis, however, was put on the role of governments to support new forms of market facilitation by market liberalisation and institutional reform. The reform

[31] E.g. Mansell /Wehn 1998, for an overview see Hanna (2003), Tipson / Frittelli (2003), Melody (2001).

model that emerged consisted of three integrated components: privatisation, competition and independent sector regulation. Market liberalisation was expected to lead to new investment in the infrastructure, and, in combination with privatisation and competition, an increase in efficiency of operations was anticipated. It is important to note, that this is not simply a neo-liberal textbook approach founded on faith in the so-called free market, but the strategic shift has been mainly a response to defunct, monopolised public telecommunications operators (PTO) in many developing countries increasingly unable to provide reliable and affordable telecommunication services to the public. The monopolies had great difficulty adapting to changing economic, political and social conditions, of which changing technology is only a part. Monopolies, operating in a protected, stable environment, in which access to telecommunication was often politically motivated and telecommunication service revenues were used as government funds, were not well-suited to adapt to a new and increasingly diversified and dynamic market place (c.f. Melody 2001, World Bank 2004). Particularly for this reason, there has been widespread recognition that an expanded role of the market in telecommunications can facilitate improved efficiency. In fact, there is a considerable amount of literature giving evidence that both privatisation and competition lead to significant improvements in performance (c.f. Wallsten 2001, Fink / Mattoo / Rathindran 2002, Noll 2000).[32]

Market liberalisation and new entry in the telecommunication supply has been one step towards achieving effective telecom structural reform. Furthermore, comprehensive reform programmes were accompanied by new regulatory measures unbundling functions of policymaking, regulation and operator management, and thereby promoting competition and ensuring that telecommunication operators provide their services at efficient prices. Regulation can be regarded as both a safeguard measure for investments of private telecommunication operators and consumers. Thus, the basic objective of regulation consists of two components, one economic and one social. "From an economic perspective the services should satisfy the full range of consumer demand and be supplied under conditions of optimal efficiency. From a social perspective, the service should be made available to everyone on reasonable terms, whether or not it is profitable to do so" (Melody 2001: 2). Effective regulation levels the playing field for all market participants by coherent, fair and transparent regulation, it encourages operators, by incentives or obligations, to extent their networks and services on an affordable basis, also including areas

---

[32] For an enlightening review of empirical literature on fixed-line telecommunications policy and performance, see Fink/Mattoo/Rathindran (2002: 29).

deemed to be unprofitable, and, finally, independent regulators have the capacity to implement and enforce the rules made to meet economic and social objectives (c.f. Melody 2001, Intven / McCarthy Tétrault 2000, ITU 2002b, World Bank 2004).

Competition and privatisation accompanied by effective regulation have been regarded as fundamental to improving telecommunication infrastructure. However, ICT promotion was never meant to solely depend on pure market mechanisms. Market forces will not always provide opportunity to all segments of society and increased market efficiency alone will not be sufficient to induce pro poor broad ICT development. Policy-makers and regulators need to go beyond the market to address market failures and asymmetric levels of access to infrastructure and content. This particularly affects rural telecommunication users and marginalised segments of society, whose inclusion might be less economically viable, but politically or socially desired. To this end, governments have to define national access targets and affordability goals, generally labelled as universal access or universal service goals (see chapter 3.3). Targeting remote and disadvantaged communities is often high on the political agenda and reflects the many efforts to attack a national digital divide that has developed alongside urban-rural and centre-periphery relations. Low demand density might be due to widespread poverty or a lack of skills and resources, but can, from a network cost ratio, always be explained by proportionally higher communications costs and lower available functionality (c.f. infoDev 2000, ITU 2003a). Expanding connectivity has been a primary goal for developing countries, but policy options focusing on complementary mechanisms beyond the market have been only developed recently. A useful model for rural telecommunications development has been introduced by Navas-Sabater et al. (2002). The authors focus on two separate 'gaps', to be addressed by different mechanisms (see figure 2.3.).

- The market *efficiency gap* refers to the difference between what markets actually achieve under current conditions, and what they can achieve if market barriers are removed. This gap can be bridged through effective competition, private provision of service, and market-oriented policies and regulations that create a level playing field for new entrants
- The *access gap* refers to the people and places which remain beyond the limits of the market unless additional investments are mobilised through intervention, in the form of subsidies to encourage service providers to enter (Dymond/Oestmann 2002: 2).

Figure 2.3: Universal Access Market Model

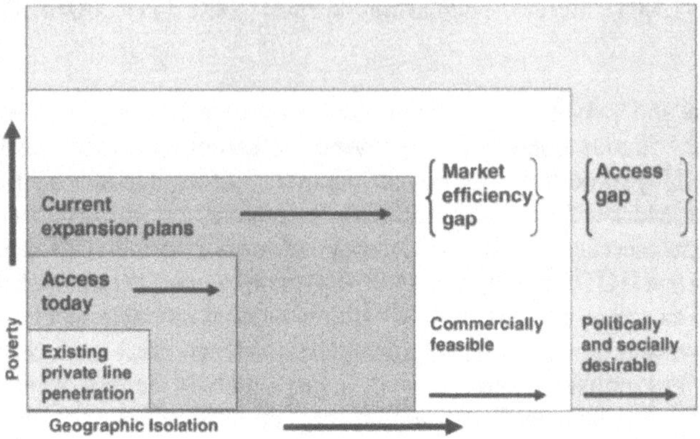

Source: Navas-Sabater et al. (2002)

Addressing the access gap involves a number of options, including direct subsidies and, for example, their competitive allocation to rural telecom operators or roll-out obligations attached to licensing procedures. Governments can impose performance requirements or levies on operators directing them to provide telecommunication services in underserved areas or governments can provide incentives, such as tax breaks or reduced licence fees offered to providers that extend their networks or improve services in target areas (c.f. ITU 2003a, Intelecon 2003a, Hudson 2001).

Furthermore, universal access issues have been closely associated with the provision of public access facilities as a tool to bring ICT services to the broader population.[33] In due course, different types of centres materialised, ranging from 'phone shops' or 'public call offices' solely providing basic telecommunication services to large-scale centres including awareness building, training, and extra provisions for specially chosen target groups. The most prominent model of donor funded community telecentres has been the so-called 'Multipurpose Community Telecentre' (MCT) advocated by the ITU, the Canadian

---

[33] Originally, the idea of public access facilities or telecentres came from Scandinavia, but subsequently has been copied successfully throughout the world. The shift towards an explicit focus on developing countries originated in a 1994 call for proposals issued by the International Telecommunication Union (ITU) that paved the way for the Buenos Aires Action Plan (1994-1998) by the ITU to "develop best-practice, sustainable and replicable models of ways to provide access to modern telecommunication facilities and information services, particularly to people in rural and remote areas" (Ernberg 1998).

International Development Research Centre (IDRC) and UNESCO.[34] In contrast to other models, MCTs highlight the 'multipurpose' aspect covering a wide variety of ICT services and applications, often including libraries, conference facilities, training programmes, as well as special provisions for small enterprises, distance education, tele-health, e-commerce and the like, and, hence, emphasising their potential to serve as a vehicle for improved integrated rural development. For these reasons, public ICT facilities have gained much attention[35] and public access remained a critical issue to extend the reach of ICT, not only in rural areas, but also in the densely populated urban centres.

However, developing appropriate instruments and policies, by addressing both the market efficiency gap and the access gap, have not been restricted to infrastructure development and improved access to ICT. Government ICT strategies also include the linkages with education policies and strategies. In a narrower sense of human capacity building, there are three key challenges in education, such as (i) increasing user knowledge, that is, enabling users to use ICT confidently and creatively, (ii) ensuring a sufficient professional ICT workforce with specialist ICT skills to meet the needs of the economy; and (iii) ensuring that the workforce is able to take full advantage of different sectoral applications and individually tailored ICT solutions (c.f. Kearns 2002: 31). In a broader sense, however, learning, education and training providing skills and knowledge are generally regarded as a key to progress. This applies for the above-mentioned three layers of ICT education, but clearly goes beyond, as prospects of the information society depend on investment in education and training (c.f. ILO 2002, TechKnowLogia 2003).

Finally, enhancing human capacities must be compatible with the mix of ICT available, and user oriented strategies are essential to building a critical mass of local users. However, beyond the general call for developing relevant content

---

[34] Promotion of MCTs is part of the ITU's programme on 'Integrated Rural Development and Universal Access', for a brief history see ITU (1998), a more up-to-date progress report on the established pilot telecentres can be found in Oestmann / Dymond (2001).

[35] For example, some studies contend with comprehensive rural access strategies (Intelecon 2003, infoDev 2000), while others investigate the grounds of rural ICT access and develop possible approaches to encourage wide-spread adoption of ICT (Chapman / Slaymaker 2002, Mayanja n.d.). Some highlight individual core issues such as finance and technology choice (Caspary / O'Connor 2003), while others are practical guides addressing key issues of successful implementation derived from a collection of case studies (Commonwealth of Learning 2001, McConnell 2001), and again others are designed as manuals to set up telecentres (Jensen / Esterhuysen 2001). Finally, there are studies that address issues of evaluation and means to assess the impact of telecentres on the local community (Scharffenberger 1999, Whyte 2000).

and statements such as the extent to which national ICT strategies offer access to information depends on the type of information that is accessible and affordable from a variety of sources (Mansell / When 1998: 106), a set of more difficult issues must be addressed. These relate to strategic choice of technical designs – ICT based or including traditional media, such as paper based, TV or radio with a potentially broader reach than solutions focusing solely on ICT (c.f. Haddad / Draxler 2002). Furthermore, choices relate to questions whether a country should attempt to develop its own content in accordance with its curricular framework, buy the material or pay a licensing fee or to adopt an open source software (OSS)[36] driven information technology strategy (c.f. Sida 2004). Thus, ICT utilisation does not solely depend on individual applications – these range from data banks, e-governance application, health management systems or environmental monitoring programmes – but on strategic choices that need to address both the learning objectives, the acceptability of the means of communication and the appropriateness of technical designs.

Thus, strategy formulation has been a key component for developing countries to prepare themselves to efficiently harness ICT for development. This chapter gave on overview on the general framework for national ICT strategies and their justification. There is little dissent on the need for a coherent framework, but, at the same time, it became also evident that proposed strategy components – mainly developed by a number of international organisations – only provide a rough orientation. Although considerable research output has contributed to a better understanding of ICT and ICT promoting policies, there still are a number of open questions and unsettled issues. The research community has not yet offered clear answers to the links between ICT utilisation and social and economic development or the exact type of skills and abilities required for today's information society. Nevertheless, there is a point of departure. Good governance and effective regulation are, and will continue to be an essential instrument for implementing dynamic markets and universal access policies as fundament of promising ICT for development strategies. The way, how this agenda was influenced by multiplying international fora dealing with ICT in Africa and, finally, adopted in the African context, will be subject of chapter 3.

[36] Open source is, in contrast to proprietary software, a model of distributed, shared, open software development. With OSS, the programming code used to create software solutions is available for inspection, modification, re-use and distribution by others. It is often assumed that open source software is free of charge, but this is not necessarily the case. "The concept of "free", in this context, emphasises what can be done with the source code rather than its cost. Because of its collaborative nature, the open source model lends itself to allow participants to be both producers and consumers/users of the software" (Dravis 2003: 5).

3. Development and State of ICT in Africa

Since the beginning of the 1990s the state of telecommunications has changed radically – in Africa and throughout the globe. In Africa, first rudimentary telecommunication networks had already been installed during colonial rule, but only since the last ten years developments have progressed dynamically. Two major reasons can be put forward: Firstly, technology got progressively more powerful. Secondly, technical innovations coincided with far-reaching transitions towards democracy and free market economy, with the latter putting new emphasis on competition and private sector participation. State-centred development strategies were replaced by open market systems; partly due to poor performances of the former, partly as a result of external pressure from the donor community. This also affected African telecommunications. Defunct Public Switched Telephone Networks (PSTN) gradually commercialised their business and became more market oriented. Particularly, the World Bank and the International Monetary Funds (IMF) advocated the rethinking. They had made the liberalisation of the telecommunication sector a condition of their structural adjustment programmes and, thus, were the most important protagonists of liberal sector reforms.

The telecommunications revolution was global, but affected the world regions in varieties of pace and scope. A global ICT ranking puts Africa, jointly with a few South Asian countries at the bottom. At the end of 2003 Africa accounted for 1.4 percent of global Internet users, 2.1 percent of fixed lines and three percent of mobile phone users (ITU 2003); on a global scale this is hardly significant. But global figures only give half the story. They do not reveal the radical changes within African telecommunications. Average African fixed-line penetration grew up to nine percent annually between 1998 and 2003 and, at the same time, became more efficient and more reliable. Similar figures appeared for the Internet, which has witnessed steady and considerable growth, though still on a low level. The African success story, however, has been mobile or cellular telephony. Only in 1998 there were a modest two million mobile telephone users. In 2003 this number reached 51 million. Since 2001 the number of mobile phone subscribers has surpassed the number of fixed-line customers and taking 2002 and 2003 alone, in both years mobile telephony attracted 14 million new subscribers. Also on a global scale, this has been the highest growth figure (ITU 2004). However, there have been considerable variations within Africa: Fixed-line telephone density per 100 inhabitants varies from 27 in Mauritius to 0.02 in DR Congo, mobile phone density from 55 in Seychelles and 0.07 in Ethiopia (ITU 2003).

These developments went far beyond mere technological changes, but touched upon the prospects of human development in general. For example, the UN Summit on the Information Society that took place in December 2003 in Geneva was the latest visible high-level event underlying the new relevance of ICT. The summit dealt with no less than discussing new ways to make ICT work for human development. In fact, there has been increasing evidence that telephones and the Internet have been utilised for development. An improved flow of information, more efficient working procedures and improved productivity, all contributed to an increasingly promising and development related record of ICT.

However, a individual examples should not be taken as a complete record of ICT as a tool for development. The research community has not yet provided a comprehensive performance record of ICT including detailed impact analysis as well as a thorough assessment of overall costs and benefits. But even without such a record, some basic developments are traceable and can be deduced from general data. Firstly, basic ICT indicators inform about the diffusion of the technology as well as about basic tendencies in their utilisation, whether it is fixed line or mobile telephony or the Internet. Secondly a closer consideration of the underlying development strategies can be helpful identifying the likely road ahead. The creation of sound political and economic frameworks, fair and transparent sector regulation, in short good governance and efficient markets are regarded to be the key to increased ICT utilisation from where further applications in different segments can develop. This chapter attempts to assess the last ten years of telecommunication sector reforms and tries to identify some same basic tendencies in African telecommunications. The first part of this chapter identifies the origin of the reforms, the second part analyses developments in the ICT core segments regulation, fixed line telephony, mobile cellular and Internet for the past decade.

3.1. Origins of Reform

Telecommunications sector reforms in Africa commenced only slowly and in the late 1980s and early 1990s, when often defunct public telephone monopolies dominated the scene. The main momentum for reform came from the World Bank and the International Monetary Fund (IMF). Since the end of the 1980s both organisations called for macroeconomic stabilisation and structural adjustment measures as a prerequisite for international assistance. From the very beginning, a particular emphasis was put on the telecommunications sector. Therewith, World Bank and IMF did not only provide the general direction towards liberal markets, they also guided the sectoral programmes. Both institutions became key actors and participated in almost all negotiations on African telecommunications and ICT (c.f. Wallsten 1999). Additional input to the liberal reform agenda came from the GATT negotiations during the Uruguay round (1986-1994) that led to the creation of the World Trade Organisations (WTO) and the General Agreement of Trade in Services (GATS). One pending issue of GATS had been the telecommunication sector that had been renegotiated in the due course and resulted in the Agreement on Basic Telecommunication (ABT) and the so-called Regulation Reference Paper.[37] The agreement took effect in February 1998 and committed the member states to a progressive liberalisation. Though the now annexed GATS provided for waivers and special provisions for developing countries, the agreement introduced the WTO principles "open markets", "most favoured nation treatment", "transparency" and "non-discrimination"[38] to the global telecommunications market, and, though many developing countries had been rather sceptical about GATS, a group of African reform protagonists emerged: Côte d'Ivoire, Ghana, Mauritius, Senegal und South Africa committed themselves to open their markets according to WTO rules, including transitional provisions for developing countries (c.f. ITU 1998: 8/9; WTO 2004).

World Bank, IMF and WTO put considerable pressure on the African countries to liberalise their economies – and their telecommunication sectors. This led to bitter conflicts between reform protagonists and opponents. Wilson (2004) described these as „one of the key battlegrounds in the history of Internet development in Africa" and accordingly, the pace of change was slow. In an attempt to maintain control of the strategic telecommunication sector and to secure resources from revenues from international telephone traffic, most

[37] The Agreement on Basic Telecommunication Services introduced competition in over 95 percent of the world telecommunications market (by revenue). It became applicable to most of the sector segments, ranging from submarine cables to satellite systems, from broadband data to cellular services, from business networks based on the Internet to technologies designed to bring low-cost access for under-served rural communities. The majority of WTO Members have also made regulatory as well as market access commitments, ensuring adherence to a multilateral framework for promoting competition in this sector (c.f. Senunas 1997, WTO n.d.). The Regulation Reference Paper outlines some basic regulatory guidelines, e.g. for competition, transparency, interconnection or spectrum management. It can be found in Intven/McCarthy Tétrault (2000: A-1).

[38] "Most favoured nation treatment" means that members are held to extend immediately and unconditionally to services or service suppliers of all other members "treatment no less favourable than that accorded to like services and services suppliers of any other country", "transparency" means that GATS members are required, inter alia, to publish all measures of general application and establish national enquiry points mandated to respond to other member's information requests, "non-discrimination" implies that the Member concerned does not operate discriminatory measures benefiting domestic services or service suppliers. The key requirement is not to modify, in law or in fact, the conditions of competition in favour of the Member's own service industry (WTO n.d.).

countries preserved the status quo. Only after external pressure increased, and, at the same time broad programs in support of sector reform had been introduced, a handful of forerunners – the WTO reform protagonists plus Tanzania and Uganda – initiated first reforms. With considerable donor assistance, first careful policy changes slowly started to pay off, and, with increasing success, the number of countries subsequently adjusting their strategies also increased. At the end of the 1990s, African countries could be divided into (i) the reform countries, such as Ghana, Mauritius, Senegal, Tanzania, Uganda and Côte d'Ivoire, which had initiated some successful liberal reform steps, (ii) a group of countries that opposed the reforms, such as Ethiopia, Malawi, Nigeria, Zambia or Kenya, which first successfully withstood the international pressure. However, since 2000 they started to open their economies, too. Finally, (iii) conflict or post-conflict states as well as marginalised small states formed a third group. Due to their size or due to high risk investments these countries hardly were relevant and investors avoided them accordingly. Only as reforms started to consolidate in the stable markets and investors gathered experience in African operations, these were increasingly targeted, too. This applies particularly to the populated and therewith economically interesting states like DR Congo and Angola (c.f. Wilson 2004).

Whereas the first part of the decade was characterised by conditionalities, positive and supportive measures came to the forefront in the second half. Thus, broad support programmes that, in turn, based on experiences made in some earlier project approaches accompanied first successful reforms. The World Bank had already started respective sector programmes in the 1980s and further significant input came from the International Development Research Centre (IDRC) that contributed to the commencement of electronic networking in Africa since the end of the 1980s. UNESCO's Rinaf programme (Regional Informatics Network for Africa), that supported Internet connectivity and IT competencies in 21 African states between 1992 und 1998 and finally UNDP and the ITU were among the first agencies who targeted ICT in Africa (c.f. Hafkin/Wild 2002).

Until 1995, to a large extent, international donors set the African ICT agenda and only slowly African actors caught up and replaced policy proposals from the outside with new approaches that came from within the continent – though still with substantial assistance from external actors. Subsequently, ICT became subject to high-level political decision-making. The process commenced at a „Regional Symposium on Access to Telematics", organised by IDRC, ITU and UNESCO at the UN Economic Commission for Africa (ECA) in Addis Ababa in 1995. Only one month later the ECA ministerial conference approved a resolution entitled „Towards an African Information Superhighway", which identified the need of good policies and a sound political and legal framework as one major condition for increased ICT utilisation. Again one year later, the ECA ministerial conference adopted an action plan in May 1996. The „African Information Society Initiative" (AISI) became a milestone in the African ICT history, because it introduced corner pillars of an African information society that gave substantial input towards the commencement of national policy processes in support of ICT. AISI assisted countries to initiate, formulate and implement coherent e-strategies to harness ICT for development (UNECA 2003, AISI 2003).[39] In recent times, particularly the creation of the "New Partnership for Africa's Development" (Nepad) in 2001 tied up with AISI became a new institutionalised focus of African efforts in technology promotion. The e-Africa Commission launched in 2002 as part of Nepad's infrastructure programme became the focal point of Nepad's ICT related activities. The e-Africa portfolio included 13 project approaches referring to the global, regional and national policy level and ranging from private sector development, education, infrastructure- and institution-building to strategies aiming to identify and develop local relevant content for Internet based applications (c.f. Bridges.org 2003 Chetty 2003, Quaynor 2003).[40]

In parallel to the African efforts, ICT also gained momentum in international cooperation and got some special attention particularly at the end of the 1990s. Most bilateral donors had included issues of ICT in their portfolio and at the turn of the century particularly Canada, USA, Great Britain, the Netherlands, Sweden, Norway and France had a considerable ICT profile in their cooperation programmes (Wild / Hafkin 2001, OECD 2003). At the same time, the issue of strategic partnership as a new instrument in development cooperation emerged and assigned the private sector a new crucial role in the development process (c.f. Nielinger 2001b). As a result, multilateral and bilateral donors, the private sector and civil society have built new coalitions and alliances and became powerful protagonists of an ICT for development debate. Among them the Global Knowledge Partnership organised by the World Bank, the Development

---

[39] Notwithstanding the broad impact of AISI, the initiative also had shortcomings, particularly the follow-up activities that took place as regular African Development Fora I-IV between 1999 and 2004 were rather weak, esp. the implementation capacity of agreed action plans. Similarly, the core issue of the initial AISI – the formulation and approval of national ICT policies – only progressed slowly (c.f. Soltane 2003: 128).
The most recent ADF IV took place in October 2004 in Addis Ababa and with regard to ICT, regulatory issues were at the top of the agenda, see ADF (2004).

[40] In addition to the above-mentioned initiatives, most regional organisations like SADC, ECOWAS, COMESA, UEMOA had also adopted some ICT promoting measures, mostly targeting joint infrastructure projects, but above all, initiatives have related to policy harmonisation (c.f. Zongo 2001, WEF 2003a).

Opportunity Initiative by UNDP, Accenture and Markle Foundation, the DotForce (Digital Opportunity Taskforce) created by the G-8 during the Okinawa summit in 2000 or finally the UN ICT Taskforce that was inaugurated in September 2001.[41] Accordingly, the number of ICT projects, programmes and initiatives mushroomed. Though these differed considerably in scope and directions, they all shared the common features of the prevalent ICT for development approaches. The consensus was reendorsed during two recent events that also became key references to the African debate: the World Telecommunication Development Conference 2002 (WTDC-02) and the World Summit on the Information Society 2003 (WSIS).

The major outcome of WTDC-02 has been the adopted Istanbul Action plan that charted a course for developing countries to "transform the digital divide into digital opportunities" (c.f. ITU 2002c). The action plan agreed on a series of six programmes to be implemented by the ITU's Development Bureau between 2003 and 2007 aiming to provide access to ICT and promote their use not only as an engine for economic growth, but also for social, educational and medical progress. The programmes comprised (i) regulatory reforms, (ii) technologies and telecommunication network development, (iii) e-strategies and services and applications, (iv) economics and finance, including cost and tariffs, (v) human capacity building and (vi) a special programme for least developed countries (c.f. ITU 2002d).

The second event was the UN World Summit on the Information Society that took place in December 2003 in Geneva (c.f. ITU 2001b). With regard to Africa, already the preparatory process to the summit gave new input to the ICT for development debate, which was mirrored in the Bamako declaration adopted by the heads of African states during a regional conference held in May 2002 in Bamako. The Bamako Declaration covered a broad range of issues, but particularly highlighted problems of access, local content and finance.[42] Notwithstanding the relevance of the individual issues, the Bamako conference took most credit from streamlining joint African positions on ICT. However, in the end WSIS failed to meet the expectations. Particularly those questions pushed by the African states such as establishing new finance mechanisms, questions of access and infrastructure development, local content, special

---

[41] For a general overview on the different initiatives see Okpaku (2003), WEF (2003) and Hafkin/Wild (2003), for more detailed information see the organisation's websites: Global Knowledge Partnership (http://www.globalknowledge.org), Development Opportunity Initiative (http://www.opt-init.org), DOT Force (http://www.dotforce.org) and UN ICT Taskforce (http://www.unicttaskforce.org).

[42] For the declaration text, see WSIS Africa Regional Conference (2002).

promotion of health or educational applications or the inclusion of marginalised groups in the information society remained largely unresolved. In addition also those issues that emerged as the most contentious ones, such as financing, Internet governance or issues of national control versus a free flow of information, could only be solved by very general compromise solutions. Both the declaration and the action plan remained incomplete and pending issues dealing with the complex technical, economic, political and social implications of ICT were transferred to the follow-up summit (c.f. Nielinger 2003). This will be held in Tunis in 2005 and African preparatory proceedings will be set at an African regional conference to be held in Accra in February 2005.

3.2. Measuring ICT

The new attention paid to ICT was accompanied by the need to build suitable indicators and reliable data to understand and to assess the role of ICT in development. Various indicators have been introduced, ranging from teledensitiy per 100 population, number of telephone lines, mobile or Internet users, number of Internet hosts or domain names to bandwidth and data capacities and other figures. Naturally, all indicators have strengths and weaknesses. For example, Internet user numbers do not reflect Internet utilisation in public access facilities, user numbers do not account for the depth and duration of usage. Similar applies to the quantity of Internet hosts, because *.com domains are not assigned to a specific country, but are quite frequent in Africa. Vice versa, country domains like *.tz for Tanzania or *.ke for Kenya can be hosted physically in the US or on European servers (c.f. Jensen 2003; Minges 2000, Quaynor 2003). Alternative approaches are in their infancy, so that measuring international bandwidth or bandwidth per capita might develop into useful indicators, but there is hardly any reliable data to make reasonable use of the newly proposed indicators. Thus, today's indicators cannot be more than a rough approximation, without doubt indispensable, but only eligible to identify rough tendencies, but not a thorough analysis of ICT diffusion or utilisation.

Accordingly, today's analyses are always based on rather weak data. This particularly applies to different attempts to aggregate accurate ICT rankings from a chosen set of ICT indicators to build a precise national ICT profile and make it comparable to other profiles based on the same ranking. This is just summing up inaccuracy. Nevertheless, there is an increasing demand for such ICT rankings and the various attempts to introduce such rankings reflect the request for quickly accessible data on the quality of ICT developments. The most well known ICT indices are the (i) Network Readiness Index of the World Economic Forum, (ii) Orbicom's Infostate Index, (iii) the ICT Diffusion Index by UNCTAD and (iv) ITU's Digital Access Index.[43] What they all have in common is that they have developed aggregated indices based on individually weighted indicators. Usually they incorporate data on infrastructure, user numbers, costs and others (c.f. ITU 2003; Orbicom 2003, WEF 2003, UNCTAD 2003). Table 3.1 gives an overview on the respective top ten and the bottom of these rankings.

Table 3.1: Country rankings of selected ICT indices

	ITU (DAI)	Orbicom (Infostate)	WEF (Network Readiness Index	UNCTAD (ICT Diffusion)
1.	Seychelles	Mauritius	South Africa	Mauritius
2.	Mauritius	South Africa	Mauritius	Botswana
3.	South Africa	Namibia	Botswana	Swaziland
4.	Botswana	Botswana	Namibia	South Africa
5.	Cap Verde	Gabon	Tanzania	Seychelles
6.	Namibia	Simbabwe	Ghana	DR Congo
7.	Swaziland	Senegal	Nigeria	Equ. Guinea
8.	Gabon	Togo	Uganda	Cameroon
9.	Simbabwe	Gambia	Senegal	Rwanda
10.	São Tomé and Príncipe	Kenya	Gambia	Namibia
	. . .	. . .	. . .	. . .
47.	Mali	CAR	Angola	Niger
48.	Burkina Faso	Ethiopia	Ethiopia	Angola
49.	Niger	Chad	Chad	Eritrea

Whereas one might identify some similarities in the top rankings mainly occupied by the Seychelles, Mauritius, South Africa and Botswana, the middle rankings are patchy at their best. Some countries such as Namibia, Senegal or Zimbabwe have a medium rank in at least two rankings, others like Tanzania or Uganda only appear once and again others like Togo or DR Congo turn up rather unexpectedly. Similar can be said for the bottom of the rankings, where Ethiopia, Chad, Angola and Niger turned up twice and thus can catch some attention. Basically, however, the indices allow only two conclusions: Firstly, the analytical tools available to date to assess different features of the information society are limited in their scope and can only deliver proven results

[43] There are other indices such as UNDP's Technology Achievement Inndex or the IDC Information Society Index. With regard to Africa, both include only South Africa and Zimbabwe (the latter only in the UNDP index) in their rankings, so that they are not eligible for an inter African ranking.

in some individually chosen areas of interest. Furthermore it is notable that the role of policy and regulation only plays a subordinate role, although those organisations, which compiled the rankings regard both as critical, elsewhere (c.f. ITU 2003, WEF 2003, UNCTAD 2000). As said above, both, the chosen indicator sets as well as the reliability of the data must be considered with caution. Nevertheless, applying the indices despite their shortcomings, they would allow a second, though temporary conclusion. If the rankings are taken only as rough approximations, it seems to be fair to assert that due to the high number of variations in the middle of the rankings, the majority of African countries might differ according to individual measures, but generally belong to the relatively crowded middle rankings, hence would differ in some of their data characteristics, but share the overall feature of only poor ICT diffusion rates and their modest utilisation.[44] This general statement is hardly satisfactory. Thus it is necessary go a little further and look at some more details of typical patterns of ICT in Africa, that is telecom regulation, fixed-line and mobile telephony, as well as Internet developments and their respective causal policies.

3.3. Regulation

Disregarding the individual motivations to opt for a more market oriented development strategy, the states were the key players to implement it. As stated above, reforms in telecommunications were normally only part of much broader reforms, including measures for macroeconomic stabilisation and structural adjustment. Thus, general adjustments like judicial reforms to guarantee the rule of law, financial reforms for economic stabilisation, investment promotion schemes or reforms in the tax administration benefited the telecommunications sector, too. Beyond such macro measures, sectoral measures gained significance. These were policy measures for general sectoral guidance and regulatory measures to create a fair and transparent playing field – both also meant to build confidence in slowly emerging markets. OleKambainei, long serving director general of the Tanzanian regulator, explained the basics of regulation with the following formula: „The most crucial action by Governments is setting up independent, autonomous and facilitative, well-resourced, credible and accountable regulatory agencies (authorities) to ensure the creation and

[44] See also the tables provided in appendix 1, which give individual, other than aggregated, data on chosen ICT indicators. The tables reconfirm the above-mentioned. Apart from the special cases South Africa and the small island states, there are no typical ICT diffusion patterns that would allow the identification of main beneficial factors in support of ICT. Turning away from the role of proactive policies in favour of ICT for a moment, the rankings show particularly those countries performing better, which are small, densely populated or have a high national income.

enforcement of conducive, consistent and transparent regulatory environment and frameworks“ (OleKambainei 2003: 162).

The creation of regulatory bodies as a step towards open markets usually had been a result of the separation of regulative and operative functions that both had been performed by the public telephone monopolies, so that the new sectoral setup comprised a public telephone operator, then ready to focus on service provision and a new regulating agency assigned the mandate to establish a sound framework for commercial telephone and data services. The first regulator in Africa was established in 1988 in Mauritius, however, the broader trend only commenced in 1992 with new agencies in Mozambique, Namibia and Nigeria. By 1995 further six independent regulators had been created and by 1997 11 more were added to the list. In 2003 overall 29 more or less independent regulating agencies came into being (see table 3.2).

Table 3.2: Countries with separate Regulatory Authority

	Regulator	est.	Legal Basis	Sources of Finance
Mauritius	Telecommunications Authority (TA)	1988	enacted	Licence fee 55% Spectrum fee 38%
Cap Verde	Direcção Geral das Comunicações – DGC	1992	enacted	Government budget
Mozambique	Instituto Nacional das Comunicações Moçambique – INCM	1992	Ministerial decree	n.a.
Namibia	Namibian Communications Commission – NCC	1992	enacted	Government budget 100%
Nigeria	Nigerian Communications Commission – NCC	1992	enacted	Licence fee 68%
Tanzania	Tanzania Communications Commission – TCC (now: Tanzania Communications Regulatory Authority – TCRA)	1994	enacted	Licence fee 6% Spectrum fee 62
Zambia	Communication Authority – CA	1994	enacted	n.a.
Côte d'Ivoire	Agence des télécommunicationes de Côte d'Ivoire – ATCI	1995	enacted	Licence fee 10% Numbering fee 7,6% Spectrum fee 57,4%
Ethiopia	Ethiopian Telecommunications Agency – ETA	1996	proclamation	Government budget 100%
Botswana	Botswana Telecommunications Authority – BTA	1996	enacted	Licence fee 12%
Eritrea	Communications Department – CD	1996	proclamation	Government budget 100%
Guinea	Direction nationale des poste et télécommunicationes – DNPT	1996	n.a.	Government budget
Sudan	National Telecommunications Corporation – NTC	1996	n.a.	n.a.
Burundi	Agence de Régulation et de Contrôle des Télécommunicationes – ARCT	1997	presidential decree	n.a.
Ghana	National Communications Authority – NCA	1997	enacted	Licence fee 16% Spectrum fee 46%

Madagascar	Office malagasy d'études et de régulation des télécommunicationes – OMERT	1997	ministerial decree	Licence fee 6,3% Spectrum fee 44,9%
South Africa	South African Telecommunication Regulatory Authority – SATRA (now Independent Communications Authority of South Africa – ICASA)	1997	enacted	Government budget
Uganda	Uganda Communications Commission – UCC	1997	enacted	Licence fee 2,8% Spectrum fee 22,5 World Bank 43,5%
Burkina Faso	Agence de Régulation des Télécommunicationes – ARTEL	1998	enacted	Numbering fee 24,26% Spectrum fee 36,72% Licence fee
Cameroon	Agence de Régulation des Télécommunicationes – ART	1998	enacted	Government budget Numbering fee Spectrum fee
Malawi	Malawi Communications Regulatory Authority – MACRA	1998	enacted	Licence fee 23% Spectrum fee 24,5%
Togo	Autorité de Réglementation des Secterus de Postes et Télécommunicationes – ARSPT	1998	n.a.	Numbering fee 3% Spectrum fee 50,2%
Chad	Office Tschadien de Régulation des Télécoms – OTRT	1998	enacted	Licence fee 20% Nummerngebühren Spectrum fee 60%
Angola	Direcção Nacional de Correios e Telecomunicações – DNCT	1999	ministerial decree	Licence fee 20% Spectrum fee 80%
Guinea Bissau	Institut des Communicationes de la Guinée-Bissau – ICGB	1999	enacted	Licence fee 7,3% Budget 42,5% Spectrum fee 34,6%
Kenya	Communications Commission of Kenya – CCK	1999	enacted	Licence fee 19% Spectrum fee 34,6%
Mali	Comité de Régulation des Télécommunicationes – CRT	1999	ministerial decree	n.a.
Mauritania	Agence de Régulation – AR	1999	n.a.	n.a.
Gabon	Agence de Régulation des Télécommunicationes – ART	2000	enacted	Government budget
Lesotho	Lesotho Telecommunications Authority – LTA	2000	enacted	Budget Licence fee
Senegal	Agence de Régulation des Télécommunicationes – ART	2001	enacted	Spectrum fee 90%
Zimbabwe	Postal and Telecommunications Regulatory Authority – PTRA	2001	enacted	Licence fee 75% Spectrum fee 25%
DR Congo	Agence de Régulation des Postes et Télécommunicationes ARPTC	2002	enacted	n.a.
Quelle: ITU 2001: 11; ITU 2004a				

Regulatory mandates were ambitious, because they had to determine a healthy mix of incentives and obligations to make the telecommunications sectors work. Thereby, four focal working areas have materialised:[45]

1. Licensing,
2. Frequency management,
3. Interconnection, and
4. Universal access.

Ad 1) The licensing of operators aims to establish a market structure, introduce distinctive market segments, determine the number of players in each segment or to attach specific service obligations to each operator. Dependent on individual market structures, licensing policies were differently motivated. For the new regulators licence policies have been a tool to control markets, often meant to improve the position of the incumbent telephone operator, so that regulators established semi-competitive regimes focusing on safeguard measures in favour of the PTO. Secondly, licences became an instrument to apply service obligations to individual operators, which underlined their role in the provision of universal access strategies (see below). Thereby, a prevalent licensing pattern emerged. Assigning competitive licences in new segments of the communication market, such as the Internet or mobile telephony, was usually less complicated than in those segments occupied by the old public telcos. Particularly the PTO's traditional core businesses like fixed-line telephony or the operation of an international gateway for international telephone calls often continued as monopolies – or were subject to preferential treatment. As a consequence, there have been only a handful competitive fixed-line markets (Angola, Ghana, Nigeria, Uganda, see table 3.3), whereas in mobile telephony, roughly three quarters of the African countries have introduced competitive markets with at least two mobile operators – with immediate effects on the user numbers as a direct comparison reveals, for example with regard to DR Congo where four operators served 560.000 users and Ethiopia where the monopoly served only 50.000 users in 2002 (ITU 2004).

In the new Internet related market segments, licences often distinguished between data carriers that is the provision of Internet connectivity, thus the

[45] Regulatory powers go beyond these four points and typically also include price regulation, particularly where a monopoly is involved, control of numbering schemes, quality of services, equipment type approval, and provision of dispute settlement mechanisms and consumer protection. Notwithstanding the challenging nature of each of the points, the four issues introduced here, have been the ones regarded as most critical, see for example Bipe (2000), Intven/McCarthy Tétrault (2000), ITU (2002b, 2003a) or Intelecon (2003, 2003a).

infrastructure, and Internet services, that is to deliver services to the end-consumer. Since the latter has to acquire connectivity and data capacities from the former, regulative focus has been on the carriers, usually with a view on coordinated infrastructure deployment. ISPs, in contrast have been a typical downstream business segment that relies on the capacities provided by the data carriers and provided services that did not overlap with PTO operations, so that, from an economic point of view, there was very little need to regulate ISP businesses. In fact, recent studies show that unregulated ISP markets tend to perform better than regulated markets (c.f. Wallsten 2002: 13). Nevertheless, there is virtually no African country that that has not required at least some formal approval for ISPs. Thus, ISP licences follow less an economic ratio, but have been merely an instrument of political control and to lesser extent also a source for revenues from licence fees for regulators (see table 3.2).

Ad 2) Frequency management relates to the allocation of radio frequencies for wireless communication, aiming at an efficient allocation of frequencies that were increasingly regarded as a limited national resource. Radio frequencies are utilised by a variety of users ranging from radio and television to weather and aircraft radar and data communication. Traditionally, spectrum allocation has been done in accordance with ITU recommendations, which divided the spectrum into different bands and services. The ITU's radiocommunication sector maintains a Table of Frequency Allocation, which identifies spectrum bands of about 40 categories of wireless services with the goal of avoiding interferences among these services (c.f. Lie 2004, ITU-R 2003, Nunno 2002).[46] Thus, regulators do not assign standards, but decide who is getting how many frequencies for what purpose. In the past, the most contentious area in frequency allocation was mobile telephony, when old analogous systems occupied the frequencies and, thus, prevented the provision of more efficient digital GSM-systems.[47] Today, as analogous systems almost disappeared from the African landscape, regulators have been occupied with monitoring the use of frequencies to eliminate harmful interferences of increasingly deployed wireless technologies. Illegal band utilisation, e.g. for wireless data transmissions recurrently interfered with wireless voice technologies so that radio signals overlapped and disturbed the smooth functioning of the licenced services.

[46] In accordance with the ITU recommendations, the Southern Africa Development Community has elaborated on the tableau and recommended a regional frequency allocation plan covering frequencies between 20 and 3100 MHz (c.f. SADC 2000).

[47] For details on mobile telephony generations, see ITU 1999a: 14.

African regulators have only gradually built the capacities to effectively monitor their frequency band and enforce regulations.[48]

Ad 3) Fair interconnection agreements are critical for effective regulation. Interconnection of telecommunication services targets the interoperability by and a fair cost distribution of interoperability among different operators. Creating a fair interconnection regime has been a key instrument in building a level playing field for different operators in the market. At the same time, interconnection has been one of the most contentious issues for African regulators and interconnection disputes have been a recurrent feature in markets with more than one service provider (c.f. MT 2002: 2343; ITU 2004: 18). Particularly affected by disputes have been fixed-line and mobile telephone operators, because these are not only the most advanced market segments, but have also touched upon the preferential treatments for PTOs. Unfair or anti-competitive interconnection policies can seriously retard competition and reinforce the old monopoly by regulatory means.[49] New market entrants had little weight in interconnection negotiations and, thus, they relied on fair terms of interconnection agreements under regulatory oversight. Interconnection principles have also been part of the above-mentioned WTO basic telecommunications service agreement and the related reference paper that calls for the availability of reasonable interconnection terms including cost-oriented

---

[48] The spectrum fee schemes applied have been relatively simple. New approaches, however, go beyond a merely static assignment approach and use more market based management approaches such as frequency auctions, spectrum pricing or trading (Neto 2004, Cave 2002). The second challenge relates to new technical improvements in interference management techniques. So-called smart or polite technologies do not interfere with each other and would necessarily question the current ratio of charging licence fees altogether. For these technologies, Wireless Local Area Networks (WLAN) or Wireless Metropolitan Networks (WMAN), licensing is regarded to be too cumbersome, so that they have been generally defined as unlicenced or licence exempt bands (c.f. Wireless Internet Institute 2003, ITU-R 2003, Lie 2004: 18). However, due to their cost structure and their flexibility, they will become increasingly interesting for commercial use in developing countries (c.f. Best 2003) and thus it will be very difficult for African regulators to find a balance between the promotion of promising technologies on the one hand, and to maintain their regulatory oversight on the other.

[49] Incumbents may try to charge excessive rates for interconnection, refuse to make available adequate interconnection capacity or refuse to unbundle network elements or services necessary for efficient linkages between different operators (ITU 2002b: 69). Furthermore, unfair practices could continue if the incumbent had also started a mobile operation and entered into preferential interconnection rates with its subsidiary, so that cost structures for telephone services within the incumbent's family would have been lower than those for other operations.

rates and unbundled access from dominant suppliers (c.f. Intven/McCarthy Tétrault 2000).

However, fairness imperatives were not always taken into account. Sometimes because they were not the primary goal, more often, however, because fair interconnection agreements are easy to stipulate, but it has proved to be notoriously difficult to determine fair interconnection rates that are truly cost based and, for example, distinguish between individual call volumes and different cost structures of originating or terminating a telephone call in telephone networks of different size and quality. In the recent past, the introduction of cost based interconnection agreements often failed to enter into force because operators could only seldom agree on a mutual basis for calculating the factual costs – aggravated by conditions specific to Africa, in which telephone bills could not be collected but related interconnection fees had to be paid (Bipe 2000: 63). Until today regulators have been overstrained in proposing and enforcing a widely accepted interconnection formula.[50]

Ad 4) Universal access policies, finally, developed into a fourth main working area of regulators. The issue of universal access is at the very core of the whole ICT for development debate, because it has addressed trade-offs between access targets that are politically or socially desired, but which cannot be supplied on a practical and economically justified basis. Thus, universal access concepts deal with potential benefits of telecommunication services for low-income groups or people living in disadvantaged and remote areas. Regulators and policy makers have the critical role of defining universal access targets and ensuring that goals are reached. Access goals have been different from country to country, but generally the objective is to provide affordable access to communication e.g. on community level by the provision of public access facilities like payphones or telecentres (c.f. Dymond / Oestmann 2002).[51] Once the targets are set, a strategy needs to be developed and implemented.

---

[50] Beyond the role of interconnection for fair and transparent markets, interconnection also can play a role in the promotion of rural telephony by introducing asymmetric interconnection regimes charging calls terminating in rural areas slightly higher in order to provide incentives to invest in disadvantaged areas. Asymmetric solutions would allow different network access charges and rural telephone networks could be installed as beneficiaries of new interconnection agreements (c.f. Intelecon 2003b: 10-11, Melody 2003:18). Though such mechanisms would extend the market efficiency, regulatory routines today are generally still preoccupied with some basic agreements on interconnection targeting fair treatment of all operators.

[51] In contrast, universal service obligations as prevalent in most developed countries refer to the objective to make facilities available to every household.

In the past, regulators often have combined their licensing power with their universal access goals and have incorporated service obligations into operator's licences, for example, network roll-out either to target a predefined number of new service lines or a specific regional coverage to be reached in a given period. Predominantly service obligations have been attached to privatised or partly privatised incumbents. The Tanzanian investor MSI / Detecon was obliged to install 660,000 new telephone lines in a four year period, France Télécom had to install 290,000 new lines in Côte d'Ivoire within five years and 300,000 in Senegal in 10 years. SBC / Telecom Malaysia, the main investor in South Africa Telkom had the obligation to increase the number of main lines to 2.8 million in a five-year period (c.f. Pyramid 2001: 8).

Only until recently, African countries started to make use of new concepts for universal access strategies, mainly touching upon the World Bank model and the Canadian telecommunication consultancy Intelecon introduced in chapter 2.3.3., which have been successfully tested in Latin America (Navas-Sabater et al. 2002, Intelecon 2003a). The concept based on the assumption that under liberalised conditions markets have proved remarkably effective, including, in many instances, poor rural and remote areas. The key instrument aiming to close the access gap has been the establishment of so-called rural telecommunication developments funds or universal access funds that make available the resources needed to extend into or subsidise markets in marginalised areas. Thereby, the establishment of universal access funds has faced several difficulties, like defining the borderline between market and access gaps and establishing accepted mechanisms to fill up the fund and to withdraw resources from the fund (c.f. ITU 2003a, 2003b, Intelecon 2003, 2003a, Navas-Sabater 2002).

At present, the World Bank supports universal access strategies in Burkina Faso, Ghana, Kenya, Madagascar, Mali, Mauritania, Mozambique, Senegal, South Africa, Tanzania, Togo and Uganda (Intelecon 2003: 25).[52] Some advanced experiences with a working universal access fund have been made in South Africa. Here the fund, administered by the Universal Service Agency has operated on relatively small scale and provided finance for community telecentres only (c.f. USA 2003).[53] The most advanced universal access fund has been established in Uganda. The Ugandan Rural Communication Development

[52] For an overview on universal access policies in Africa see Navas-Sabater (2002: 67).

[53] The difficulties define target areas of subsidisation (and, thus, potential market reach) become evident, when the South African model of telecentre deployment is compared to Senegal, where the incumbent Sonatel has introduces a fully commercial telecentre franchise scheme which did not only reach a considerable coverage of ICT services, but, at the same time, also accounted for some 30 percent of Sonatel's revenues (c.f. Intelecon 2003b: 12).

Fund, administered by the regulator Uganda Communications Commission (UCC) and financed by one percent service levy by operators, government allocation and donor assistance, has targeted the broad provision of telephone and Internet services in the country. UCC mapped out 154 sub-counties eligible for tendering of communications services to private sector operators. Though not all sub-counties could be covered in the reverse auction process, that is the lowest subsidy demand wins a licence for service provision, there were considerable achievements: Each successful bidder received a subsidy amount ranging from 30 to 50 percent of the investment costs and until 2003 the fund supplied 69 public payphones, 20 Internet points of presence (PoPs), development of web portals for the 26 districts, establishment of five Internet cafes and one training facility (UCC 2003, see also BA 220/2004).

In retrospect, there can be no doubt, regulatory tasks have broadened and deepened in the past and will continue to do so in the future. Taking into account the nature and the complexity of the assigned mandate, and also the modest capacities regulators have at their disposal, African regulators, in fact, have gained a considerable profile in the recent years. On the other hand, there have also been differences regarding the scope of the assigned mandate, regulatory independency and implementation power. The most frequent critique was raised with regard to the regulators independence. Even if a regulator was installed as a formally independent entity, independence was de facto undermined by political interventions or ministerial prerogatives that weakened the process of confidence building in the newly emerging markets. Furthermore, regulators often lacked the capacities to fulfil their complex mandates and this has been the reason why agencies were often accused of solely being administrators and collectors of licence fees. Thus, a lack of independence, lack of competencies and limited resources have been the major problems for African regulators.

3.4. Fixed-line Developments

Despite increasing liberalisation rhetoric in the beginning the decade, public telephone monopolies have come to an end only slowly and sectoral reforms as demanded from World Bank and IMF were the exceptions rather than the norm. This was due to several reasons: Some states opposed the new liberal approach to telecommunications in principle, others resisted reforms due to specific considerations, such as they feared a loss of jobs, loss of control or, particularly, a loss of revenue that foremost came from international settlement payments, i.e. agreements on interconnection charges for international telephone calls, that, for example, only until the mid 1990s accounted for over 15 percent of the national income in Ethiopia, Gambia, Senegal or Sierra Leone (ITU 1998: 11, c.f. Africa Internet Forum / UNECA / infoDev 1999: 15).[54]

Basic rethinking commenced in the mid 1990s and the credo of a private sector led development strategy gained increased acceptance – caused by continuous pressure from Washington, withdrawal from an accounting rate system benefiting the African states by the United States government and above all the growing number of successful telecom liberalisations particularly in Latin America (c.f. Wallsten 1999). Chad and Djibouti had sold off their public telephone operators (PTO) already in the mid 1970s, however the first wave of telecom privatisation in Africa commenced in 1995 with PTO privatisations in Guinea, Ghana and Madagascar. In 1997 Côte d'Ivoire, South Africa and Senegal followed. All in all, the first wave between 1995 and 1997 comprised of privatisations or part privatisations in ten countries. Main investors were France Télécom and Telecom Malaysia. Whereas the investment of France Télécom followed France's colonial ties – the French acquired 40 percent of the Central African Socatel already in 1990, in 1995, 34 percent of the Madagascan Telma, 1997 51% of the Ivorian CI Télécom and also in 1997 61 percent of Sonatel from Senegal – the Telecom Malaysia strategy was shaped by a new Asian self-confidence and explicitly meant to increase the role of South-South cooperation. Telecom Malaysia took over 30 percent of Ghana Telecom shares in 1996, 60 percent of the Guinean Sotelgui in 1997 and jointly with a US American co-investor 30 percent of the flagship of African PTOs, Telkom South Africa. In addition, Portugal Telecom played a minor role which was only later extended. At this time Portugal Telecom had shares in PTOs in Cap Verde, Guinea-Bissau and São Tomé and Príncipe, that, like France Télécom investments, were characterised by old colonial ties. Table 3.2 gives an overview of PTO privatisations and progress in the installation of a second national operator (SNO) in Africa.

---

[54] These resulted from settlement payments from international telephone traffic, which reflected the fact that African subscribers, proportionally, made a much higher percentage of international rather than domestic calls and that African countries received much more traffic than they sent out. Consequently, African countries were net recipients of settlement payments. This scenery has gradually changed since 1997, when the United States, the major contributor to international settlement payments, introduced price caps for international settlements, so that international traffic lost its role as a major source of revenue (c.f. ITU 1998: 11; ITU 1999a:12).

Table 3.3: Privatisation of African PTOs and installation of SNOs

	PTO	Status	Investor	SNO	Investor
Angola	Angola Telecom	state-owned		Mundo Startel 2002	Telecom Namibia
Benin	Office des poste et télécommunicationes (OPT)	state-owned			
Botswana	Botswana Telecommunication Corporation	state-owned			
Burkina Faso	Office Nationale des Télécommunicationes ONATEL	Ongoing privatisation	Interested parties i.a. Maroc Telecom, Telenor, Detecon and France Telecom		
Burundi	Office Nationale des Télécommunicationes ONATEL	state-owned planned privatisation			
Cameroon	CAMTEL	state-owned			
Cap Verde	Cabo Verde Telecom	partly privatised 1992	Portugal Telecom (40%)		
CAR	Socatel	partly privatised 1990	France Télécom (40%)		
Chad	Sotel Chad	state-owned			
Comores	Société nationale des postes et télécommunications, SNPT	state-owned			
Congo	Société des télécommunicationes du Congo (Sotelco)	state-owned			
Côte d'Ivoire	CI Télécom	privatised 1997	France Télécom (51%)	Arobase Télécom	Banque Atlantique
Djibouti	Societe des Telecommunications Internationales de Djibouti	state-owned			
DR Congo	OCPT Renatelsat	state-owned			
Equatorial-Guinea	Sociedad Anónima de Telecomunicaciones de la República de Guniea Ecuatorial (Getesa)	partly privatised	France Télécom (40%)		
Eritrea	Telecoomunication Services of Eritrea	state-owned			
Ethiopia	Ethiopian Telecommunications Corporation (ETC)	state-owned			
Gabon	Gabon Télécom	state-owned planned privatisation			
Gambia	Gamtel	state-owned			

	PTO	Status	Investor	SNO	Investor
Ghana	Ghana Telecom	partly privatised 1996	Telecom Malaysia (30%)	Westel 2000	Ghana National Petroleum Corporation, Adesemi
Guinea	Sotelgui	privatised 1996	Telecom Malaysia (60%)		
Guinea Bissau	Guiné Telecom	privatised 1989	Portugal Telecom (51%)		
Kenya	Telkom Kenya	state-owned privatisation postponed		bids invited	prequalified: China Network, Telenor
Lesotho	Lesotho Telecom	privatised 2000	Econet (70%)		
Liberia	Liberia Telecommunications Corporation	state-owned			
Madagascar	Telecom Malagasy (Telma)	partly privatised 1995	France Télécom (34%)		
Malawi	Malawi Telecommunications	state-owned ongoing privatisation (30%)	Prequalified: Press (local consortium), CDC (Commonwealth), Detecon		
Mali	Sotelma	state-owned planned privatisation			
Mauritania	Mauritel	partly privatised 2001	Maroc Telecom		
Mauritius	Mauritius Telecom	partly privatised 2000	France Télécom (40%)		
Mozambique	Telecomunicações de Moçambique	state-owned			
Namibia	Telecom Namibia	state-owned			
Niger	Société Nationale des Télécommunications Sonitel	partly privatised 2001	ZTE / China RAILCOM / LAAICO		
Nigeria	Nitel	state owned		Globacom 2002	Coneptro (local oil company)
Rwanda	Rwandatel	partly privatised 1998	MTN 31% 1998 +9% 2003		
São Tomé und Príncipe	Companhia Santomense de Telecomunicações	Privatised 1989	Portugal Telecom (51%)		
Senegal	Sonatel	partly privatised 1997	France Télécom (61%)		
Seychelles	Cable & Wireless	privatised	Cable & Wireless (100%)	Airtel	Bharti Gruppe (Indien)
Sierra Leone	Sierratel	state-owned			

	PTO	Status	Investor	SNO	Investor
Somalia			Saudi Telecom Company (STC) Telecom Somalia Somatel		
South Africa	Telkom South Africa	partly privatised 1997	SBC Communications Telecom Malaysia (30%)	In preparation	Prequalified: Two Consortium (Swedtel/Telenor) CommuniTel (T-Systems, Telecom Namibia), Transtel, Eskom, Nexus Connection
Sudan	Sudatel	state-owned			
Swaziland	Swaziland Posts and Telecommunication Corporation	state-owned			
Tanzania	Tanzania Telecommunications Company Ltd, TTCL	partly privatised 2000	MSI Detecon (35%)	In preparation	Zantel announced to be likely SNO
Togo	Togo Télécom	state-owned			
Uganda	Uganda Telecom	partly privatised 2000	Telecel Detecon (51%)	MTN Uganda 1998	MTN
Zambia	Zambia Telecommunications Company	state-owned			
Zimbabwe	Tel One	state-owned privatisation postponed		TeleAccess 2003	

Source: ITU (2004b), MT, ARB, ION, AA, BA (multiple issues).

Despite commencing liberalisation, the states usually maintained a majority share in their PTOs. Only the Seychelles have a fully privatised fixed-line network. However, the general tendency could not be reversed, all the more that at the end of the 1990s – the boom phase in international telecommunications – privatisation proceeds and revenues from licence fees reached maximum values and African governments until 1998 sold licences that added up to 1.3 billion US$ (c.f. Pyramid 2001: 3). The prospective cash flow also persuaded previously indecisive and hesitant governments. In 2000/2001 a second wave of privatisation commenced and six more PTOs in Mauritania, Niger, Mauritius, Tanzania, Uganda and Lesotho were privatised. The privatisation of Nigerias incumbent Nitel has been cancelled, so that advanced plans of an international consortium of Portugal Telecom, Korea Telecom and Deutsche Telekom subsidiary Detecon to take over 51 percent of the Nigerian PTO never materialised. The deal would have been the biggest telecom privatisation deal in sub Saharan Africa to that date, worth the record sum of 1.3 billion US$ (c.f. PWC 2001).

In contrast to basic patterns of the first wave, the second wave had some new qualities: Firstly the background of investors diversified. Apart from new investors from emerging countries like the Chinese ZTE[55] in Niger and Maroc Telecom in Mauritania, the German Detecon gained a significant profile in the African Telecommunication market with investments in Nigeria, Uganda, Tanzania and Mozambique.[56] Similar applied to Portugal Telecom that started to change its strategy from merely administering PTOs in its former colonies to transform them into responsive market players. The second distinction relates to the increasingly successful mobile operations that also started to invest in fixed-line operations, often as part of a consortium like Telecel in Uganda, MSI in Tanzania or ZTE in Niger, rarely also as sole investor like Econet in Lesotho or MTN in Rwanda. Thereby, investments hardly targeted the fixed-line network, but the mobile licence that was usually attached to the PTO privatisation package. Investing in a PTO was seen as an appropriate strategy to get a mobile telephone licence in an increasingly competitive environment for mobile telephony.

Recent tendencies in fixed-line operations vary. Even in monopoly markets PTOs have commercialised their operations and have introduced a more market oriented business approach. Furthermore, PTO privatisations have become an accepted strategy of telecommunication development. With the exception of Kenya and Zimbabwe where privatisations were due to political rejections recurrently postponed, all economically important African countries have privatised their incumbent. In due course also smaller markets were seized by the general trend, so that at the end of 2003, 19 former public monopolies had been privatised or partly privatised. Even the Ethiopian ETC has announced privatisation plans and further transactions are in preparation or close to finalisation in Burkina Faso, Mali, Malawi, Gabon, Burundi (see table 3.2).

Further steps of liberalisation, however, have gained ground only rarely, so that monopolies – public or private – have dominated the scene. The introduction of a second national operator (SNO) to strengthen the competitive environment for fixed line telephony has been rather exceptional, though a first trend in this direction has been traceable. Uganda and Ghana introduced an SNO in 1996, a special case has been the quasi autonomous Zanzibar where Zantel was installed

---

[55] Over the past years, ZTE has established a predominant position in North Africa – in 2003 the company was selected to build Africa's largest CDMA (3rd generation wireless) network in Algeria. ZTE has also built large-scale GSM networks in Nigeria and Ethiopia (BA 226/2004).

[56] The investment in Mozambique has been a 26 percent share in the mobile operator M-Cell, which is affiliated to the PTO Telecomunicações de Moçambique (TDM), (CN 17.6.2002)

as a second operator on the Zanzibar islands.[57] In 2002, Globacom in Nigeria and Mundo Startel in Angola became SNOs. In 2003 TeleAccess started as SNO in Zimbabwe but failed to roll-out its project due to a lack of funds (The Independent [Harare] 17.9.2004) and the SNO process in Côte d'Ivoire, where Arobase Télécom was licenced as SNO in 2004, has been constricted due to political unrest. Tenders for a second operator are also on their way in Kenya and South Africa. But the same way as governments were not ready to give away all authority in their PTOs, they were neither willing to abstain from all their influence in approving an SNO licence. SNOs were only installed with a considerable public share in the new operator, often realised by large parastatal corporations like the Ghana National Petroleum Corporation in Ghana or Conpetro in Nigeria, which also came from the oil and petrochemical sector. Thus, the SNO process was less an attempt to seek an investor with a high profile in telecommunication, but more to seek sufficient national control of the SNO.

African fixed-line telephony has changed in the last ten years and today is characterised by a mix of commercialisation and competition on the one hand, and adherence to the old monopoly structure on the other. Commercial, but non-competitive fixed line structures of partly privatised monopolies are dominating the scene. Thus, removing imprints from the past and building a competitive advantage in a climate of intensifying competition are the main challenges for fixed-line telephone operators. Main competitors will be mobile operators. Already in 2001, mobile density surpassed the fixed-line density and mobile telephony has developed rapidly ever since.

3.5. Mobile Cellular Developments

Mobile telephony has been an impressive African success story. Only in 1998 there were just two million mobile subscribers on the continent. At the end of 2003, the number has risen to roughly 51 million subscribers (ITU 2004). Even on a global scale, with regard to subscriber numbers, Africa has become the fastest growing market in the world. Main engines in this dynamic are South Africa, today's largest market on the continent with 18 million subscribers and a 71 percent mobile coverage,[58] and Nigeria, the today most dynamic market, in which mobile telephony (GSM) started only in 2001 and subscriber numbers grew to 1.3 million in 2002 and 3.1 million in 2003 (c.f. ITU 2004: 3). However,

[57] Zantel will become the SNO in Tanzania in early 2005, when the exclusivity period granted to the incumbent TTCL will end.

[58] Figures for June 2004 taken from the South African information service Cellular Online, www.cellular.co.za.

mobile telephony in Africa was no automatic success, but developed gradually due to a mix of several factors, such as technical innovations, market liberalisation, and private sector initiative that together built the fundament of mobile telephony in Africa.

First generation mobile networks, based on analogous technology, have been deployed since the beginning of the 1990s, in some cases, like Gabon or Mauritius, even at the end of the 1980s. But penetration rates remained low, because coverage was limited and costs high. This started to change with the introduction of digital or second generation networks based on GSM standards (= Global System for Mobile Communication[59]). In Africa, GSM networks were first installed in Senegal and Cameroon in 1994 and Malawi, Namibia and Uganda in 1995. Eight more countries followed in 1996 (ITU 1998: 5), but the technology remained limited in scope. In order to understand the commencing success story, it makes sense to make two basic distinctions: firstly with regard to different mobile operators pushing the market and secondly to differentiate between three phases of mobile telephony development. To start with the first distinction, in the beginning of liberalised cellular markets, the following companies can be labelled as pioneers of mobile telephony in Africa (c.f. ITU 2002a: 16, ITU 1999a: 64 Pyramid 2001: 15):

(1) *Developing country and emerging market specialists* without a traditional home market. This group comprised of the Luxembourg based Millicom International Cellular (MIC) with mobile operations in Asia, Latin America and Africa, and Telecel international, which later got insolvent, but, at that time, was part of the Egyptian Orascom Group. Telecel International operated in 13 African countries in 2001. The third group representative was the Dutch Mobile Systems International (MSI), later renamed in Celtel International, that pursued a pan African growth strategy.

(2) *Expanding mobile telephone companies from developing countries,* which had started business in their original home market and expanded their business into other developing countries. This group was made of the globally operating Telecom Malaysia or the TRI Group originating from Malaysia, too, with operations in Asia and Africa, both pursuing a South-South strategy. Furthermore, two African companies fell under this category, Mobile Telephone Networks (MTN) from South Africa and

[59] Originally, back in 1982, GSM stood for Groupe Speciale Mobile, only since 1987 it was changed to Global System for Mobil Communication – though the acronym remained the same.

Zimbabwean Econet Wireless. Both started as national companies and developed into pan African players.

(3) *International mobile telephone operators*. This included the South African Vodacom, today the largest operator in Africa. Vodacom is a joint venture between Telkom South Africa (51%) and the global cellular phone market leader Vodafone (35%).[60] In the early phase the French conglomerate, Vivendi also belonged to this group, but later withdrew from Africa.

(4) *Traditional telephone companies from developed countries*, which subsequently developed into global operators. This group covered companies like France Télécom, which has operated under the brand Orange, and, with some restrictions, Portugal Telecom and the Deutsche Telekom subsidiary Detecon, but both generally focused on the fixed-line market. Until recently, the Swedish Telia had some operations in Africa, but refocused their business on the Baltic region.

In a second distinction, the African mobile telephone market can be divided chronologically into the three phases of (i) entry, (ii) development and (iii) consolidation:

- The *entry phase* started in the early 1990s and lasted until 1998. Commencing sector reforms included the application for new mobile telephony licences and the first pioneers of the African cell phone market emerged, particularly the developing country specialists Millicom and Telecel. At the same time, the first PTO privatisations brought a second opportunity to enter into the mobile telephone market, because usually a mobile licence was attached to the operations of the incumbent to make the monopoly fixed-line operator a lucrative and promising investment destination. France Télécom started its African mobile telephone strategy with investments in fixed-line operators. However, in retrospect, there had not been a distinctive investment pattern. Companies invested wherever an opportunity arose. With the exception of France Télécom that focused on colonial ties and Vodacom applying a South African centred strategy, the early mobile operators' investment map had been patchy.
- The *development phase* lasted from 1999 to 2001. In this phase distinctive investment patterns became visible, mainly as a result of new tenders for

[60] Therewith, Vodafone also shared some preferences with the expanding companies from developing countries, because Vodacom only operates in Africa and developed its business from its home market. Furthermore, Telecom Malaysia is shareholder in Vodacom's main shareholder Telkom South Africa, so that a clear distinction faded away.

mobile licences which brought new leeway for the operators to build their profile. The phase also coincided with the commencing boom in mobile telephony, in which the operators could gradually develop pan African strategies. Early adopters were MSI, Telecel and MTN, who together built a presence in 22 African countries (c.f. ITU 2002a: 16).

- The *consolidation phase* set off in 2002. The investment patterns stabilised and due to strategic expansions, a group of seven main operators emerged. These were Vodacom, Celtel International (formerly MSI), MTN, Orange, Millicom, Econet and, with some restriction, Telecel. The profiles of these operators are given in table 3.4.

Table 3.4: Profile of main African Mobile Telephone Operators (2004)

Operator	Vodacom South Africa		Celtel International (formerly MSI)		MTN – Mobile Telephone Networks		MIC – Millicom International		Orange		Telecel International		Econet Wireless	
URL	www.vodacom.co.za		www.msi-cellular.com		www.mtngroup.com		www.millicom.com		www.orange.com		www.orascom-telecom.com		www.econetwireless.com	
Share-holders	Telecom South Africa (50%), Vodafone (31,5%)		Widespread shareholdings		M-Cell South Africa (100%)		Widespread shareholdings		France Telecom (100%)		Orscom Telecom (100%) (liquidated in 2003)		Altech	
Opera-tions / Sub-scribers	South Africa	7.874.000	Tanza-nia	140.000	South Africa	6.050.000	Tanza-nia	168.800	Bots-wana	150.000	Benin (sold off 2003)		South Africa	
	Tanza-nia	541.285	Burkina Faso	107.000	Came-roon	538.000	Ghana	117.800	Came-roon	500.000	Burkina Faso (sold off 2003)		Simbab-we	153.167
	Lesotho	77.474	Chad	27.000	Nigeria	1.650.000	Sene-gal	206.500	Senegal	576.000	Burundi (sold off 2003)		Nigeria	900.000
	DR Congo	600.000	Congo	180.000	Rwanda	134.000	Mauri-tius	136.600	Côte d'Ivoire	600.000	CAR (sold off 2003)		Lesotho	20.000
	Mozam-bique	100.000 (since 2004)	DR Congo	700.000	Uganda	468.000	Sierra Leone	31.700	Mali	206.300	DR Congo (sold off)		Botswa-na	310.000
	Failed acquisi-tion of Econet Nigeria early 2004		Gabon	120.000	Swazi-land	88.000			Mauriti-us	326.000	Côte d'Ivoire (sold off 2003)		Kenya	Start in 2004
			Malawi	45.000					Mada-gascar	140.000	Chad	599.000		
	Kenya (Voda-fone shares in Safari-com)	750.000	Niger	31.000							Gabun (sold off 2003)			

Operator	Vodacom South Africa		Celtel International (formerly MSI)		MTN – Mobile Telephone Networks		MIC – Millicom International		Orange		Telecel International		Econet Wireless	
			Sierra Leone	58.000							Congo			
			Sudan	255.000							Uganda (sold off 2003)			
			Uganda	61.000							Zambia (sold off 2003)			
			Zambia	92.000							Zimbabwe			
			Kenya								Togo (sold off 2003)			
Brands	Vodacom, Safaricom (Kenya)		Celtel, Mobitel (Sudan) Kencell (Kenya)		MTN		Mobitel (Tanzania, Ghana), Millicom (Sierra Leone), Sentel (Senegal), Emtel (Mauritius)		Orange Sonatel (Senegal) Ikatel (Mali) Cellplus (Mauritius)		Telecel		Econet	
Source: Company websites, MT, ARB, ION, AA, BA, CN (multiple issues)														

The gradual move to market consolidation was the result of different corporate strategies. A typical expansion pattern has been the development of some core operations in advanced reform countries and from there expanding into lucrative markets with different risk readiness according to the operator. Thereby, basic considerations were related to the size of the market and weak fixed-line structures (c.f. ARB 2002: 15230). Vodacom predominantly applied this model strategy.[61] The South African company only expanded slowly, firstly developed its home market South Africa and adopted a policy of wait and see. The company entered fewer, but densely populated markets starting with Tanzania as a test case for suitable development strategies in a fairly stable environment and subsequently entering Mozambique and DR Congo, in which particularly the latter has been regarded as a potential, but much riskier endeavour.[62] The attempt to move into Nigeria, the second most important market after South Africa, however, could not be realised, despite several attempts to do so.[63] Celtel International made a mark with investments in smaller and riskier markets.[64] MTN focused on selected prime target markets like South Africa and Nigeria. All in all, the environment for mobile telephony became more competitive and inevitably smaller operators got under pressure. Econet Kenya almost failed to start operations in Kenya due to capital shortage (ION No. 1104) and could only start after a significant US$ 70 million capital injection by South African Allied Technology (Altech) in the mother company Econet Wireless.[65] Tritel Tanzania ceased operations completely. The largest liquidation, however, was related to Telecel, a full subsidiary of the Egyptian Orascom group. Due to its many and often unfocused operations, Telecel accumulated high debts and individual operations were sold off. Only Chad, Congo and Zimbabwe remained under the Telecel roof, all others were taken over by local rescue companies (ITU 2004: 7,

---

[61] Via its mother company, Vodafone, it has an indirect presence in Kenya, too.

[62] Apart from applying for a new license, typical market entrance patterns included the acquisition of smaller operators. For example Vodacom started operations in the DR Congo by entering a 51 % Joint Venture with a local operator Congolese Wireless Networks (CNW) – who had at the time of acquisition in December 2001 20,000 subscribers and a 20-year licence (ARB 2001: 15048).

[63] Vodacom's 600Mio. US$ acquisition of Econet Nigeria was already finalised in October 2003, but, after all, failed due to interventions of the parent company Econet Wireless International in mid 2004 who objected to the acquisition (ARB 2004: 16126, AA 2004 No. 449, East African [Nairobi] 28.6.2004).

[64] After numerous unsuccessful attempts, Celtel managed to enter the Kenyan market, finally. The company took over Kenya's Kencell, recently, for a total of US$ 250 million (ION No. 1090/2004, AA No. 449/2004).

[65] Therewith Altech also took over the majority shares in Econet Wireless (AA No. 440/2004, MT 2004: 211, BA 228/2004).

AA 422/2003, ARB 2003: 15657). These developments show that the consolidation in the African mobile market is not yet finalised. New tenders are expected in Angola and Liberia, however new growth opportunities will particularly derive from future privatisations and possible takeovers, particularly from the still numerous smaller and independently operating mobile telephone operators. Thereby, Vodacom, MTN, Orange and Celtel at present act from a stronger position, while Millicom, Econet and the former Telecel operations from a weaker one.

The history of African mobile telephony has been, above all, a history of successful companies that capitalised on open markets created by politics. The success of the companies, in turn, based on a further factor of success: the introduction of prepaid cards. Prepaid solutions quickly became accepted due to their functionality and their cost structure, so that prepaid mobile telephony could develop into a mass product. Previous strategies of mobile operators to attract only the high end postpaid, customer segment failed and remained an absolutely niche market. Today, 80 – 98 percent of national cellular markets are covered by prepaid (c.f. ITU 2004, Vodacom 2004). The concept was first successfully introduced in South Africa in 1999. Cellular telephones were sold without contract and the customer bought airtime for telephone calls in advance. Though operators praised the advantage of prepaid with the argument that customers had full control over their telephone costs, the true beneficiaries have been the operators, themselves, because prepaid prevented bad debts and upfront cash flow could easily be reinvested. Recharge cards which had a minimum denomination of 20 US$ in the early times, went down to denominations of two US$,[66] increasingly available in local currency and usually include one month of airtime, i.e. readiness to receive calls. The broader population only became involved as recharge cards denominations decreased significantly.

To assess the performance of mobile operations, the number of subscribers is usually put in relation with the average revenue per user (ARPU). On average, ARPU in Africa has been 27 US$ (ITU 2004). However, regional figures have varied considerably, e.g. for MTN South Africa the figure is 30 US$, for Vodacom Tanzania 18.5 US$ and for Vodacom DR Congo 20 US$ (Vodacom 2004, CN 9.2.04). The young Nigerian market, in contrast, has an ARPU of relativly high 55 US$, because new operations tend to attract the wealthier customer segments first. South Africa and Uganda, which can be regarded as consolidated markets, are well within the African average of 24 and 23 US$ (ITU 2004). In comparison to the more advanced markets in Egypt or India,

[66] Kenyan Kencell has introduced the lowest denomination. Since August 2004 recharge cards have been available for one US$ (East African [Nairobi] 9.8.2004).

which have an ARPU of 17 and 11 US$, and thus have a higher proportion of low usage subscribers, average Africa still has more room for mobile telephony expansion. At the same time, costs for infrastructure deployment decreased due to savings from economies of scale. In addition, infrastructure builders like Nokia, ZTE and Interwave Communications increasingly offer low cost and light GSM components and, thus, have become serious competitors to the well established market leaders Siemens, Alcatel and Motorolla. The overall decreasing cost ratio for mobile telephony and the high number of potential subscribers are a clear indicator, that the full potential of the African cellular market is not yet achieved.[67]

From today's point of view, the market for cellular services has been much bigger than previously expected. A mix of liberal policies, good company performances, technical innovations and the strength of prepaid services have extended the market reach of mobile telephony. Costs have decreased and increasingly cellular telephone usage has decoupled from formerly typical features like income and an urban bias.[68] This also means that other factors will become more relevant. Foremost, speed has been the name of the game. The faster a network could be established the better the operator's competitive advantage. Additional factors like service quality, customer care, product innovations such as per second billing and the like would give an additional competitive edge. Beyond the rules applicable today, the increasing convergence of mobile and fixed line telephony, whose impact is hardly foreseeable, will also determine the way forward in African telecommunications markets.

---

[67] This applies more because, in contrast to other mobile markets, Africa has been unencumbered from high debts of third generation mobile licences and thus belonged to the few profitable mobile operations. For Africa, mobile technology will continue in the foreseeable future to predominantly rely on voice services. Nevertheless, the upgrading of GSM, so that it will also be capable of data services will be a challenge for African operators, too. Some commercial GPRS networks (2.5 generation networks) have been installed in South Africa by MTN and Vodacom. Also, a few third generation CDMA networks (UMTS) have been deployed, but mainly for fixed-line wireless operations in urban centres in Angola, DR Congo, Mauritius or Zambia (ITU 2004, ITU, 2003).

[68] It is also noteworthy that actual spending on communication or surveys to assess the willingness to pay for communications services indicate that the de facto income situation is much better than given in, for example, World Bank statistics (BA 206/2004). In a similar way the correlation between urbanity and mobile telephone use is losing its former significance. This means that mobile telephone services have been increasingly used by those, whose home settlement is not covered by a mobile network, but who are willing to walk or take public transport to the next telephone facility (BA 206/2004; Apoyo 2003).

3.6. Internet Developments

Internet developments have been notably less dynamic than those in the mobile sector. Nevertheless, the sector witnessed some stable growth, too. First predecessors of the Internet as it is known today, had been deployed since the late 1980s and based on simple store and forward email only systems. The most common systems were FidoNet and UCCP. The first FidoNet installation in Africa was set up in Kenya in 1989 and until 1992, all in all, 16 African Fido nodes were in operation. Therewith, FidoNet had been the pioneer of electronic networking in Africa. Particularly between 1990 and 1997 it developed into a cost-effective and efficient tool of data communication and was especially utilised among the research and the NGO community. Whereas the first Fido nodes connected to Fido's main hub in London (see chapter 4.1.2.3., box 4.2), the intra-African networks expanded gradually, too. ESAnet (Eastern and Southern African Network) was the first African network. It was built in 1991 as a research network of five universities in Uganda, Tanzania, Kenya, Zambia and Zimbabwe. It later merged with Healthnet, a health information portal run under the aegis of the US-American NGO Satellife in close collaboration with the same five universities (c.f. Levey / Young 2002).

At the same time as electronic networking advanced due to the information need of research institutions and NGOs, the public telephone companies subsequently started to modernise the African telecommunication infrastructure. In 1993 South Africa had been the first African country that had established a commercial permanent Internet connection. In 1994 Zambia and Mozambique followed and in 1995 eight more countries permanently connected to the Internet. Until 1998 all African countries, except Eritrea and Somalia, had been connected and since March 2000, when Eritrea hooked up to the Internet, all countries maintained a permanent connection – notwithstanding the occurrence of temporary outages (c.f. Jensen 2001). For the time being, the technology was limited to the main cities, only gradually including the secondary cities, too. However, broad Internet development strategies to move beyond the secondary cities like the introduction of national access numbers or within universal access strategies hardly emerged (Jensen 1998). Only the rapid increase of public Internet access facilities, particularly small scale Internet cafes induced a widespread and sustaining Internet dynamic. Public access points emerged in many variations ranging from small private telephone kiosks, Internet cafes, franchised telephone shops to large scale donor financed telecentres offering a variety of services from Internet to training.

In retrospect, Internet diffusion in Africa had a clear pattern. South Africa had been the engine and the neighbouring states with some minimum stability benefited from a vibrant South Africa. With only a few exceptions, e.g. Senegal in Francophone Africa, this had been the prevalent pattern until 1998. Namibia, Botswana or Zimbabwe developed in a South African slipstream (see figure 3.1). Until 2001 South African Internet user numbers constantly accounted for half of all African Internet users, for 2001 e.g. 250,000, and the bordering countries took advantage from South African infrastructure and first expansions of South African IT companies in neighbouring states and proximity markets. Though this pattern has been valid until today – geographic proximity to South Africa is still one of the most benefiting factors of promising Internet development – it is not any more the exclusive configuration for Internet success. Overall increasing user numbers and clearly diversified usage patterns revealed that new factors determined the pace and scope of Internet diffusion in African countries, too. Consequently, South Africa today just stands for only one quarter of all African Internet users (ITU 2004). Evidently, the other countries caught up and coped increasingly well with the challenges of Internet development including hardware and software, connectivity and human capacities that all together have considerably burdened countries to make use of the Internet effectively and efficiently. Due to this complexity there is hardly a uniform pattern of Internet diffusion traceable. However, factors like high national income levels, small geographical size, access to the sea, good politics and potential markets and a competitive environment seem to be in favour of good Internet developments.[69]

The example of Internet service providers (ISP) might help to show some typical features of the Internet. The public telephone operators were the first larger ISPs. Apart from universities, which often restricted services to members of the academia, PTOs were often the only providers of public Internet services between 1995 and 1998. Commercial ISPs could only develop in open markets and, because initial investments needed to start an Internet business were relatively low, smaller companies could also enter the market – often as spin offs of universities or from the universities' environment. However, the business was cumbersome and margins low, so that the ISP segment market never became as attractive as the mobile segment and international or pan African ISPs analogous to mobile telephone companies hardly emerged. The history of ISPs has predominantly been national histories of small and medium companies. Only three ISPs have built an interregional profile. Of these, Africa Online has been the biggest endeavour with branches in Côte d'Ivoire, Ghana, Kenya, Namibia, Swaziland, Tanzania, Uganda and Zimbabwe. The second example has been the South African M-Web that has operations in South Africa, Namibia, Zimbabwe and Nigeria, as well as shares in overseas operations in

[69] See also the tables given in appendix 1.

Thailand und China. The third company Mediapost, also from South Africa, has been active in South Africa, Rwanda, Congo and Tanzania and is expected to expand to Kenya, Uganda, Senegal, Nigeria, Cameroon, Malawi and Zambia in the near future (Business Times [Dar es Salaam] 3.5.2002). Other attempts to build an extended Internet presence failed like in the case of UUNET, who merged with Africa Online or Swift Global who retreated into their Kenyan home market (c.f. Nielinger 2001: 59). On the national level, however, a vibrant mix of generally locally rooted ISPs emerged.

Figure 3.1.: African Internet Density 1997 (per 10,000) and 2003 (per 100)

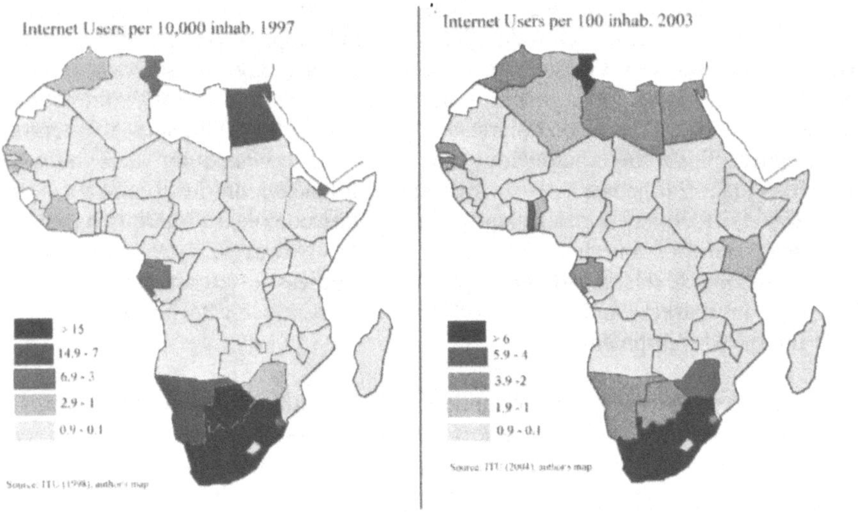

Costs for Internet connectivity soon materialised as one major problem for ISPs and other data communication providers. Costs accounted for up to 50 percent of ISPs overall revenue to connect the Internet (Jensen 2000). In countries like Nigeria, Tanzania or Kenya monthly connection fees exceeded the 10,000 US$ mark in 1998 and only very gradually, coinciding with commencing sector reforms, the prices started to decrease. New technological solutions also contributed to this trend. Though the price level was still high, the times of exorbitant connectivity fees had come to an end. Open markets and technological innovations, together with a high demand for Internet services induced a stable growth of the Internet in Africa, in which approximately 14 million Africans make regular use of the Internet and the trend continues to rise.

New opportunities for scalable and cost-effective networks rise from new technologies that offer promising solutions to replace costly cable solutions for the so-called last mile of connectivity, like 30-mile range Wireless Metropolitan Area Networks (WMAN) that have already been developed into marketable commodities. Furthermore, the international links to the Internet have improved considerably in recent times with according effects on costs and the likelihood that these will decline further. In the past, international connectivity was provided by the national telephone operators which got their bandwidth, apart from very few exceptions, over satellite: mainly via Intelsat, in some cases via Arabsat (Djibouti, Mauritania) and Intersputnik (Madagascar) (CIA Factbook 2003). Internet connections via satellite have been a booming business, and providers like Intelsat , Panamasat, SES-Global, Eutelsat and New Skies have launched or will launch new satellites to increase their capacities to meet the demand (c.f. Imfundo 2003: 133, Moroney/Hamilton 2004).[70] New liberal environments have ended the PTO monopolies for international traffic and new providers have broadened the supply, and, thus, a number of new carriers apart from the PTOs were able to deliver more cost-based direct satellite links to the Internet (Imfundo: 2003: 56).[71] It is also noteworthy that a number of African satellite providers, resalers of capacities provided by satellite operators, have been able to build a strong presence in this market segment, among them Gateway Communications, Afsat Communications, SatCom, Networks Africa, Paconet, ComCarrier or Alledan (ARB 2004: 16009).

Furthermore, the implementation of a fibre project, formerly known as Africa One targeting a fibre completely encircling the continent, has additionally improved the African ICT infrastructure. Since 2001, some parts of the project in Southern and Western Africa have started operations. In 2003 the complete west surroundings have been installed and connected to the individual project components SAT-3 / WASC / Safe comprising 28.000km fibre linking Europe with West and South Africa and Asia. Project shareholders come from, among others, Senegal, Ghana, Benin, Côte d'Ivoire, Nigeria, Cameroon, Gabon, Angola, South Africa and Mauritius (ARB 2003: 15675, MT 2002: 2560). These components have already led to a notable price reduction in the provision of bandwidth. The lacking Eastern route of the fibre, that would close the gap from South Africa to Djibouti is in advanced planning. A memorandum of understanding for the realisation of the US$ 230 million East African Submarine System (EASSY)[72] had been signed by the PTOs from Botswana, Kenya,

---

[70] For a map on satellite access in Africa, see IDRC / GVF (2003).

[71] Further cost relief has been due to the growing number of Internet exchange points in Africa to facilitate direct peering between African ISPs – to avoid the costly satellites routes via a downlink in the US or Europe if one African ISP communicates with another African ISP.

[72] A map of EASSY can be found in chapter 4.3.5.3.

Mozambique and Rwanda. Zantel from Zanzibar and MTN-Uganda had been non PTO signatories of the agreement (MT 2004: 376, ARB 2004: 15935).[73] But even without EASSY, according to Telegeography (2004), cable based Internet capacities have increased to a significant 75 percent in 2003. Even though this has been less than in Asia or Latin America, the short supply of bandwidth and international connectivity, which have been the largest cost factors for Internet operations, is targeted and numerous individual projects will alleviate Internet businesses in the near future.

3.7. Concluding Remarks

The landscape of telecommunications has changed radically in the last ten years. Technical innovations, liberal policies and committed entrepreneurs have been the corner pillars of dynamic ICT developments. These positively affected the individual segments – fixed-line telephony, mobile telephony and data communications – that have emerged as more or less vibrant economical sectors. The developments also had a positive impact on numerous downstream markets that benefited from new ICT applications. Tools of modern communication came into the reach of many Africans for the first time and new technologies have been increasingly utilised in many areas, such as governance, health, education, private sector or civil society. Political and economic actions taken by governments and broad private sector participation led to the advancement of ICT and facilitated ICT for development applications.

However, ICT diffusion rates and the ability to utilise technologies have not been evenly distributed and sectoral developments were neither linear nor always consistent. A detailed cost benefit analysis of African telecommunications is still missing; so far studies can only deliver some rough approximations of the net costs and the impact of ICT. Nevertheless, two major tendencies can be deduced: Firstly, telephone services grew considerably more dynamic than advanced and value added services. Especially mobile telephony due to appropriate products, a favourable prepaid cost structure and an increasing coverage quickly became accepted and marked the success story in African telecommunications. In contrast, the Internet has grown less dynamic and usage has been mainly restricted to urban areas. The second observation can be made with regard to competition. Those countries that opened up their economy and introduced some level of competition have performed significantly better than those countries that stick to the monopoly status. Thus a first

[73] Connecting to EASSY, mobile operators from Kenya, Uganda and Rwanda also plan to build a fibre-optic link from Kigali via Kampala and Nairobi to Mombassa (ARB 2004: 16200).

approximation to assess ICT performance records in Africa would include a distinction between reform protagonists and opponents that, to a large extent, can explain the progress made in the telecommunications sector. A second step, however, that would take into consideration the increasing number of reform countries, would also include the quality and scope of the reform process, because often reforms were ostensible, ineffectual or inappropriate. Particularly the mixed record of fixed-line developments pointed towards an unfinished business of reforms in many countries and also highlighted the fact that reforming segments embedded in old political and economical structures has been considerably more challenging than introducing new structures to new businesses like mobile telephony or the Internet. However, even if sound reforms are on track, the existences of good policies do not provide sufficient explanations for a good Internet performance record. Other beneficial factors were market size, estimated potential of the market, but also factors relating to geography and demography.

Still in the early 1990s, African countries resisted reforms and objected to new development models. Only first successful model examples in some countries – Uganda, Ghana, Mauritius, Tanzania – leveraged a private sector led growth strategy and induced a process in which a broad reform agenda coming from the outside was subsequently adapted into an African agenda with many stakeholders and various positions. To the extent as private sector growth strategies became accepted, companies started to build a presence in stable reform countries where they gathered experience and substance, and, on this basis, subsequently also expanded to more risky countries dependent on the companies' expansion strategy. The individual target markets, in turn, were characterised by different political, economical and regulatory environments aiming to increase market efficiency to bring ICT opportunities within the reach of the broader population and in some cases also introduced additional targeting schemes like the establishment of universal access funds for the marginalized population to deliberately narrow the access gap between urban and rural or the low and high ends of the market. Though this general outline might mirror the basic tendency in African telecommunications, it also reveals a basic dilemma: If markets are intended to reach the broader population, they have to be poverty oriented markets, because otherwise they fail due to limited scope and scalability. Small markets characterised by prevalent poverty are difficult to serve, all the more that knowing how to make markets work for the poor[74] is still limited. In order to understand the challenges of upscaling telecommunications, the nex chapters will focus on the country case Tanzania

[74] This limited knowledge has been the reason why the World Bank devoted the 2004 world development report to this issue, see World Bank (2004a).

4. The Case of Tanzania: Creating an Environment for ICT – Policy, Regulation, Markets and Education

This chapter is about creating a sound environment for information and communication technologies (ICT) in Tanzania. It is not about ICT applications and their impact in development, but solely focus on those factors identified as being key enablers for increased ICT diffusion – policy, regulation, markets and education. The country case Tanzania is at the centre of this analysis, because the East African country is very well suitable for a more detailed discussion for several reasons: it has not only been a stable and peaceful environment[75] with a considerable record of economic reforms that create a good starting point for the analysis, Tanzania was also among the first African countries that was committed to telecommunication sector reform and ever since has conducted a proactive ICT promoting policy. Thus the article highlights the role of policy, the creation of efficient institutions and the role of different players to provide efficient market mechanisms in a given country context, covering a time span between the early days of informatics in the 1960s and 1970 and today. It analyses the viability of the approach, discusses the potential scope of a private sector led strategy and also seeks to identify its limitations, in which politics needs to go beyond markets to a more inclusive set of broader development objectives.

For a better orientation, the Tanzanian ICT history is divided into three phases. The year 1994 was a turning point of telecommunications in Tanzania. After the socialist period between 1961 and 1985 the Tanzanian government adopted a new development model. A comprehensive Economic Recovery Programme (ERP) was introduced in 1986, followed by an Economic and Social Action Programme (ESAP) in 1989, both based on the strategy to shift from a centrally planned system to a market-oriented economy.[76] Beyond general policies applied under the structural adjustment programmes, the government also declared the rehabilitation of key infrastructure components a priority.[77] In this context, the telecommunications sector became subject to reform since 1992 and

[75] This applies particularly if compared to other African countries. However, it should be noted that Tanzania is not conflict free, and in some areas conflict intensity, in fact has grown in recent years; not only on Zanzibar and Pemba, but also on the mainland, c.f. Hirschler (2004).

[76] An analysis of the economic transformation process is given in Wobst (2001: 7-24).

[77] Initially, infrastructural rehabilitation efforts focused on transportation facilities like roads, railways and ports. It was not before the economy responded positively to the initiated reform steps that the telecommunication sector was targeted (c.f. Wobst 2001: 12).

respective legislation took effect in 1994, both considerably pushed by the World Bank and the International Monetary Fund. However, reforms and adjustment processes were not lineal, but consistently monitored and readjusted. One of the main reform readjustments appeared in 1999, when the regulative set-up was modified and the process to privatise the public telephone operator TTCL was accelerated. On the basis of these landmarks, the following phases of ICT in Tanzania can be identified:

- *Recognition* (until 1993): Introduction of telecommunication (1951) and informatics (1965), rigid framework with some minor sectoral growth;
- *Orientation* (1994 – 1999): Commencement of market reforms; separation of operative and regulative responsibilities, liberalisation in various market segments;
- *Departure* (since 2000): Increased proactive sectoral policy, improved profile of the regulative authority and a noticeable diversification in the market.

These phases comply with the basic structure of the chapter. It is organised in chronological blocs, in which policy, regulation and market developments are assessed for the given period. The arrangement allows both, to identify the main developments in each of the analysed areas and also to show the interlinkages. This is clearly advantageous in contrast to a strict sectoral assessment, in which policy, regulation and market developments would have been analysed one after the other. That would have inclined an underestimation of increasing interlinkages (or would have been at the expense of a coherent reflection). The phase structure shall make it easier to understand the reciprocity of ICT development in Tanzania. It is the objective to show the dynamics, the cause and the impact of increasing sectoral differentiation. It will be demonstrated how market players have reacted to policy and regulation, and how politics as a facilitator of a sound business environment has managed to cope with sectoral challenges, adjusted to new conditions and gained new functionality with regard to both: reactive capabilities to adjust to maturing market conditions and proactive to provide orientation and guidance to all actors involved. Hence, basic tendencies of policy, regulation and market developments are given for each phase trying to identify strengths and weaknesses of each phase and also draw upon other factors that have positively or negatively influenced ICT development in Tanzania.

4.1. The First Phase until 1993: Recognition

4.1.1. Telecom until 1993

Telecommunication is much older than just recent history. The early phase can be subdivided in three phases: (i) a precolonial period until 1961, (ii) a postcolonial period mainly determined by role of the first East African Community (EAC) as joint infrastructure provider for the three member states Tanzania, Kenya and Uganda until 1977 and (iii) a national period covering the creation of a national Tanzanian telephone operator after the collapse the EAC. As figure 4.1 shows, the development of the telecommunication sector corresponded to the major changes in the political structure.

Telecommunication arrived in Tanzania with the beginning of colonialism. When colonial administration, settlers, trading companies and missionaries set up their centres, they also started to build first rudimentary postal, telegraph and telephone networks that followed the colonial patterns. But only 1951, with the creation of the East African Posts and Telecommunications Administration (EAPTA), service provision was put under an administrative roof. As an instrument of colonialism and corresponding to British colonial territory, EAPTA provided postal and telecommunications services to mainland Tanzania, the former Tanganyika, Kenya and Uganda (TCC 2001b: 5). In these early days the telecommunication network was restricted to the absolute minimum. Just before the arrival of independence in 1961 the network covering the then Tanganyika territory deployed a modest quantity of 66 telephone exchange lines. A number that equalled a teledensitiy of 0.10 telephones per 100 population (see table 4.1).

Figure 4.1: Schematic overview on telecommunication sector development 1951 - 1993

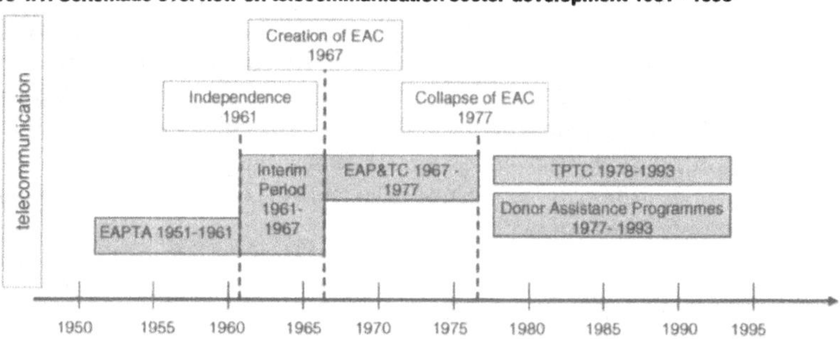

After independence, Tanganyika, Kenya and Uganda inherited the EAPTA, but the consecutive achievement of independence[78] and the turning away from the colonial past led to a suspension of most EAPTA's activities. For an interim period between 1961 and 1967 national telecommunications continued to rely on the leftovers of the colonial infrastructure and international communications for the three East African countries was provided by the UK based Cable & Wireless. It was only in course of the formation of the first East African Community (EAC) in 1967, that the old EAPTA structures had been revived. The remains of the old telephone operator were transformed into the East African Posts and Telecommunication Corporation (EAP&TC) serving as the PTO for the three EAC member countries. But also EAP&TC only had a short lifetime and performance indicators remained low, too. It lasted until the collapse of the first EAC in 1977. All in all the post colonial period between 1961 and 1978 resulted in a very modest increase of the existing network capacity to an overall 99 telephone exchange lines, whereas the low teledensity, due to the population growth rate, remained the same (see table 4.1).

Table 4.1: The state of telecommunication between 1961 and 1993

	1961	1977	1993
Population	**12,329,372**	**17,019,057**	**26,735,254**
Number of Exchanges	**66**	**99**	**230**
Total Number of Exchange Connections	**9,079**	**15,849**	**85,005**
Public Coin Telephone/ Payphones	**93**	**834**	**663**
Teledensity (Telephone per 100 population)	**0.10**	**0.10**	**0.32**

Source: TCC (2001b)

With the collapse of the EAC the Tanzanian holdings of the EAP&TC were subsequently transferred to the Tanzania Posts and Telecommunications Corporation (TPTC) that came into being in 1978 and took over the mandate to deliver postal and telecommunication services in Tanzania. The starting position for the first Tanzanian telephone operator was anything than easy, because the failure of EAP&TC led to a loss of several shared facilities, like the gateway for international telephone calls, accounts, supplies and training centres. Furthermore, a slow transfer of administrational procedures from the EAPTC headquarter in Kampala to the new Dar es Salaam based TPTC contributed to TPTC's humble performance.[79]

[78] Tanzania gained independence in 1961, Uganda in 1962 and Kenya in 1963.

[79] The main causes of the poor performance was due to (i) weak finance management as expressed by inadequate tariffs – local tariffs were very low (equivalent of US$ 0.02 per minute), while international tariffs were some of the highest in the world (equivalent of US$ 6.40 per minute) –, poor billing and collection performance; and poor audit controls, and (ii) a

In accordance to the then common approach to telecommunication, TPTC was not only the state owned telephone operator[80] it also ruled the sector. Accordingly and as illustrated in figure 4.2, the sectoral setting was very rigid. The Ministry of Communication and Transport (MoCT), the central government authority dealing with policy and regulation, and TPTC shared regulative competencies. The ministry provided general guidance, for e.g. investments or tariffs, and executive principles, such as service obligations and operative performance targets, inter alia for rural services, and TPTC concentrated on operational issues and related regulatory functions like the numbering administration, type approval of equipment or radio frequency spectrum management. However, neither the ministry nor TPTC fulfilled their mandate adequately. The ministry delayed necessary sectoral adjustments and TPTC struggled to provide the most basic services: waiting time for a telephone was up to 10 years and only 26% of all national phone calls could be completed without failures (c.f. TCC 2001b: 7).

Only as external and internal pressure to adjust the sectoral settings grew, the picture started to change. The sector became target of donor assistance programmes from 1977 onwards. Especially the Swedish International Development Agency (SIDA) and the World Bank started to support the expansion of the telecommunication network and until the early 1990s modest achievements had been made. A new telephone gateway to Tanzania, an earth satellite station for international calls, was set up, the number of telephone exchanges was raised to a total of 230, capable of 85,000 telephone connections, and teledensity increased to 0.32 telephones per 100 population (see table 4.1).

---

poor operational record such as unclear lines of authority; weak management information systems; inadequate training; lack of customer orientation; and poor staff compensation (c.f. UNECA 1999; TCC 2001b: 5).

[80] In addition, the military and other security services operated private telecommunication networks. In the the public domain, the Tanzania Electric Supply Company (TANESCO), SITA (provider for a airline reservation system), Tanzania Railways Corporation (TRC) and Tanzania Zambia Railways Authority (TAZARA) also operated private telefone networks. Especially TANESCO's network had a fairly good reputation based on more sophisticated traffic handling facilities than those available on the TTCL network (c.f. Luhanga 1994).

Figure 4.2: Tanzania Telecommunication Sector until 1993

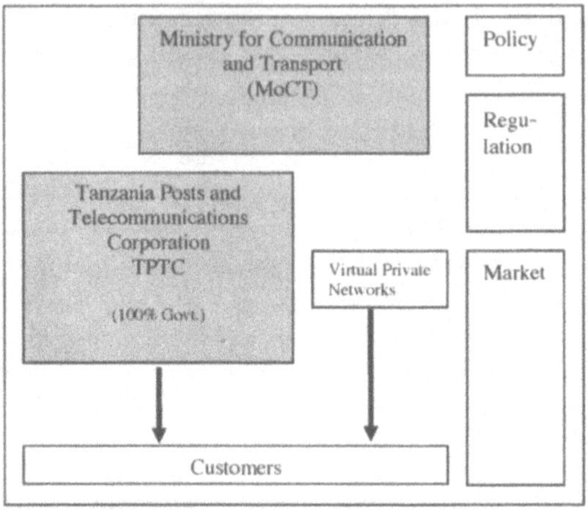

Only as external and internal pressure to adjust the sectoral settings grew, the picture started to change. The sector became target of donor assistance programmes from 1977 onwards. Especially the Swedish International Development Agency (SIDA) and the World Bank started to support the expansion of the telecommunication network and until the early 1990s modest achievements had been made. A new telephone gateway to Tanzania, an earth satellite station for international calls, was set up, the number of telephone exchanges was raised to a total of 230, capable of 85,000 telephone connections, and teledensity increased to 0.32 telephones per 100 population (see table 4.1).

Thus, after some rudimentary service provisions in the colonial and post colonial period, the early phase of telecommunications in Tanzania was induced by the collapse of the first EAC, when Tanzanian authorities were forced to reshape the institutional setup of and transfer the sectoral responsibility to the national level. The chosen sectoral design was limited to the absolutely necessary and offered some minimum provisions with regard to policy, regulation and network operation. Sectoral changes always coincided with major incisions in the political structure – the coming of independence and the period of the EAC. Therefore all approaches to telecommunication were reactive developments caused by superordinate factors rather than on deliberate attempts to adjust the sectoral policy. Accordingly, the overall performance had been weak, with some additional input since the late 1970s, when particularly the World Bank and SIDA started to assist the telecommunications sector. Further signs of change appeared with a shift to a more market oriented approach to development in the mid 1980s. Therewith, the whole sector was due for reform and the drafting of a new legal basis begun in the early 1990s.

4.1.2. Early Informatics

Before the 1980s, telecommunications and computerisation were completely distinctive. The advent of information technology in Africa followed a very similar pattern throughout the continent. It was not long after the wider introduction of mainframe computers in the developed world[81] that large scale computer units also had been introduced in most African countries by the end of the 1960s.[82] The then systems were designed to suit large-scale centralised computing purposes, and, consequently, new systems were mainly deployed in headquarters of ministries (finance, defence), central statistics offices and major public utility agencies like telecom, railways, harbours, or electric power. Therewith utilisation was limited and the sales quantity of large computer systems in Africa remained marginal. This changed not until the invention of the micro- or personal computer in the early 1980s that offered new opportunities at more reasonable prices.

Figure 4.3: Schematic overview on informatics sector development 1951 - 1993

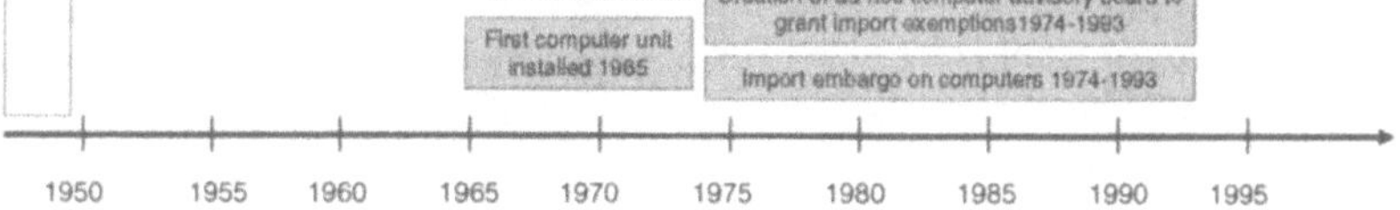

Figure 4.3 gives an overview of the beginning informatics development in Tanzania. In contrast to the developments in the telecommunications sector, the changes were not influenced by the political structure, but can be explained by intrinsic patterns of computerisation and the related dynamics of an emerging

[81] For a typology of different computer types, refer to box 4.1.

[82] E.g. Ethiopia: 1960, Zimbabwe and Zambia: 1961, Cote d'Ivoire, Nigeria and Kenya: 1962, c.f. Adam (1999); Kyarie (1995).

global IT market. In Tanzania, an import embargo on computers imposed in 1974 structured the national history of computerisation and divided the early phase in two parts, before and after the ban. Figure 4.3 also shows that the main activities had been introduced at the same time as the ban had been issued, and that the activities comprised a peculiar mix of positive and negative actions.

4.1.2.1. Banning Computers from the Country

The early discussion about ICT had been conceptualised under the term of *informatics* and started in 1965, when the first computer had been imported in the country[83] – a large-sized IBM Hollerith ICT 1500 –, deployed in the Ministry of Finance. By 1974 the number of computers subsequently rose to a modest seven computers (see figure 4.4). These were mainframes and they were both maintenance intensive and costly, so that access was restricted to some government agencies and large parastatals. But even those, who could afford the technology, soon struggled to run their machines properly and problems occurred in almost all installations.

These shortcomings became most visible in the Treasury, itself, when the ministry had started to computerize the whole government accounting system in the early 1970s. The system included data exchange flows from the regional, and partly district level, to the ministry headquarters. However, as regional administrators started to implement different codes for their budget figures, it became increasingly difficult for the headquarters to consolidate national budget figures. The project failed completely and caused heavy losses. From 1972-1974 the government could not tell how much money it had spent or collected as revenue (c.f. Mgaya 1994). In retrospect, the failure of the project seemed inevitable. The accounting system was designed as a centralised card-based process, but incompatible data collection systems impeded smooth information flows resulting in devastating effects. Thus, with the exception of payroll and pension systems the government ceased all computerisation processes and got back to manual systems.

These events and some similar experiences in other areas rendered most of the applications developed obsolete, and for the time being this meant the end of computerisation in Tanzania. Especially the financial losses exposed the government to criticism from members of parliament and the general public

[83] Some observers even refer to an earlier date. For example David Sawe identified 1956 as the beginning of information technology in the region when so-called hand-punch input devises for computer cards were installed for the British colonial government (interview with David Sawe in: The Express [Dar es Salaam] 18.03.1999).

raising the question if "computers are relevant and necessary at all in Tanzania?" (Shila 1994). Finally, the government concurred with the general feeling that the country was not yet ready for computerisation. On the 14th June 1974 a government order was issued banning the importation of all computers and peripherals in the country. Although never strictly executed, the ban remained in force until 1993.

The decreed embargo implied a radical solution to the perceived failures. But the ban also marked a new beginning, because the government started to address questions of viability and utilisation of computers more systematically. It was no complete rejection of computer technology, but an attempt to commence a new strategy with an emphasis on planning, coordination and control. Thereby, the government pursued two intentions, firstly to avoid the mistakes of the past and secondly to provide some guidance for the road ahead. In fact, government authorities quickly resumed their work. Shortly after the ban had been issued, the Ministry of Finance appointed two teams to study the prospects of informatics for the country. Despite ambitious mandates[84] the final outputs were rather modest and only resulted in the creation of a new body. The Computer Advisory Committee was assigned the task to advise the Ministry of Finance on all computer issues and, more importantly, define exceptions and providing rules to circumvent the embargo and function as an importation control system to prevent the uncontrolled spread of incompatible computer systems. So the policy largely met the first objective to avoid the mistakes of the past, but failed to introduce a deeper discussion on the role of computers and on public interest criteria for computerisation.

4.1.2.2. The Impact of the Ban

The embargo of computer equipment has served as a besetting argument for the poor Tanzanian IT record until today.[85] But this statement must be qualified.

---

[84] The first team comprised experts from the National Institute for Productivity and the International Labour Organisation (ILO) and advised the government to formulate a national policy on computer technology focusing on acquisition, utilisation and a national training programme in computer science. The second team was called Government Computer Task Force and recruited most of its members from the Government Computer Service Centre and the Ministry of Finance, the two agencies in charge of all matters of informatics coordination and planning. The tasks assigned to the second team comprised issues of computer utilisation in existing installations, applications running on the computer, and an evaluation of the manpower base in the country (c.f. Shila 1994; Mgaya 1994).

[85] See for example David Sawe's statement: [...] because of its prohibitive policies, Tanzania was subjected to status of a spectator to revolutionary changes that swept across IT soon after 1974. [...] The country sat idly as around the world educational systems geared up towards

Firstly, ban enforcement policy did not change between 1974 and the mid 1980s. The technically biased import provisions imposed some bureaucratic hurdles for computer importation, but they did not put an insurmountable burden on purchasing a computer. Though with difficulties, computers continued to be imported under temporary permits. Secondly, the numbers of installed computers show significant variations between 1974 and 1986, a period in which the importation procedures did not change. Figure 4.4 shows that computer utilisation declined until the late 1970s and increased from the early 1980s onwards.[86] This suggests that it was not the policy that changed the landscape, but other reasons must have influenced the importation pattern. So, what explains the IT-stagnancy after 1974?

Three related answers can be put forward. Firstly there were operational reasons. The old systems in place proofed to be insufficient and alternatives to mainframe computers had not yet entered the scene. Predominantly high costs and a lack of skills limited the spread of further computer technologies. Secondly, any constant IT-utilisation in the early 1970s would have implied the search for new solutions after the old failures. This would have increased the cost-value ratio and thus presented a prohibitive burden for further computerisation. Thirdly, these inherently arguments were closely tied to a psychological momentum. Failures, losses and broad criticism created a climate of IT-disaffirmation, in which it was simply inappropriate to promote any new computer based solutions – computers simply had a bad reputation.

The scene changed radically in the mid 1980s with the invention of mini- and microcomputers (see figure 4.4 and 4.5). Increased computerisation was not due to policy changes, but can be explained by the availability of new and more appropriate technology solutions. The beginning era of the microcomputer, the predecessor of today's PCs, was a major boost for the computer market. They became affordable for an increasing number of users, and, due to their small size, they could be transported and deployed easily, and combined a plus of user-friendliness with a better computing capacity at declining costs. The technological quantum leap brought the computer within the reach of many

---

promoting the emerging technology and responded to gigantic leaps in IT between 1974 and 1984 (in: The Express [Dar es Salaam] 18.03.1999).

[86] The total number of mainframe computers dropped from seven to five in 1974, then remained stagnant until 1981 before it increased, though from a very low level, to 51 if adding the number of mainframes and minicomputers in 1986 (see figure 4.4).

more possible users and many private companies started to import and utilised them.[87]

Figure 4.4: Number of Large Computer Systems in Tanzania 1968-1986

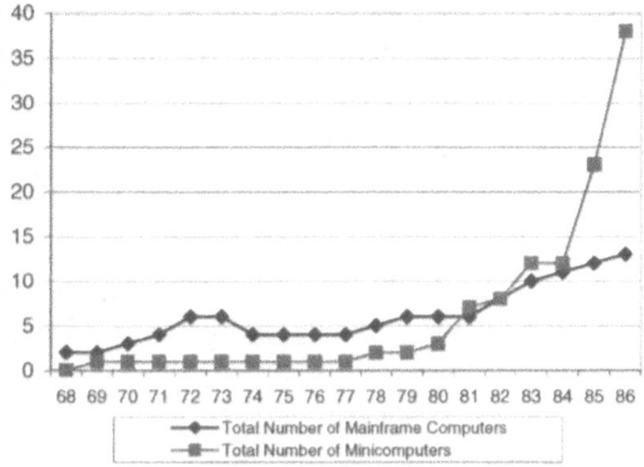

Source: Sheya / Koda (1987), in: Mgaya (1994)

Figure 4.5: Number of Microcomputers 1980-1986

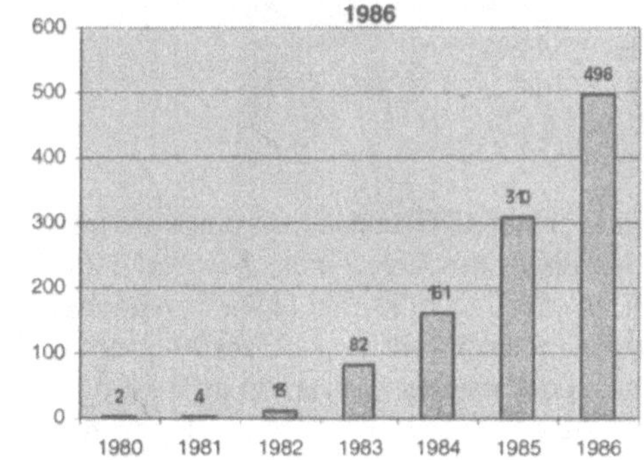

Source: Sheya / Koda (1987), in: Mgaya (1994)

The scene changed radically in the mid 1980s with the invention of mini- and microcomputers (see figure 4.4 and 4.5). Increased computerisation was not due to policy changes, but can be explained by the availability of new and more

[87] In contrast to the early days of computing the user basis diversified considerably. With regard to the 1986 sectoral distribution of deployed computers, government presented 30,5 percent, parastatals 35,7 percent and the private sector 33,8 percent of the overall computer imports. Data taken from from Ndamagi (1988, in: Mgaya 1994).

appropriate technology solutions. The beginning era of the microcomputer, the predecessor of today's PCs, was a major boost for the computer market. They became affordable for an increasing number of users, and, due to their small size, they could be transported and deployed easily, and combined a plus of user-friendliness with a better computing capacity at declining costs. The technological quantum leap brought the computer within the reach of many more possible users and many private companies started to import and utilised them.[88]

Box 4.1: Types of Computer Technology in the early 1980s

The early history of computing distinguished between different types of computers. Historically, a useful distinction can be made between:

- **Mainframe** is an industry term for a large computer for the commercial applications of businesses and other large-scale computing purposes. Historically, a mainframe is associated with centralized rather than distributed computing
- **Minicomputer**, a term no longer much used, is a computer of a size intermediate between a microcomputer and a mainframe. Typically, minicomputers have been stand-alone computers (computer systems with attached terminals and other devices) sold to small and mid-size businesses for general business applications and to large enterprises for department-level operations
- **Microcomputer** is a complete computer on a smaller scale and is generally a synonym for the more common term personal computer or PC, a computer designed for individual usage.

Source: Tech Target Network (n. d.)

The above factors have changed the pace of computer diffusion in Tanzania. With regard to all available computer types the total number roused from a neglecting 11 units in 1980 to 550 in 1986.[89] Although the government continued to monitor the importation, it soon faced a increased demand for the new technology and as a consequence, government's importation policy became even laxer and opened the door for a first low-level growth period of computer

[88] In contrast to the early days of computing the user basis diversified considerably. With regard to the 1986 sectoral distribution of deployed computers, government presented 30,5 percent, parastatals 35,7 percent and the private sector 33,8 percent of the overall computer imports. Data taken from from Ndamagi (1988, in: Mgaya 1994).

[89] Numbers based on Sheya/Koda (1987, in: Mgaya 1994). The given numbers must be seen as a rough approximation. A second count conducted by Ndamagi (1988, in: Mgaya 1994) gives a total of 778 computers in the country in 1986. Despite this variation, both numbers show the same basic pattern: Nearly 90 percent of the installed units were microcomputers, which again highlighted the importance of the technology shift as the main reason for the growth rate of computer importation.

acquisition in Tanzania in the late 1980s.[90] Thus, the importance of the ban on the IT record of Tanzania must be downgraded. A strict compliance never existed and never was intended. Certainly, the agencies in charge never were computer promoting agencies, but an instrument of political control. Thus, issues of information technology kept a low profile and generated a restrictive environment that inhibited any further growth in computer utilisation. Further and even stronger explanations of a low computerisation profile related to general cost considerations – acquisition costs, high after sales service and spare parts cost –inappropriateness of the technology and a lack of IT-awareness and skills. Imprudent and hastily implemented donor assistance programmes played a crucial role, too.[91] The ban never had the devastating effect as sometimes put forward to explain the modest Tanzanian IT record. This can be shown with respect to ban enforcement, the number of importations and market developments, but also in comparison to other African countries that had a similar IT history without a ban burden.[92]

4.1.2.3. The Arrival of Email Service Provision

After the growing acceptance of microcomputers, the arrival of electronic networking, introduced as email-only systems in the late 1980s, marked a second milestone of broader computerisation. The most important system was the FidoNet (see Box 4.2), an easy to use store and forward email systems making use of international telephone calls to exchange emails and other information. The first African system installed was ESAnet (Eastern and Southern African Network), a network of five university computer centres in

---

[90] The successive change also became noticeable in growing number of computer vendors in the country. Up to the end of the 1970s there were only two registered computer companies in the country, ICL and NCR, in the successive years the number of computer selling agents grew noticeable, generally as a sales office for one computer brand, exclusively. The first private vendors were Computer Corporation of Tanzania (CCTL) founded in 1980 as a distributor of Wang computers, Computer and Telecom Systems Limited (CATS), founded in 1987 as agent for ICL computers and Business Machines Limited (BML) with a history that dates back to 1960, but only since the mid 1980s with a record as computer selling agent. Company information was collected during a vendor survey in late 2002 within the scope of this research project. For details see section 5.1. of this text. A summary of survey results can be found in Nielinger (2002).

[91] This often has been neglected. Considering the figures quoted in Odedra-Straub (1995), that half of the computers in Africa acquired in the early 1980s were 'aid donated', donors contribution to the bad IT-performance would deserve a closer reflection.

[92] Taking 1980 figures, Kyarie (1995) counted a neglecting number of microcomputers in Kenya and Odedra-Straub (1995) noted that Zimbabwe had less than ten and Ghana none microcomputer in 1980. Similar findings are quoted in Nwachuku (1994) and Adam (1999).

1991.[93] In a parallel effort, the same universities also pursued another project focusing on satellite-based communication to exchange health and medical information via HealthNet, a communication service operated by Satellife, a Boston based NGOs. Because both projects, ESAnet and HealthNet, were funded by the International Development Research Centre (IDRC) and conducted by the same universities, they finally merged in 1991 into one HealthNet project and, thus, became the "father of email communication in the country" (AAAS 1996; see also Miller 2003, Levey/Young 2002). HealthNet operated both satellite and telephone based systems and the installation of satellite connectivity also brought up first regulative issues, because satellite operations rely on available frequencies, and therefore the operators needed to apply for an operating license and frequencies on the frequency band. In the late 1980s and early 1990s wireless communication was not yet an overcrowded business, and so, getting a license was less a technical, but a political challenge. Two issues were particularly critical: firstly security concerns needed to be cleared to avoid frequency interferences with the security forces. Secondly, and even more important the telephone operator, TPTC needed to make far-reaching concessions with regard to its monopoly on data communication, because HealthNet would operate its own independent satellite link. Of course, TPTC was concerned on its market position and on potentially losses in revenue,[94] but in the end some pressure from the Ministry of Health, backing from the international donor IDRC and the negotiating skills of Tanzanian Internet pioneer Bill Sangiwa (see below) brought first lower cost connectivity solutions and the needed satellite frequencies for the HealthNet project.

Despite some satellite-based solutions, FidoNet – and its smaller pendant UUCP – and international telephone calls remained the major tool for connectivity. The Fido system was installed since 1992 and the then very rudiment network became part of a global system that comprised 16 African e-mail nodes, from Algeria to Zimbabwe, that served mainly the electronic communication needs of research organisations and NGOs,[95] connected via GreenNet, a London based

---

[93] These were centres at Makerere University, University of Dar es Salaam, University of Nairobi, University of Zambia, and the University of Zimbabwe (c.f. Levey / Young 2002).

[94] Miller (2003) describes the negotiations behind the scenes that resulted in low cost connectivity for the HealthNet project. There was not much experience and especially in the beginning the telco was not very supportive to accept the loss in telecom revenue that would come with the satellite connection.

[95] Fatma Alloo, a founder of the Tanzania Media Women's Association (TMWA) recalls in an interesting report her first contact with ICT through the APC in the late 1980s: "They put us in touch with a Mr. Sangiwa who, based in our local hospital, was operating an e-mail link in collaboration with GreenNet. So I paid him a visit just to see for myself and took along my 12-year-old son. To be honest it is his [Suhail's] excitement that drew my involvement, for I

NGO (see box 4.2). The first FidoNet node in country outside UDSM was established at the Tanzania Commission for Science and Technology (COSTECH), and subsequently five additional nodes were introduced between 1992 and 1995.[96] In that period COSTECH developed as the most important national FidoNet node and email provider.[97] The emerging FidoNet system initiated a development, in which an increasing number of email pioneers prepared the ground for accelerated ICT diffusion in the country and reflected the crucial role of research institutions and NGOs as facilitators of early electronic networking.

---

was scared of this technology myself. Eventually, I became the catalyst in TAMWA to get women enthused about this technology and to invite Ms. Karen Banks of GreenNet to come down and train us....I still chuckle when I think of the huge size of the modem that Karen carried to Dar es Salaam for us from APC.

Initially, the women of TAMWA, like myself, were a bit apprehensive about this technology....Understanding this technology and its potential for empowering ourselves also formed the basis of our discussions with the donor community. I remember arguing with our donors to give us the 'means of production,' the desktop publishing tools, rather than funds to produce our magazine *Sauti ya Siti, Voice of Women*. There was a little company across our office called Computer Centre, and I walked in and convinced the young man that we would get the computer for him provided he trains seven TAMWA members on the production of the magazine. He agreed! Years later he admitted to me that I was his first customer, and having just returned from studies abroad, the prospect of meeting women thrilled him. Little did he realize that this encounter would play a role in his personal and commercial growth. Today he is one of the major dealers of computers in Dar es Salaam." (quoted in: Esterhuysen 2002)

[96] In chronological order these were COSTECH (March 1992), HealthNet node at Muhimbili University College of Health Science (August 1993), Centre for Energy, Environment, Science and Technology (CEEST), an international environmental NGO (April 1994), Arusha Node Marie (April 1994), UDSM (December 1994), and HealthNet node Mwanza (c.f. GreenNet (n.d.), a brief description of FidoNet nodes in Tanzania can also be found in AAAS (1996).

[97] In the beginning COSTECH provided email links to the Economic and Social Research Foundation (ESRF), Sokoine University of Agriculture (SUA), Tanzania Food and Nutrition Centre (TFNC), National Institute of Medical Research (NIMR), Tanzania Industrial Research Organisation (TIRDO), and Institute of Finance Management (IFM) (c.f. AISI 1999). The number reached its highest level in 1995, when all in all 60 organisations were connected through COSTECH. However, COSTECH lost its role as service provider as commercial ISPs entered the market. In addition, COSTECH also introduced training courses on networking targeting the research, government and NGO sector. All measures were meant to transform COSTECH into a national IT centre of excellence and to meet the objective it received considerable support, e.g. by UNECA's CABECA programme (Capacity Building for Electronic Communication in Africa) programme or later by UNESCO's RINAF (Regional Informatics Network for Africa).

Box 4.2: The origins of electronic communication in Tanzania – FidoNet
Between 1990 and 1997 FidoNet provided means of cheap, efficient electronic communications. It based on simple 'store-and-forward' technology and proved to be very appropriate for electronic communication in developing countries, where phone line quality was poor, reliable electricity supply problematic and costs of communications expensive. The store and forward system allowed to write emails offline. For distribution international phone calls were made on a daily basis from the FidoNet gateways to over 50 small hosts in Africa, Asia, Latin America and Central and Eastern Europe, delivering mail from, and collecting mail to, their respective user communities. Between 1990 and 1997, it is estimated that somewhere between 2 and 5 million messages were sent across the Fido gateways, at a cost of about $0.30 per message. FidoNet gateways were installed at nodes of the Association for Progressive Communications (APC), and international NGO attempting to promote ICT for social justice and development, for example at Web Networks (Canada), IGC (USA), GreenNet (UK), Laneta (Mexico), Comlink (Germany), Nordnet (Sweden) and Worknet / Sangonet (South Africa).

The APC network provided networking and information resources to over 17,000 subscribers in 94 countries. Among them the Fido gateway for African networks. In Africa, FidoNet started in Kenya in 1989. Tanzania got its first connection in 1992. All FidoNet hosts were connected through the FidoNet gateway GreenNet in London.
(Source: GreenNet n.d.)

Behind the figures of growing email utilisation, first electronic networking and the diversification of organisations, which made use of the new technologies, the role of personal champions became a crucial engine for progress. In the case of Tanzania, this role can be credited to Bill Sangiwa, a bio statistician at the University of Dar es Salaam. Most of the institutional diversification in email and Internet usage could be traced to Sangiwa's work. He played a decisive role in the development of ESAnet and HealthNet, he was in charge of the FidoNet node at COSTECH, he negotiated the satellite frequency spectrum for HealthNet and he also founded Tanzania's first commercial Internet service provider Cyber Twiga in 1996 (c.f. Miller 2003; NSRC 1999; Int. Sangiwa 22.7.2002). It has not only been the availability of services that led to increased usage; technological innovations always needed to be phased in and this process, to a large extent, depended on personal commitment and dedication of individual champions.

4.1.2.4. Informatics Education

In the 1960s informatics was in its infancy and computer skills were scarce in Tanzania like in most African countries. As the new technologies gained ground, education became a critical issue and subject to political decision-making, too. The provision of training facilities was part of the political process, all the more because Tanzania still was under socialist rule and educational matters and curriculum development were centralised in the Ministry of Education. Early approaches to informatics education and training were in line with early policy approaches, and not unexpectedly, the ban on computers also changed the approach to computer education. In 1974, the year of the ban, the University of Dar es Salaam (UDSM) introduced a master degree course in computer science. As mentioned above, the ban also marked a new beginning and this respect, the master course was an attempt to build a new resource pool of computer experts. An overview of early informatics education could be divided in three phases:

- Until 1974: dominance of individual application courses that reflected the uncoordinated ICT approach at that time;
- 1974 – 1983: new attempts to strengthen computer training in higher education, the introduction of a master degree course in computer science and its subsequent failure to meet the expected output of trainees; and
- a standstill in university computer education, coinciding with a gradual diversification of offered courses and emergence of alternative training institutions since the early 1980s. A new beginning of higher computer education only commenced in 1991 (see also figure 4.3).

The first computer training was offered as early as 1965 at the University College Dar es Salaam as optional courses in the mathematics department and later extended to some other faculties of the college (c.f. Mgaya 1994). However, none of the courses aimed to train computer specialists, but focused more on application training in a given field of interest, e.g. statistics, accounting or economic forecasting. This partial approach was revised in 1974, when the Ministry of Education approved a two-year MSc programme in computer science at the University of Dar es Salaam. The programme was attached to the Computing Centre, which already came into being in 1968 as a unit within the Mathematics Department of the Faculty of Science, but made its first noteworthy appearance only with the introduction of the new MSc course. From the very beginning, the programme failed in providing the most basic prerequisites – trainers and instructors, books and teaching materials and a sufficient quantity of hardware equipment. The course was abandoned in 1983 after only twelve students had been trained (Mgaya 1994).

At a time when computerisation gained some momentum, the public education system withdrew almost completely from computer education and the local education system was in no position to respond to the needs of a specialised informatics workforce. Higher computer education was not resumed before 1991, when the Computing Centre of the University of Dar es Salaam was able to launch a new BSc programme. Learned from past mistakes, the new bachelor programme was designed shorter and more flexible than its predecessor and had an input of 25 students in 1992 and 40 in 1993 (Shila 1994). The BSc was the first lasting attempt to introduce a university degree in IT training in the country. However, only with the rebuilding of the University Computing Centre (UCC) in 1995 and the formation of the Department of Computer Science in 1999, both at UDSM, computer education at university level gained some new significance.

At the same time as the university abandoned the Masters programme in computer science in 1983, increased computer diffusion led to a largely unmet, but growing demand for IT education. In the absence of any alternatives, several government-owned institutions started to introduce their own computer courses.[98] Although these institutions faced similar problems as the university courses, i.e. a lack of trainers, teaching materials and equipment, they soon became accepted IT training facilities. But they also repeated old errors. Every institution introduced individual courses designed to meet individual needs and granting individual certificates. Although they alleviated computer training in a difficult environment, they did not manage to go beyond isolated applications and seek a bigger impact by introducing some common standards and certificates of computer education. Remembering that all institutions were state owned, these shortcomings highlight the fact that the government still was not able to coordinate approaches to computer education and, hence, could hardly fulfil its role as facilitator of the technology.

In conjunction with the arrival of economic reforms in the mid 1980s, the first private computer training centres entered the scene. The first was the Dar es Salaam based International Computers Limited (ICL) that started to offer courses in computer studies. Others followed, such as the Dar es Salaam Institute of Computer Science (ICSM), the Institute of Information Technology (IIT) or the Institute of Management and Information Technology (IMIT). They offered a variety of courses from short to full-time and from application courses to diploma programmes in computer science. However, quality often was poor and common standards not yet in place. Nevertheless, the private centres stimulated the IT environment in the country and together with a number of new small training centres mainly run by computer vendors offering basic introductory courses,[99] they facilitated some increased computer utilisation and became early signs of an emerging small IT market in the country.

[98] For example, the Eastern and Southern African Management Institute (ESAMI), the Institute of Development Management (IDM), the Institute of Finance Management (IFM) or the College of Business Education (Mgaya 1994, Shila 1994).

[99] Shila (1994) estimated that 55 of such small centres were launched in the early 1990s.

4.1.2.5. Concluding Early Informatics

Informatics in Tanzania has a surprising long history, but from a broader perspective neither informatics nor telecommunication were priority sectors, not politically and not economically. From a global point of view the Tanzanian record was hardly sizeable. On a global scale, microcomputers just had entered a first phase of accelerated growth in the early 1980s. The then most successful minicomputer, the Commodore-64,[100] had started to occupy the children's rooms and sales figures were well in the millions in the industrialised world.[101] The contrasting Tanzanian figures are telling. In 1986 Tanzania had a computer density of 0.002 per 100 or one computer per 46,000 population.[102] Though such comparisons fall short due to many respects, the numbers very well highlight the status quo and the difficult and challenging starting position of early informatics in the country.

Moving the perspective to an African level, apparently, and despite Tanzania's unique history and a 20-year ban burden, the Tanzania IT diffusion pattern to a large extend coincided with the performance record of other African countries. This applies for the arrival of the technology in the 1960s, the growth pattern, in which for some 15 years a handful of individual systems were primarily used for planning and administrative functions mainly in the public sector, and then an increasing computer awareness throughout the continent following the advent of micro- and minicomputers. And this also applied for computer education, where only a handful of countries such as Kenya, Malawi, Nigeria and Zimbabwe had universities that offered computer science degrees (c.f. Odedra-Straub 1995; Drew / Forster 1994).

Tanzania has been a typical rather than special case of early computing in Africa. Notwithstanding the existence of numerous particularities, the basic IT diffusion pattern was similar throughout the continent. But this also cut down the role of policy. Basically, early developments in the informatics sector were due to the dynamics of computing rather than a result of deliberate decision-making. Even bad policy, if not in shape of a complete rejection of the technology, could not reverse the trend of a commencing IT sector in most

[100] History hides an interesting side aspect. One of the first pioneers of home computing, Shiraz Shivji, who developed the Commodore-64 and later the Atari ST, the first best selling microcomputers, was Tanzanian by origin (c.f. Happy Computers 1986).

[101] In Germany, for example, 3 Mio. Commodore-64 were sold in the three year period between 1983 and 1986 (c.f. Hoogen 1997).

[102] On the basis of an overall population of 23 Mio. in 1986, as given by the World Bank (1988).

African countries. However, when computerisation gained momentum in the mid 1980s, market induced changes increased the necessity to frame the new developments politically. Basic rethinking became evident in 1989, when the then Tanzanian finance minister declared "computers are essential for the future development of the country" (quoted in: Mgaya 1994). This introduced a fundamental new approach to informatics. The focus shifted from control to promotion and the government started to restructure the sector and respective legislation came into force in 1994.

4.2. The Second Phase 1994 - 1999: Orientation

The telecommunication sector reform did not occur in a vacuum but was affected by the overlaid transformation towards a market economy and democracy. The reform process also concurred with a decade of rapid technology changes. These trends also affected Tanzania, so that a whole set of factors paved the way for a new approach to telecommunications. Firstly, the World Bank and the IMF pressed for radical policy changes, and e.g. the privatisation of the incumbent telephone operator was one condition for the release of additional financial or technical assistance, among others under the IMF's enhanced structural adjustment facility ESAF (c.f. East African [Dar es Salaam] 26.4.1999, Wilson 2004). Furthermore the WTO agreement on basic telecommunication was in the offing, in which WTO members have agreed on liberalised telecommunication markets and bilateral donors started to advocate the new approach, too. They pursued a stick and a carrot approach, demanding substantial economic reforms and in turn offering extensive support to get the reforms on track. Thus, pressure from the international community was strong. On the other side, a largely defunct TTCL and other sectoral deficiencies strengthened the proponents of policy changes, too. In the early 1990s there was hardly a realistic alternative to reforms. The international pressure was too high and on the national level the liberal reform idea also gained ground.

4.2.1. Background and Legal Basis of Telecommunications Sector Reform

Tanzania was among the first African countries that restructured the sector[103] and there were hardly models and only little experience to build on. The expected outcome of reform matched a positive sum game, in which all parties should win. The overall objective was the creation of a sound policy framework

[103] In fact, the Tanzanian introduction of full or partial competition in all telecom market segments with the exception of basic services was widely recognised in many African countries (c.f. TRASA 1999).

conducive to development and increased private sector participation. The government maintained control by carefully allotted means of control and guidance meant to induce market based development dynamics. Therefore, new laws needed to be passed, new policies needed to be developed and new regulatory frameworks were required to provide a stable and predictable legal environment, to guide the sector and to maintain order.

However, announcing some inexpensive advocacies has been one thing, developing an appropriate implementation plan that is politically accepted and technically sufficient quite another. Tackling policies had been a learning process. This is why the period between 1994 and 1999 can be labelled as an orientation phase. The general direction was fixed, but the technical road to implementation was not. All major stakeholders recurrently redefined their agenda: government as policy maker, regulator as supervisor, operators as service providers and the public as consumers. The new approach called for radical rethinking and demanded careful guidance and a deliberate sequencing of reform steps. The new telecommunications development strategy comprised four steps:

i. Establishment of a market-oriented regulatory and policy framework;
ii. Formation of a regulatory body to license private operators to provide non-basic telecommunication services;
iii. Commercialisation and corporatisation of the state-owned telephone company TTCL;
iv. Rehabilitation and expansion of the basic telecommunications network.

Three pieces of legislation passed in 1993 and took effect on the 1st January 1994. They provided the legal ground for restructuring and liberalisation:[104] (i) the Tanzania Communications Act, (ii) the Tanzania Telecommunications Company Incorporation Act and (iii) the Tanzania Posts Corporation Act. Law enactment led to the dissolution of the former Tanzania Posts and Telecommunications Corporation and the creation of the new Tanzania Telecommunications Company Limited (TTCL) as public telephone operator, the Tanzania Posts Corporation (TPC) as postal service provider and the Tanzania Communications Commission (TCC) as regulating agency for the post and telecommunication sector. The Tanzania Communications Act also outlined

[104] Already in 1992 the Ministry of Communication and Transport and TPTC signed a Memorandum of Understanding clarifying the role of the ministry and outlining the elements of the commercialisation of TPTC (c.f. UNECA 1996).

the responsibilities of government vis-à-vis the new organisations. The new responsibilities included:[105]

- the government provides policy framework, national security, law and order,
- TCC became solely responsible for a majority of regulatory functions, including licensing, monitoring, tariff approval and technical standards,
- TTCL was incorporated as a wholly state-owned, but autonomous company operating as a commercial enterprise.

The changes affected all actors in different ways. But before the experiences of each stakeholder in the new environment will be reviewed, it makes sense to dwell on the overall approach a little longer. Although the sector has been segmented in different sub functions, still, the process was under strong guidance and government control. Part of the national strategy was the introduction of a Telecommunication Restructuring Programme (TRP) designed as a general construction plan aiming to induce and support the reform process.

4.2.2. The Telecommunication Restructuring Programme (TRP) (1993-2000)

The Telecommunications Restructuring Programme (TRP) was the first comprehensive programme aiming to substantially reform the telecommunication sector in Tanzania. It paved the way for the separation of regulative and operative functions, development and implementation of a legal and regulatory framework and for the privatisation of the incumbent telephone operator TTCL. As a result considerable improvements could be observed in both the governance of the telecommunication sector and in the provision of basic telecommunication services. The TRP was delivered in three phases (TRP I – TRP III), covered the time period between 1993 and 2000 and highlighted issues of:

- policy and regulation,
- institution-building, and
- network modernisation.

[105] The Tanzania Post Corporation (TPC) is not subject to telecommunication and, thus, not discussed in this chapter. See MoCT (2001) for some TPC performance indicators. For a recent ICT profile of TPC, see chapter 4.3.5.4.

4.2.2.1. Donors and Commitment

Though the project was conducted under the main responsibility of the World Bank as the prime donor agency, the TRP was a joint effort brought together eight different donors.[106] At the time of inception the TRP had a volume of a little less than US$ 190 million. Together with CIDA and KFAED, who joined later, and TTCL's local contribution of about US$ 30 million, the overall project portfolio summed to US$ 250 million. The largest parts were made available from the World Bank (US$ 75 million) and SIDA (US$ 40 million) (World Bank 2000; Int. Karhammar 23.07.2002). Consequently, both partners had the highest profile within the programme, with SIDA as the evident junior partner. In the definite scope of duties, only the World Bank appeared in all three major programme windows, whereas SIDA focused on institution-building and network extension. All other partners solely focused on infrastructure.

4.2.2.2. Objectives and Project Components

The TRP addressed the main actors and issues in the Tanzanian telecommunication sector. regularly Because it was designed as a long-term project, the project components were revised,, so that project adjustments could be included as the sector matured and new institutions and responsibilities came into being. First project focus was on separating TPTC into the Tanzania Postal Corporation (TPC) and Tanzania Telecommunication Company Ltd. (TTCL) as commercially oriented units on the one hand and Tanzania Communication Commission (TCC) as regulatory body on the other hand. With completion of the restructuring process, institutional and corporate performance targets of the newly created organization came to the fore of the programme and resulted in new objectives. These were: (i) to establish a market oriented policy and regulatory framework, to introduce private sector participation in non-basic services, (ii) to facilitate commercialisation and corporatisation of TPTC, and (iii) to eliminate existing bottlenecks in the availability of telecommunication services (World Bank 2001). Figure 4.6 shows the scope of the project and illustrates the various addressees and different project components – MoCT, TPTC, TTCL, TPC, TCC and the Parastatal Sector Reform Commission (PSRC) as the agency in charge of privatisation process in the country became subject to the programme.

[106] These were: the World Bank, the African Development Bank, (ADB), the European Union (EU) as multilateral agencies and the bilateral donors Swedish International Development Agency (SIDA), Danish International Development Agency (DANIDA), Japan International Cooperation Agency (JICA), Kuwait Fund for Arab Economic Development (KFAED) and Canadian International Development Agency (CIDA).

Figure 4.6: The Telecommunication Restructuring Programme (1993-2000)

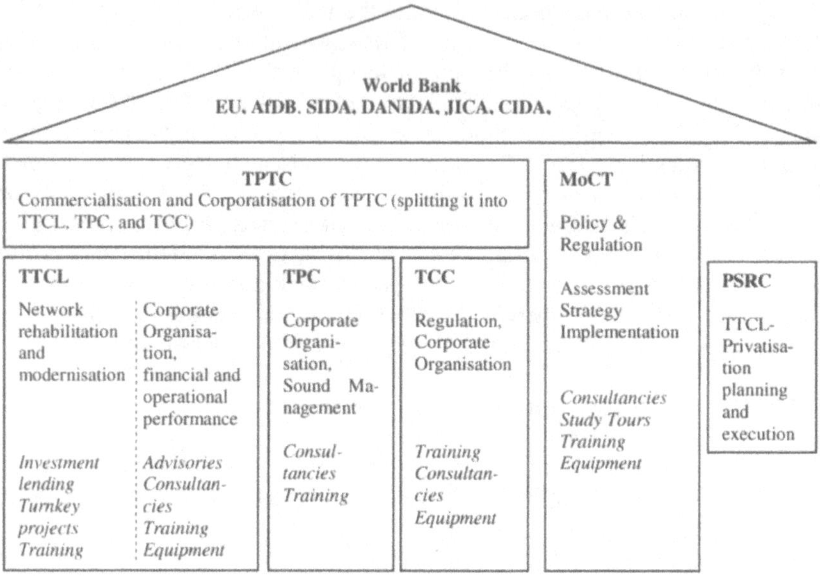

Figure 4.7: Disbursement by destination and source under the TRP (in %)

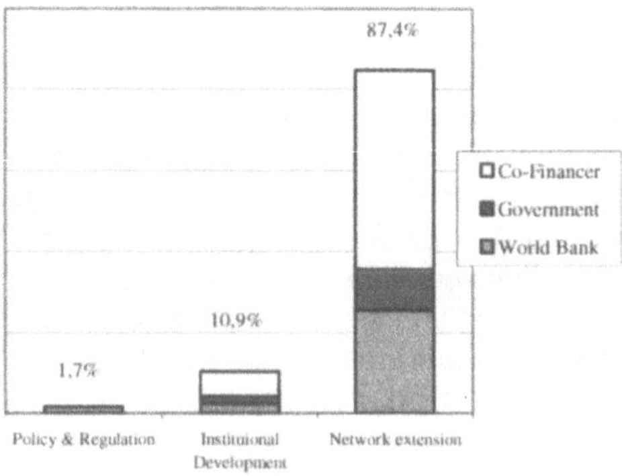

Source: Author's calculation based on World Bank (2000, 2001) and UNECA-NICI (1998) figures

Thus, TRP had various sites under construction, but as the programme progressed, TTCL's poor performance was increasingly identified as the major bottleneck to reform process. The incumbent operator had already received considerable assistance in the early phase of the TRP, especially to strengthen its corporate organisation with respect to financial management, planning and marketing activities, but still in the mid nineties, the incumbent was not able to respond to technological and market developments. Accordingly, TTCL became a prime target of the programme, so that the majority of projects appeared in the rehabilitation and extension of TTCL's existing telecommunication network. Overall expenditures on infrastructure covered almost 90% of the overall TRP budget (see figure 4.7) and covered numerous individual infrastructure projects (see box 4.3).

Despite the indicative ratio of TRP disbursement it would be wrong to solely value the programme as a pure network rehabilitation project – a prevailing misinterpretation until today.[107] This would unduly cut short the scope of the project and ignore that the other components remained equally important. This applied all the more because policy, regulation and institution building measures had been implemented to support the efforts to improve TTCL's performance record and construction components were more costly than consultancies or training measures.[108]

[107] In fact, the TRP was mainly perceived as an infrastructure project. Many articles highlighted the role of the TRP, especially in the early phase of the project and almost exclusively with regard to infrastructure improvements. This has been an astonishing and prevailing ignorance of the comprehensive character of the programme (c.f. africaconnection (1999); TRASA (1999); UNECA-NICI (1998); AISI (1998); Luhanga (1994)).

[108] All in all, the TRP comprised a diverse mix of different funding and assistance arrangements. Issues of policy, regulation and corporate restructuring were approached by advisories, consultancies or study tours and training. The Ministry of Communication and Transport (MoCT) received assistance to assess policy needs, in particularly the current sector structure, review the government sector strategy, identify options for private sector involvement and define a respective implementation plan. Furthermore, the World Bank reallocated funds to assist in the formulation of a National Telecommunication Policy, which was finally approved by the government in August 1997. The PRSC was supported in the conduct of the preparation of the TTCL privatisation and TCC received extensive training of staff and consultancies on specific regulatory issues. However, as mentioned above the most perceptible addressee of the TRP was TTCL. Apart from infrastructure project measures addressed TTCL's corporate capacity by means of introducing contemporary management principles procedures, financial strengthening, i.e. compliance to internationally accepted accountancy rules, international audit, pre-privatisation restructuring study and financial and legal advice with regard to TTCL's sale to an international strategic investor (Int. Karhammar 23.07.2002; see also World Bank 2000a, 2001; ÅF International 2002; Mallango 2000).

Box 4.3: Infrastructure related TRP components

- *Digital Backbone Microwave System*
  The project became operational in 1997 and solved congestion problems between Dodoma-Dar es Salaam-Zanzibar-Pemba-Tanga-Moshi-Arusha.
- *Provision of Telecommunication Materials (Cables) for Tanzania*
  The DANIDA funded project has provided new telephone distribution facilites at Tanga and Moshi. The installation of cable networks was completed by TTCL in 1997 and has connected a total of 13,163 customers to the telephone network.
- *JICA phase 2A and 2B*
  The project has provided new cable networks in Dar es Salaam. Phase A was completed in 1995 (coverage of the industrial areas of Nyerere Road and Tabata), phase B was ready for operation in 1997 (coverage of Dar es Salaam Central areas). As a result customer numbers rose from 15,863 to 43,186.
- *ADB-1: DSM Telecommunication Project & ADB-2: Dodoma-Mwanza Telecommunication Project*
  Completed in 1997. ADB-1 provided a digital broadband backbone for Dar es Salaam installing fibre oprtic cables with a capacity of more than 22,000 digital telephone lines between Dar es Salaam Central, Upanga, Kariakoo and Mogomeni. The ABD-2 provided considerable telecommunication network equipment for Dodoma, Shinyanga, Mwanza and Musoma extending the upcountry digital backbone, e.g. adding over 15,000 digital telephone lines.
- *The Southern Highland Telecommunication Project (SHTP)*
  The EU-funded project was completed in late 1998. It covered the remote regions of Iringa, Mbeya, Ruvuma and Rukwa and added 16,700 new telephone lines.
- *DSM Telecom Project/2*
  The SIDA-funded project was successfully completed in 1998 and extended the Dar es Salaam digital backbone to the northern city areas Kijitonyama, Mbezi and Kimara (among others: 16,700 digital telephone lines).
- *Zanzibar and Pemba Telecommunication Project*
  Provision of 10,000 direct exchange lines covering Zanzibar, Bububu, Chake-Chake and Wete. The project was completed in early 1999.
- *Kagera Region Telecommunication Project*
  The CIDA-funded project was successfully completed in early 2000. Kagera region is now fully automated and modernised.
- *IDA-Wireless Local Loop (WLL)*
  The IDA-funded Integrated Telecommunication Project provided WLL connections and started in 1998 and was completed in early 2001. Part A covered Arusha and Morogoro, part B parts of Kilimanjaro, Tanga, Dar es Salaam, Coast, Mara, Dodoma, Zanzibar and Pemba. In total the project provided roughly 67,000 new telephone lines. The IDA Integrated Telecommunication Project was the first of its kind in the whole of Africa
- *IDA-New Service Order and Billing System (SOBS)*
  The Service Order and Billing System was approved in 1997 and was approaching its final stage in the end of 2000. SOBS targeted TTCL`s basic business procedures like regular billing, smooth application, provision and collection services, and also the enhancement of management capacities through improved data management, i.e. the provision of non-payment lists, calls in error list and adherence to internationally accepted rules of accountancy.

Source: Project descriptions base on World Bank (2000, 2001), Int. Karhammar (23.07.02) and Int. Egawa (17.07.2002). Some additional information can be found in: Luhanga (1994); africanconnection (1999); UNECA-NICI (1998).

4.2.2.3. TRP Results

A critical appraisal of the TRP needs to highlight the major input of the programme as a starting point for far-reaching sector restructuring. The project was a major catalyst for telecommunications development in Tanzania. It guided the course and initialised fundamental changes towards a market-friendly and demand driven environment. It facilitated the privatisation and rehabilitation work of TTCL, it supported a regulatory framework that in effect had invited new players to participate in the emerging telecommunication market and it considerably contributed to the improved provision of new telecommunication services. Of course not all achievements were solely a result of the TRP and also shortcomings and project deficiencies appeared in all targeted areas – technology, design, system operation, business procedures, regulative approaches and policy. But taking into account the scope of the project, these were almost inevitable, due to an inexperienced environment, lack of skills and knowledge, but also due to considerations beyond project design such as political interference.

4.2.3. Tackling policies

4.2.3.1. Drawing the Bow: The General Policy Framework

As indicated earlier, changes in the telecommunication sector were embedded in a broader reform process initiated in 1986 as economic and later political reforms. The first democratic multiparty elections took place in November 1995 and marked the beginning of a new era after the socialist development approach from independence to the mid-eighties. Ever since Tanzania has made notable progress transforming the basis of the economy and introducing multiparty democracy. The departure of the reform process was closely related to the political changes in 1985, when Ali Hassan Mwinyi became president after the 24-year rule of Julius K. Nyerere, whose reign ended in an economically ruinous record. However, reforms were not without throwbacks. In 1986 government reached its first agreement with the IMF on support for a three-year economic recovery programme (ERP I), but progress of the macroeconomic stabilisation programme had been erratic. Growth rebounded to 3-4 percent in 1985-1991, but inflation remained above 30 percent (c.f. EIU 2003: 23; Naschold / Fozzard 2002: 3). The second ERP initiated in 1989 reinforced the reform efforts. It continued the ongoing monetary stabilisation, but additionally drew on structural adjustments aiming a general reduction of state controls and the promotion of private sector activities, inter alia in the telecommunications sector, calling for the privatisation of TTCL. But the Mwinyi administration increasingly lost its commitment to reform, and particularly for the second term of the Mwinyi presidency, macroeconomic management and fiscal balance deteriorated. As a result, IMF, the World Bank and most bilateral donors suspended their aid programmes and growth rates dropped to an average of 1.3% for the period 1993-1995 (Wobst 2001: 12; Naschold / Fozzard 2002: 3).

Reform only got back on track during the first term of the current president, Benjamin Mkapa, who gained office in the first multiparty elections held in 1995. A new period of concord between the government and the Bretton Woods institutions not only brought the country an image of a well performing reformer, it also introduced a prolonged period of economic stability.[109] The reforms were pursued with the understanding that such stability was necessary to achieve sustained growth, which is required to reduce the pervasive poverty in the country. Therewith, the government gradually managed to restore the confidence of the donor community and appeared to be committed to intensify the reforms by consolidating and deepen the gains of macroeconomic stability and structural adjustments. Within the general policy frame, the promotion of good governance and a sound business environment became key objectives of the agenda (c.f. URT 2000: 4). Some of the major achievements of the Tanzanian reform process have been:[110]

- Adoption of a new National Investment Promotion Policy in 1996 followed by the Tanzania Investment Act No. 26 of 1997. Therewith the 1990 established, but insufficient Investment Promotion Centre (IPC) was transformed into the Tanzania Investment Centre (TIC), assigned the mandate to create a positive private sector investment climate and serve as a one-stop agency for all investment matters, hence providing assistance to all investors.
- Establishment of the Parastatal Sector Reform Commission in 1992 responsible for privatisations. Privatisation measures were intended to reduce the public sector's dominance of economic activity and to pave the way for private sector development. Under the adopted privatisation programme, more than half of the 400 public entities had been divested between 1993 and 1999.

[109] For example GDP growth accelerated to 6 percent, domestic inflation maintained at approximately 4 percent annually, the exchange rate for the Tanzanian shilling continued to be market-determined and gross official international reserves were maintained at the equivalent of approximately 4 months of imports of goods and non-factor services (URT 2000: 15).

[110] Achievements are compiled from: BOT / TIC / NBS (2001); World Bank (2002) and Madete / Aiko (2003). For a summary of sectoral policies, see also Wangwe / Semboja / Tibandebage (1998).

- Review of a number of sectoral policies to facilitate private investment, such as the telecommunication policy (see below), the mining policy in 1997, and tourism policy in 1997.
- Reform of the tax system by rationalizing the tax structure and establishing the Tanzania Revenue Authority (TRA), which started operation in July 1996 out of the former tax departments of the Ministry of Finance. Introduction of value added tax (VAT) in July 1998.
- Enactment of the Banking and Financial Institutions Act, 1991 (amended in 1995), which strengthened the supervisory of the Bank of Tanzania in managing monetary policy, and liberalised the financial sector in order to increase efficiency in the provision of banking services. Following liberalisation, the number of commercial banks increased from two state-owned banks in 1993 to 19 banks (17 private) and 12 non-bank financial institutions as of end 2000.
- Establishment of commercial courts in 1999 to speed up the settling of commercial disputes, thus building investor's confidence.
- Commencement of the Civil Service Reform Program meant to enhance the government's efficiency. Measures included downsizing of staff and operations to alleviate undue pressures on the government budget. While staff numbers have been considerably reduced (77,000 workers were retrenched between 1992 and 1998, reducing the work force to around 270,000), the envisaged efficiency has yet to be realized.

Thus, the main emphasis of reform was on credibility and reliability of market institutions and though the achievements were considerable by the mid 1990s – markets were relatively free, the parastatal sector and the civil service became considerably smaller, and a significant improvement in fiscal discipline took place – the country continued to face multiple difficulties. Despite enormous efforts, reform strategies turned out to be imperfect, technical implementations were weak or even contradicted the spirit of their original intention. As a result, markets remained fragile and investor's confidence low.

Apparently, success was not guaranteed and policy was not self-explanatory, too. In this situation the government soon realised that market-oriented reforms were short of an appropriate fundament fitting to explain the reform process, justify the intentions and set well-defined goals of the whole endeavour. Formally, the Arusha declaration of 1967 was still in place. It articulated the great vision of Tanzania's development philosophy based on socialism and self-reliance as the long-term goal of development.[111] But as reality long got ahead of the principles of the declaration, Tanzania found itself "without a meta-policy to serve as the basis and guide for the development effort" (Mallya n.d.). For that purpose, a new basic concept of politics had to be put in place and in 1995, the government appointed a group of experts to formulate a policy document to that end. The outcome was the Tanzania Development Vision 2025,[112] which was presented to the public in August 1998.

The vision was explicitly put in the context of the major previous guiding concepts – the quest for independence as the first superior, and the Arusha Declaration as the second fundamental political concept in the country's history (c.f. Planning Commission 1998). Therewith the Vision 2025 became the key reference for Tanzania's present and future development strategy. It envisioned that "Tanzanians will have graduated from a least developed country to a middle income country by the year 2025 with a high level of human development" (Planning Commission 1998). As specific goals for the year 2025 the document states:

- High quality livelihood,
- peace, stability and unity,
- good governance and the rule of law,
- a well educated and learning society; and
- a competitive economy capable of producing sustainable growth and shared benefits.

Though if considered as an isolated text, the vision might have some weaknesses such as being rather general, reformulating commonplaces of the modern development discourse, providing a rather general notion of progress or assuming rather unrealistic performance benchmarks,[113] its undoubted strength derived from its dedication to democracy, sound economic management and human development. Furthermore, the Vision 2025 makes an explicit reference to the role of information and communication technologies (ICT). Section 4.2 .(iv) reads as follows:

[111] For the declaration, see URT (1967) a summary of basic principles is given in Wenzel / Wiedemann (1989).

[112] The Tanzania Development Vision 2025 is quoted as Planning Commission (1998).

[113] Among others e.g. "absence of abject poverty", "absence of corruption and other vices" or an expected growth rate [of GDP] of "8% or more" (c.f. Planning Commission (1998).

> "Advanced micro-electronic information and communication technologies (ICTs) are central to competitive social and economic transformation. ICT costs are continuing to fall while their capabilities and resultant profitability enhancements are increasing.
> These technologies are a major driving force for the realization of Vision. They should be harnessed persistently in all sectors of the economy and should be put to benefit of all social groups with a view to enabling the meeting of basic needs of the people, increasing productivity and promoting competitiveness.
> The new opportunities which the ICTs are opening up can be harnessed to meet the goals of the Vision. However, appropriate skills and capabilities would have to be put in place. This task demands that adequate investments are made to improve the quality of science based education and to create a knowledge society generally".

The Tanzania Development Vision 2025 – in conjunction with the actual reform record at the end of the 1990s – did not only provide the basic principles to guide all Tanzanian development efforts, it also, for the first time, highlighted in prominent position the role of ICT as an important area in the development process. This claim was further vitalized in the due course. However, in the given context it is sensible to leave the big policy concepts for the moment and turn again to the sectoral policy approaches and their implementation from 1994 onwards.

4.2.3.2. The Role of the Ministry of Transport and Communication (MoCT)

As in the early phase of telecommunications, the Ministry of Transport and Communication maintained its role as the leading government authority in the sector – in a new self-conception acting mainly as organiser and facilitator. The ministry took action on several levels to implement the reform programme. It drafted the new legislation to establish an independent regulatory body and controlled the commercialisation and corporatisation of TPTC as decisive step for the creation of TTCL.[114] With the finalisation of the new sectoral setting the ministerial responsibility was (a) define the sectoral policy and ascertain that the policies are implemented; (b) incorporate the government's economic and social objectives into the telecommunications policy, (c) interact with other relevant government departments on all questions in the area of telecommunications; (d) represent the government within regional and international telecommunications bodies and conclude treaties, agreements, conventions and international regulations in the area of telecommunication (c.f. UNECA 1996; MoCT 2001).

[114] E.g. prior the formal establishment of TTCL the ministry signed a Ministerial Directive and Memorandum of Understanding in March 1992 between the Ministry and TPTC outlining the elements for the commercialisation of TPTC as main means to overcome service insufficiencies and make TPTC financially self-reliant in its operations ready to be transferred into TTCL (UNECA 1996).

The MoCT also was significantly involved in the legislative process drafting the Tanzania Communications Act and the Tanzania Telecommunications Company Incorporation Act (1993) and it assisted the newly created bodies to fulfil their new mandates as regulator and service provider. Thus, the MoCT successfully redefined its new role as sectoral facilitator, but beyond the mere words, the ministry remained a powerful player. It still possessed far-reaching and expansible core functions and continued to act as a supervisor and management recruiter of TCC and, in some cases, also intervened directly and revised TCC's directives (see below). In that sense, the ministry was both, a reform protagonist, but also eager to secure its own interest in the telecommunication sector.

4.2.3.3. The National Telecommunication Policy

Under the auspices of the MoCT, the policy drafting process of a National Telecommunication Policy (NTP) was already drawn up in 1992, but it was not adopted before 1997. Partly because the adopted legislation already had a strong policy element and served as a plausible foundation of government's intentions,[115] and hence, to a large extent, anticipated the policy.[116] But more important, it turned out a difficult task to draft an appropriate policy linking the technical implementation measures given in the adopted legislation to a comprehensible policy outline comprising foundation, goals and strategies of the new approach. Early versions equalled a long and partly spongy wish list including e.g. unrealistic scope of service and vague universal access provisions, or feeble definitions of scheduling, goals and strategies (c.f. UNECA 1996). So it took time to finalise the document, partly due to limited resources and capacities partly due deliberate delays. However, in the end the ministry drafted a comprehensive policy document that was adopted in October 1997.

The cardinal objective of the telecommunications policy has been to ensure that telecommunications services are provided in a liberalized and competitive manner. Its specific target is to achieve a telephone density of 6 telephones per 100 population by the year 2020.[117]

[115] E.g. defining the function of TCC as "to promote competition" and that the Commission "shall have regard to efficiency and economy" (URT 1993), or transforming TTCL in a corporate body and take measures "designed to ensure that business is conducted according to commercial principles" (URT 1993a)

[116] Furthermore, reform progressed lapsed in the last years of the Ali Hassan Mwinyi presidency who ruled until 1995, c.f. Naschold / Fozzard (2002: 3).

[117] In 1997 the country's telephone density stood at 0.3 telephones per 100 population, well below the average for neighbouring Kenya (0.8), and also below the African average (1.9).

> The "document underlines the objectives and strategies of the National Telecommunication Policy (NTP), for the period 1997 through 2020. The NTP aims at ensuring the accelerated development of an efficient telecommunications network that can provide an info-communication infrastructure and universal access to telecommunications services by all sectors of the national economy and segments of the population. The NTP underscores the development of the telecommunications sector as a guideline to all stakeholders in line with national macroeconomic, legal and regulatory regimes." (URT 1997)

The main government strategy in its endeavour to achieve the stated target based on market oriented approaches. Hence strategies geared towards encouraging investments in the sector and emphasising the role of the private sector, who shall ensure the provision of "adequate, sustainable, and efficient telecommunication services in all sectors of the economy" (URT 1997).

In addition to the merely economically based core statements, the policy also covers a limited number of complementary issues such as restating the role and obligations of the government, regulator and operators and also encourages the development of human resources, the adoption of industry standards, and the development of local manufacturing capacities. Furthermore the NTP highlights the need for development of telecommunications services in rural areas and states the target to provide telecommunication facilities in each village by the year 2020. To this end NTP called upon the establishment of a rural telecommunication development fund (c.f. URT 1997).

What has been the impact of the policy? At first glance, the NTP was little more than a reaffirmation of previously adopted strategies and, as it will be shown in section 4.2.6. it apparently resulted in an only modest impact on the sector. In 1997 the market already took shape and market participants have already rolled out their businesses. A second view, however, shows some strengths, too. The policy has provided a condensed and coherent orientation, it has been balanced and placed emphasis on some viable focal points that clarify the government's commitment and strategy. Therefore the NTP became an important reference for sectoral development goals, e.g. with regard to the intended teledensitiy of 0.6 and the installation of telecommunication facilities in each village by 2020. Though not legally binding, the government accepted an obligation to continue the reform process and, therewith, improved the planning reliability for the private sector to build and maintain the network infrastructure.

For comparison, the Asian average at that time accounted for a teledensity of 6.61 (c.f. ITU 1999: A-4, A-7).

4.2.4. Building Regulatory Capacity: The Creation of Tanzania Telecommunication Commission (TCC)

TCC came into being on the 1st January 1994 after the government of Tanzania enacted the Tanzania Communications Act. No.18 of 1993 in December 1993 and assigned TCC the mandate "to be responsible for the regulation of postal and telecommunications services in the United Republic of Tanzania and to provide for the designation of public postal and telecommunications licensees and to provide for all matters pertaining thereto" (URT 1993).

TCC was established under the assumption that credible and transparent regulation is primary to competitive markets. To enable TCC to fulfil its mandate, it received substantial support under the TRP, such as assistance in form of in depth consultancies on a variety of regulatory aspects or comprehensive staff training including study tours in various parts of the world.[118] Apart from the details to be applied, some general considerations guided the establishment of TCC. These were that:

i. the Commission shall be independent in all aspects, except the legislated mandate to the minister to intervene through specific directives,

ii. the adopted regulatory framework shall be flexible, allowing room for development,

iii. the administrative procedures shall be as simple as possible and allow appeals; and

iv. the Commission shall be self-financed through license fees, royalty charges as percentage of revenues of operators, and fees for services rendered (c.f. UNECA 1996).

With regard to its organisation, TCC has been composed of a chairman appointed by the President, and six Commissioners appointed by the Minister of Communications and Transport. The Commission was supported by a secretariat of experts headed by a Director General who was appointed by the President. The Commission recruited all other staff. In 1995, with the onset of reforms, this was a total of 16 senior officers organized in six departments: Telecommunications Services; Frequency Management; Postal Affairs; Legal

[118] The regulatory framework was prepared by an inter-ministerial committee, which carried out extensive consultations, including visits to a number of countries for first-hand experience on the functioning of their regulatory framework to the USA, UK, Malaysia, Singapore, Sweden and Finland (c.f. TCC 2001a: 7).

Services; Consumer and Industry Affairs; and Finance and Administration (c.f. URT 1993; TCC 2000).

Box 4.4: The Mandate of TCC

The Tanzania Communications Commission was established by the Tanzania Communications Act No. 18 of 1993 to regulate postal and telecommunications in Tanzania. Its functions include:

- Licensing telecommunications and postal operators, equipment vendors and contractors and monitoring their performance
- Ensuring quality Postal and Telecommunications services
- Allocating and managing the radio frequency spectrum
- Type approval and standardization of telecommunications equipment
- Regulating telecommunications and postal tariffs
- Numbering administration
- Law and Policy enforcement
- Arbitration of disputes between operators and between operators and customers
- Promoting competition in the postal and telecoms industry
- Advising the Government on postal and telecommunications matters
- Monitoring quality of postal and telecommunications
- services
- Research and Development

Source: URT (1993)

TCC was installed as the main implementing agency for the government's telecommunications policies and as an instrument to build confidence in the market. These were high standards, all the more because TCC needed to balance short-term regulatory issues and at the same time seize the long-term developments anticipating major tendencies in a rapidly changing environment.[119] The full range of tasks and responsibilities is given in box 4.4. Notwithstanding the relevance of each of the functions and their respective interlinkages, some regulatory core responsibilities emerged and became the focal points of TCC's regulative approach:

[119] In order to ease TCC's introduction, the new regulative approach was not only based on a deliberate set of rules and procedures, it was also supported by recruitment procedures. Especially Professor Awadh Mawenya, the first chairman of the commission, and Emmanuel N. OleKambainei, the first Director General, were recruited not only due to their professional qualification, but also due to their high personal integrity meant to be a clear indication and confidence building measure in support of the regulators independence (c.f. Miller 2003).

- *Licensing policy* aimed to establish a market structure introducing market segments and the number of players in each segment.
- *Interconnection policy* to ensure the interoperability of services and a level playing field for the operators.
- *Frequency management*, that is the allocation of frequencies to ensure compliance with the given rules avoiding electromagnetic interferences for failure free wireless data or voice services.
- *Universal access policy* aimed to provide accessibility and affordability of telecommunication services wherever they can be supplied on a practical and economically justified basis.

Among these, the Commission's powers to license service providers and establish performance standards to the licensees became TCC's key activity in the given phase. Figure 4.8 shows the different types of introduced licenses and the scope of allowed competition in each segment that predetermined the structure of the emerging telecommunication market.[120]

Figure 4.8: TCC license structure 1994 - 1999

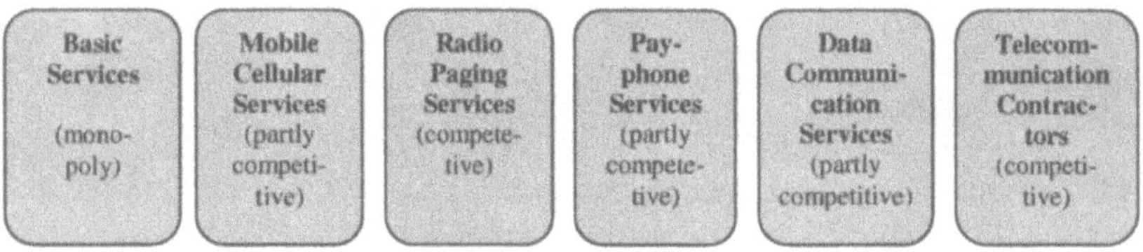

All approved licenses came together with guidelines and procedures, introducing performance benchmarks and, for example, roll-out obligations and other procedural guidelines for the operators.[121] All market segments, except the monopoly for basic services on mainland Tanzania – including the gateway for international telephone calls (fixed and mobile) – had been fully or partly liberalised. The fixed line monopoly was in accordance with the TTCL corporatisation process meant to phase out the monopoly. Mobile services were partly competitive and the same applied for radio paging, though this service

[120] According to TCC (2001a: 6-7), for an overview on licensed operators, see also figure 4.9.

[121] E.g. the following guidelines had been issued in the first two years of operation: (a) procedures for radio communication services licenses; (b) procedures for terminal and network telecommunication equipment type approval; (c) procedures for applying for contractor's installation and maintenance license; (d) procedures for importation and distribution of telecommunication and radio communication equipment; and (e) procedure for licensing land mobile cellular telephone operators (c.f. UNECA 1996; TCC 2001a).

never matured. Payphone services were actually not approved as stand alone licenses, but fell under the fixed line monopoly, in which TTCL was obliged to provide resale agreements to allow new competitors to enter the market. In data communication three providers were licensed forming a partly competitive environment and the telecommunication contractors segment (importers, distributors, wholesalers, equipment installation and maintenance) was fully liberalised.

This license setting laid the basic structure and segmented the sector. All segments required some higher-level ruling to ensure a coordinated and efficient sector development and, therewith, the basic setting was in line with the original intention to phase in a competitive environment. On the other hand licensing policy also had some shortcomings, particularly with regard to the partly liberalised segments, in which the approved operators failed to deliver the expected services. TCC issued three licenses in the data communication segment in 1996, but two of them – Wilken and Sita – almost solely focused on a premium business segment, so that their impact on a broader ICT diffusion was only marginal. The role to serve a broader customer segment was deliberately reserved for the TTCL subsidiary Datel. Thus, competitive provisions were de facto minimal and since Datel was not yet ready to offer competitive service packages, the regulator missed an opportunity to introduce capable players in the field and promote a market dynamism in favour of broader ICT diffusion (see also section 4.2.6.).[122]

The fact that TCC measures resulted in a rather imperfect liberalisation revealed a second motive for regulation and a closer inspection beyond pure economic considerations, indicate that TCC's introduction of data communication licenses was rather meant to carefully phase in the Internet than to provide for competitive markets. TCC's decision to install the TTCL subsidiary Datel as the main operator for most customers, among them the emerging Internet Service Providers (ISPs), was a safeguard measure to phase in the Internet without loosing political control. However, the potential advantage of controlling the flow of information was soon outweighed by a very low Internet utilisation in the country. Particularly the ISPs complained about prohibitive high costs and insufficient services. The poor record of the data communication providers and increasing pressure from the ISPs led TCC to reconsider the license setup in 1999. As a result, it approved new data carriers licenses to stimulate competition

[122] Similarly, TCC fixed exorbitant data communication license fees at 100,000 US$ plus a royalty fee of 5% on annual turnover (c.f. AISI 1999) and thereby forced the operators to charge high access fees on the end-consumers in order to recover costs. The high cost-structure thus incriminated the business of electronic data communications since the onset of regulation and made access to data lines affordable to a minority only.

in the sector and also introduced stand-alone licences for Internet service providers (see chapter 4.3.3.2.).

Apart from licensing TCC approved basic guidelines for interconnection and frequency management, the other main focal points of TCC's regulative approach. With regard to interconnection, TCC stipulated interoperability of all services, but left the commercial agreements of interconnection rates to the mobile and fixed-line telephone operators. TCC would only intervene by appeal (Int. Medeye 10.6.2002; Int. Mruma 10.6.2002). Although there were a number of interconnection disputes between TTCL and the first mobile operators (c.f. Miller 2003), the full complexity of the issue hardly emerged before 2001, when new players entered the scene (see chapter 4.3.3.4.). The same applied for frequency management. The full relevance of the issue only appeared later, when frequencies and radiocommunication channels were increasingly understood as a limited national resource. However, the issue started to loom as TCC rashly allocated frequencies for data communication services that interfered with voice communication leading to an outage of services. Hence, both issues interconnection and frequency management were on the agenda and repeatedly concerned TCC in individual cases, but till then hardly as comprehensive and coherent strategic approach.

The last main working area related to TCC's approach to universal access and the promotion of rural telecommunication. Here, TCC's early profile remained low, too. Though TCC had announced the plan to set up a rural telecommunication development fund to which all licensed operators would have to contribute already in 1994 (c.f. Luhanga 1994), a systematic implementation strategy did not materialised before 2003. However, this did not mean that universal service approaches dropped off TCC's agenda. Firstly, TCC, after all, channelled resources replenished by annual license royalties to support rural telecommunication development (Int. Medeye 10.6.2002, Int. Mruma 10.6.2002), but never disclosed the procedure explicitly as a working rural telecommunication fund.[123] The largest proportion was invested in the Sengerema telecentre, a pilot project for rural Internet connectivity in Mwanza region as part of an international approach advocated and supported by the ITU, UNESCO and the Canadian International Research Centre (IDRC) to promote telecentres in rural areas.[124] Secondly, TCC's efforts to promote rural

[123] An officially launched fund also would have implied rules and conditions under which funds could had been withdrawn and this might have supported the decision to run a rather low profile scheme.

[124] Sengerema was on of 6 African pilot telecentres, for details see Ernberg (1998) and ITU (1998). Initial steps to establish the telecentre in Sengerema date back to 1997, but it was only in 2000 that the telecentre became operational, c.f. Nielinger (2003a).

telecommunication was embedded in the regulator's licensing policy. Though concrete figures of service obligations remained undisclosed and became public only in individual cases as within the TTCL privatisation agreement, the general intention of the regulator's approach can be deduced from the overall licensing setup: Performance targets were set for basic services as stipulated in the National Telecommunication Policy and apecial attention was paid to the provision of payphone services as indicated by the resale agreement between TTCL and other payphone providers.

In mobile telephony, TCC had introduced service zones – coastal, northern, central, and southern – to ensure the broadest possible service availability. Only as operators almost exclusively operated in the coastal zone (=Dar es Salaam) and were reluctant to apply for other zones regarded as largely unprofitable,[125] TCC replaced the service zone approach with countrywide licenses for mobile telephony in 1999. Instead of zones to be served, it included service obligations, e.g. for remote or disadvantaged regions. Between 1994 and 1999 service obligations and performance targets were at the centre of TCC's rural telecommunication development strategy.

Evidently the scope of the assigned tasks and also pure organisational problems inherently limited TCC's working capabilities. This was further exacerbated by the fact that ministerial directives continued to overrule TCC's decisions in several cases and culminated in a suspension of TCC's duties on behalf of the MoCT in May 1997. The ministerial decision was caused by a dispute between the regulator and Mobitel following Mobitel's unlicensed upcountry rollout and the regulator's attempt to prevent it. The actual issue was later solved by a slightly opaque out of court settlement resulted in an apparently politically backed national operating license for Mobitel (c.f. Miller 2003). However, even more important, the regulative suspension lasted more than a year from May 1997 to mid 1998. It produced a regulative standstill and deprived TCC of its statutory independence. This amplified as it did not remained an individual case, but was echoed in a series of disputed decisions[126] and a further expulsion of

[125] This applied with the notable exception of Arusha and Mwanza, the most important secondary towns in the country, in which Mobitel started to rollout mobile communication services in 1997. This rollout was also an indicator of the immature institutional framework, because Mobitel started its upcountry operations without an operating license approved by the regulator, who, as a reaction, gave Mobitel formal notice to close their operations (c.f. Miller 2003).

[126] For example, ministerial interventions vetoed TCC decisions on the subject of local shareholding in any investment in the sector that should not be less than 35% to ensure that the sector would not collapse when foreign investors pull out (c.f. The Express [Dar es Salaam] 9.12.1999). Another example related to TCC's denial of a mobile telephone license

disagreeable TCC commissioners in 1999.[127] Henceforth, the openly perceivable prerogative of the government in telecommunication matters was a major setback for building confidence in the regulatory regime. Thus, in addition to merely procedural and technical inefficiencies,[128] TCC shortcomings can be summarised as follows:

- *Credibility:* The greatest concern with regard to regulation touched upon the scope of the regulator's independence from political interference. Though the Communication Act of 1993 clearly delineated policy and regulatory functions between the line ministry and TCC, it vested the ministry with far reaching competencies and stated that regulatory functions are to be exercised in accordance with government policies. Therewith, legislation provided the ministry with substantial control and extensive regulatory oversight. Doubts on the regulator's independence were exacerbated by recruitment procedures, especially after the regulatory suspension period between 1997 and 1998. From the very beginning this burdened the reputation of the regulator, especially since some of the commissioners came directly from the MoCT or had close ties to the state owned TTCL.[129]
- *Consistency of regulation:* The regulatory setup was heavily TTCL biased. Even if the TTCL monopoly for basic telecommunication services – with the exception of Zanzibar – was as a conscious safeguard to smooth the progress of TTCL's corporatisation, the regulatory provisions

---

for TTCL – under the brand Cellnet – whilst still joint venturing with Mobitel, which was overruled by the ministry. As an interesting side fact, it was mainly the World Bank who pushed the regulator to reject a TTCL mobile telephone license until TTCL was privatised. Therewith, the minister also enforced a rule against the resisting World Bank, c.f. The Guardian [Dar es Salaam] 20.4.1999, The Express [Dar es Salaam] 3.6.99.

[127] c.f. Daiy Mail [Dar es Salaam] 16.7.1999.

[128] A survey conducted by Dietrich Müller-Falcke summarised the main operator complaints. These were permanent absence of staff, a lack of timeliness and internal coordination, arbitrary license application procedures and culminating in reproaches of unequal treatment and red tape drawbacks. The interview results here base on an unpublished draft report by Müller-Falcke, which later has been published in an abbreviated version as the Tanzania chapter in GTZ / ZEF (2002).

[129] Despite the close relations between TCC, TTCL and the MoCT it would be incorrect to present TCC solely as a henchman of politics. TCC always was liable to its own mission and the relation between the three was not without friction. For example in 1999 TCC granted a license for mobile telephone services to Vodacom without any prior consultations. This caused large irritations on side of the government, namely the mighty PRSC, because this action had slowed down the TTCL privatisation and reduced TTCL's attractiveness (c.f. World Bank 2001: 11).

went much further and granted TTCL a considerable competitive advantage. This was a result of deliberate decision-making, and thus TCC was determined to install the incumbent as the dominant market player in the country, often accepting suboptimal market conditions for broader ICT utilisation.

- *Transparency:* Regulative asymmetries in favour of TTCL, often accompanied by undisclosed agreements between TCC and the PTO, made the regulator a less predictive and a less reliable partner.
- Furthermore, weak public communication ability limited TCC's interaction with the industry and with customers. In individual cases guidelines came into force, without any prior private sector consultations or even notifications (c.f. ÅF International 2002). This disrupted not only the predictability of the business environment, it also weakened the operator's opportunity to lobby TCC in favour of their own interests.[130]
- *Capacity:* TCC' rather weak capacities to ensure appropriate regulation were largely due to the small number of assigned personnel of only approximately 25 skilled employees,[131] even though in individual cases weak qualification profiles of staff members might had contributed to regulative or operative inconsistencies. Apart from political interference, most of the shortcoming could be explained with limited resources. For the same reason, TCC also had an enforcement problem. Neither necessary working tools nor needed staffs were available to adequately monitor compliance with the license conditions assigned to each license holder. This weak enforcement record was further exacerbated by deliberate toleration of non-compliance with regulation in some areas, for example with regard to Internet Telephony or VoIP (Voice over the Internet Protocol).

All in all TCC had a mixed record during the given period. It clearly set the course of the telecommunication sector, but was not yet in the position to ensure a level playing field and build the confidence that is primary for effective market operations.[132] This was partly due to political interference, partly due to

[130] It must be emphasised that it was a learning process on both sides. Like the regulator, the business community had severe difficulties to define their position vis-à-vis TCC (see chapter 4.4.)

[131] Int. Mruma (10.06.2002).

[132] It should be noted, that a TCC self-assessment covering the given time period draws a friendlier image. From the commission's perspective, itself, TCC fulfilled its mandate fairly well, especially in comparison to other African countries. A summary of achievements is given in TCC (2001a: 7). The point of a comparatively good record was also emphatically underlined in interviews with TCC staff conducted by the author (Int. Medeye 10.06.2002;

implementation deficits. The period between 1994 and 1999 marked a largely unfinished business. Rule inconsistencies, low credibility, intransparent decision-making and weak working capacities characterised the overall performance of the regulator. For that reason and as a TRP follow-up activity, TCC continued to receive considerable (SIDA-funded) technical assistance to overcome these shortcomings. However, with the growing market, regulatory issues also multiplied and posed new challenges to TCC as discussed in chapter 4.3.3.1.

4.2.5. The Story of TTCL Part I: The Road to Privatisation

The Tanzania Telecommunication Company Limited came into being on the 1st January 1994. Under the Tanzania Telecommunications Company Incorporation Act of 1993[133] it was incorporated as a wholly state-owned, but autonomous company operating as a commercial enterprise that took over the telecommunication service provision mandate of the defunct TPTC. The government appointed a Board of Directors, which was fully responsible for the management of the company and, in order to operate as a commercial enterprise, the new management embarked on a major restructuring process.[134] As described above, this happened with significant resources from the TRP. Though not every targeted figure under the TRP had been met,[135] TTCL noticeable improved for instance with regard to service quality and exchange capacity and

Int. Mruma 10.06.2002). However, a balanced view must take into account both, the basic achievements, and the technical and structural deficits.

[133] To be quoted as URT (1993b)

[134] The Tanzania Telecommunication Company Incorporation Act avoided to explicitly touching upon the issue of privatisation, though it had provisions therefore. E.g. in part II, section 6.(3) it is stated that "the total foreign investment in the Company shall not at any time exceed such percentage of the total shareholding in the Company as the Minister responsible for finance [sic] shall from time to time determine." (URT 1993b). The act provided other rather hidden provisions and this very well reflected the governments opinion during the first years of TTCL, in which it faced a strong public opinion against any privatisation deals, especially when foreign partners were involved. Therefore the government adopted a stance in which it welcomed "a friendly partner with capacity to inject both capital and expertise" (Mallango 2000), but at the same time announced that TTCL would not be privatised at that time (c.f. PSRC 1994). A good summary on the general bad attitude to privatisation is given in Augustin (1998: 76).

[135] As expressed in the Memorandum of Understanding signed by the TRP-contractors in 1992 (c.f. World Bank (2000a); Luhanga (1994).

investments made under the TRP led to an increase of 81 percent of the total number of TTCL subscribers. Other performance indicators were:[136]

- Teledensity (telephone per 100 population) gradually improved from 0.3 in 1992 to 0.46 in 2000.[137]
- Exchange capacity increased by about 67% from 128,000 in 1995 to 215,000 lines in mid 2000, and was expected to reach 247,000 lines by the closing of the project (end of December 2000).
- Exchange connections increased by about 81% from 91,000 connected direct exchange lines (=subscribers) in 1995 to 165,000 connected DELs in mid 2000.
- Call Completion Rate (CCR) for national calls reached 58% in 1999 compared to 36% in 1995.
- Incoming CCR for international calls increased to 41.5% in 2000 compared to only 23% in 1994.

Though the record was fairly well, improvements were hardly sufficient. All performance targets were related to the old and defunct TPTC, but regarding the needs of the sector, TTCL failed to cope with the rapid changes in the industry and the growing demand for telecommunication services in the country. Difficulties appeared in different areas, such as deployment of inappropriate or an incompatible mix of technology, weak system maintenance or underutilisation of installed equipment – though it shall be noted that many project components were the first of their kind in the country and, therefore, did not gain much field experience. However, the most sorrowful developments appeared with regard to weak financial and corporate management capacities. Delays in connecting up new capacities, poor billing capacity, and a high levels of accounts receivable resulted in non-compliance with some legal and financial covenants during the first years of TTCL's corporate performance (World Bank 2001: 7). On the ground of these persistent problems, the government decided to accelerate the privatisation of TTCL and it became a key objective to make the incumbent fit for privatisation.[138] This was made with continuing efforts to

---

[136] The figures are taken from World Bank (2000a 2001); ITU (2002a), ITU (2001)

[137] According to a common misunderstanding, the overall objective of the TRP was to achieve a telephone density of 6 per 100 population by the year 2020 (as compared to 0.29 in 1990). However, the target never was an explicit objective of the TRP, but a specific sector target of the National Telecommunication Policy of 1997 (c.f. URT 1997; MoCT 2001).

[138] In first range the privatisation of TTCL was among the conditions the International Monetary Fund (IMF) had laid down to enable the country to receive the second instalment of its 38.2 million US$ loan under the enhanced structural adjustment facility (ESAF), see East

modernise the infrastructure and also improving TTCL's management capacities, e.g. to introduce international auditing standards,[139] recording of revenue and debtor accounting, and the like.

The tender for the divestiture of TTCL was launched in June 1999. Invitation letters to participate in the privatisation through an issue of new shares representing 35 percent[140] of the enlarged share capital were sent to 120 organisations and in addition, advertisements asking for expressions of interest from telecommunication companies and financial organisations were placed in the international and local press and on the PSRC website. All in all 29 telcos expressed their interest, nine of them submitted pre-qualification documentation and six telcos pre-qualified (c.f. Mwandosya 2001). The process of TTCL's privatisation, in which finally a Dutch / German consortium of MSI and Detecon had been picked as investor and which was until early 2004 characterised by a lasting dispute between the investor and the government is further discussed in section 4.3.4.

In the meantime pro-privatisation measures installed and anchored TTCL as the main market player. In basic telecommunication TTCL was granted a monopoly – except for Zanzibar – and the PSRC announced that an exclusivity period for basic telecommunication would be included in the privatisation package. TTCL was also granted a full cellular license in June 1999 after withdrawing its 29% shares in MIC Tanzania (Mobitel). The joint venture was merely meant as a temporarily solution, anyway. It guaranteed TTCL's participation in the first cellular operation in a time, when cellular technology was just arriving and TTCL did not have any experiences in this area. The new mobile license was strongly backed by the government and against the original intention of the regulator and the World Bank (c.f. Guardian [Dar es Salaam] 20.4.1999, The Express [Dar es Salaam] 3.6.1999). The approval, in fact, highlighted the government's and TTCL's room for manoeuvre to put the incumbent in the position regarded as the most favoured one. A similar pattern reoccurred with

---

African [Nairobi] 26.4.1999. In due course, the government decided to include the strategic parastatals in the parastatal sector reform programme in 1996 and after the Cabinet approved the TTCL privatisation in June 1997, TTCL was handed over to the Parastatal Sector Reform Commission (PSRC), c.f. Mwandosya (2001), see also The Express [Dar es Salaam] 1.7.99).

139 Since 1997 TTCL was subject to external auditing, conducted by PriceWaterhouseCoopers (PwC).

140 The International Finance Corporation of the World Bank with support from the PSRC strongly had advocated to sell 51% of TTCL's share, the government rejected this proposal and enforced no more than 35% of shares would be sold to an strategic investor, c.f. The Express [Dar es Salaam] 24.6.1999, 1.7.1999.

Datel, TTCLs brand in data communication services. Though the Datel license remained attached to TTCL in the pre-privatisation phase, shares were quickly sold off as the new investor repelled Datel as its data communication arm. In the six years between 1994 and 1999 TTCL changed considerably in both service provision and management. But the company continued to face difficulties in all areas of business operations and TTCL remained in intransparent proximity between the state and the regulator. Nevertheless, TTCL was installed as market leader with a presence in each market segment and thus well positioned and "marketable".[141]

4.2.6. Market Developments between 1994 and 1999

As a result of the described reforms and adjustment processes, the telecommunications sector developed and diversified in the years 1994 to 1999. A schematic overview of the sectoral setting is provided in figure 4.9. The new policy and the new regulative framework laid the foundation for private sector participation in various market segments and several new operators entered the market to provide services in competitive and partly competitive environments. TTCL was installed as the main market player, which is indicated by the dashed line in figure 4.9. The period between 1994 and 1999 can be characterised as the introduction of limited competition and, to a degree, limitation were meant to be a safeguard measure for TTCL to give the incumbent the opportunity to adjust to the new environment and change into a profitable business. With regard to the individual segments, the new sectoral setting resulted in the following developments.

4.2.6.1. Fixed and Basic Services

Recalling that the primary goal of the TTCL restructuring process had been its commercialisation and the improvement of service delivery based on the assumption that both goals will aid one another in the long run, the short-term consequences also showed some trade offs. Rehabilitation and modernisation of the out-dated network also entailed the closing of existing, but unreliable or insufficient network components. Furthermore, rehabilitation was a time consuming task that was delivered in a series of individual projects. The restructuring affected the market in different ways. Figure 4.10 summarises the development of some selected telecommunication performance indicators for the given period.

[141] These were the word of the external auditor PWC, quoted in: The Guardian [Dar es Salaam] 24.11.1999.

Figure 4.9: Tanzania Telecommunication Sector 1994-1999

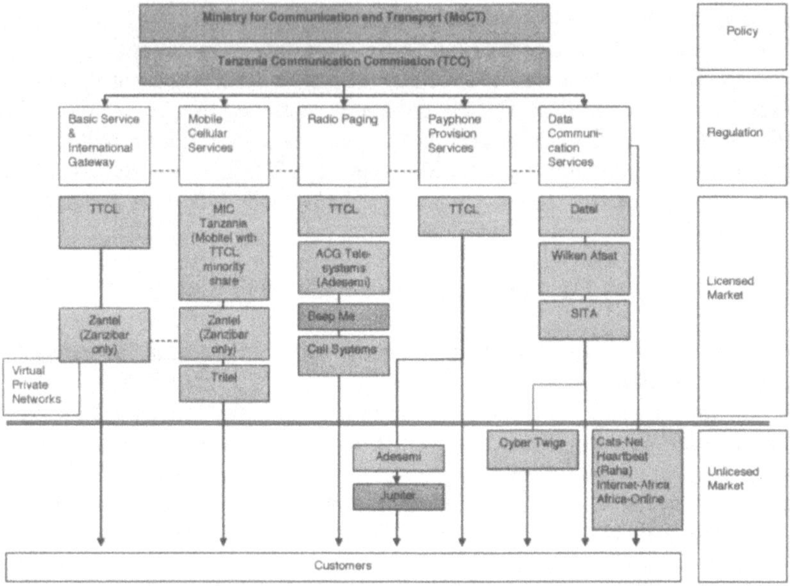

One major change was a radical change in the tariff structure. Whereas local telephone calls had been heavily subsidised by high international call charges during the TPTC phase, this subsequently had been altered. Still, until 1990 TPTC charged 3 Tanzanian shillings (TSh.) for a 3-minute local call; with the inception of TTCL, rates increased to 12 TSh. in 1994 and at the end of 1999 rates were fixed at 63 TSh. (ITU: 2001:158), after all an increase of total 525 percent between 1994 and 1999. The ratio of the new tariff structure was laid down in the Tanzania Telecommunications Company Incorporation Act to "ensure that charges for services are kept at commercial rates" (c.f. URT 1993a). Naturally, this affected the end-consumer market. The number of local calls decreased from 8,890 in 1993 to 6,947 in 1999 (figure 4.10), but service provisions for international calls improved and grew at roughly 230 percent between 1993 and 1999. This tariff adjustment was more viable, but notably burdened those customers – and this was the vast majority of the population – who relied on the accessibility and affordability of local calls. In this respect, new tariff structures forcibly induced a new attitude towards telecommunication services, and it was not without friction that commercial principles had been introduced in the telecommunication end-consumer market.

Figure 4.10: Selected TTCL performance indicators 1993-1999

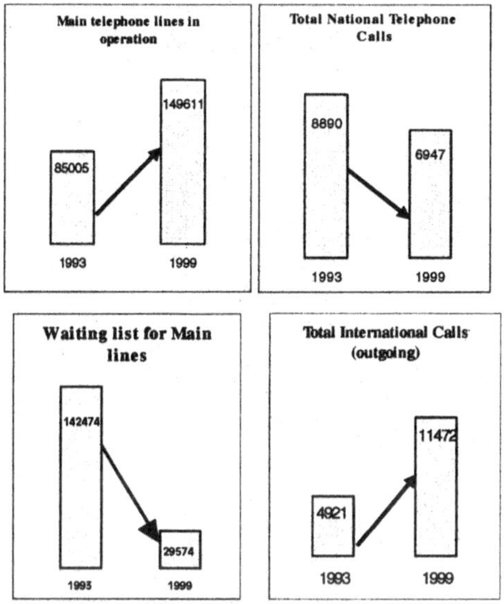

Source: Author's compilation. data taken from ITU 2001a: 158

Starting from a low level, the overall balance of TTCL's performance in the given period was modest, but clearly positive The number of main telephone lines in operation gradually increased from 85,000 in 1993 to roughly 150,000 in 1999 and, at the same time, the waiting list for main line subscription sharply declined from high 142,474 to 29,574 (figure 4.10). Therewith TTCL's improved performance record was the main contributor to an overall increased availability of telephone services in the country.

The second operator Zantel was licensed in early 1996 to provide telecommunication services on the Zanzibar islands Unguja and Pemba. The creation of Zantel followed a direct intervention of the Government of Zanzibar, intending to overcome TTCL's poor service delivery on the islands and meant to promote local economic development (c.f. TRASA 1999; SATQ 1997). In 1994 the Zanzibar Telecommunication Corporation – Zantel was created as a government joint venture (18%) with Emirates Telecommunications Corporation (Etisalat) (34%), UK based Kintbury Investment (24%) and Meeco International (24%) as local investor. Zantel received the TCC license for basic and mobile services on Zanzibar and in 1997 started the operation of its international gateway and a GSM network in Unguja. At that time, Zantel has operated on the islands only and simply because the limited scale of operation it had only little significance for the Tanzanian market as a whole. The World

Bank estimated the number of subscribers at roughly 2,000 (World Bank 2000). However, apart from end-consumer services, Zantel capitalised on international gateway operations and developed into a serious competitor to TTCL in this regard targeting international gateway customers.

4.2.6.2. Mobile Services

As in many other African countries mobile telephone services in Tanzania also developed as the most dynamic growth segment in telecommunication. First operations started in 1994. Millicom Tanzania – a joint company between Millicom International Cellular (MIC) (51%), a Luxembourg based international cell phone company and TTCL (49%) – introduced a mobile analogues network (TACS)[142] under the brand name Mobitel. MIC Tanzania was one of the pioneers in the African mobile phone history. Jointly with Ghana and Mauritius, Tanzania was the first location from where the company developed their presence in the African cell phone market. In fact, MIC already had entered the country before the sector reforms commenced in 1994. In 1993 it won a tender issued by TPTC who seek a partner to roll out mobile services. As the regulator came into being, MIC had to apply for a new license as required by the Communication Act of 1993. In the new licensing arrangements the country was divided in four service zones – coastal, northern, central and southern highland zones – and Mobitel was licensed to provide service in Dar es Salaam and Zanzibar (= coastal zone). As second operator TRI Telecommunications Tanzania Ltd (Tritel), a joint venture between Technology Resources Industries (TRI) Malaysia (65%) and VIP Engineering, a Tanzania based company (35%) was licensed in 1995 to provide services in the coastal and northern zones, but started operation in the coastal zone, only. Tritel was the first GSM operator in the country. Zantel operated a GSM System on Zanzibar. In November 1999, Zantel also got a license to operate on the mainland, but never started operations on the mainland (The Guardian [Dar es Salaam] 23.11.99).

In the first years of operation mobile phone services remained a small business. Mobile subscriber density per 100 population was between 0.01 in 1996 and 0.16 in 1999, which equalled a total of 3,500 (1996) or 50,950 (1999) subscribers (ITU 2001a: 158). Mobitel and Tritel were the main players and they managed to attract some critical mass of customers, but mainly providing services in the coastal zone, that was mainly greater Dar es Salaam. Later, the

[142] TACS = Total Access Communications System. An analogue cellular communications system, which were later displaced by digital GSM networks (=Global System for Mobile communication). For a discussion on mobile technology see ITU (1999: 14).

major secondary towns Arusha and Mwanza were covered, too.[143] Therefore, Dar es Salaam, to a lesser extent Arusha and Mwanza benefited from mobile services. The market entrance of Tritel as the second operator brought some further improvements, particularly for Dar es Salaam customers. With only two operators, market penetration of mobile services remained low – Mobitel accounted for roughly 14,000 and Tritel for 6,000 subscribers in 1997.[144] This penetration pattern was similar to other African experiences, where mobile phone services started in the second half of the 1990s.[145] Early mobile telephony faced similar problems throughout the continent. Tariffs were high, coverage limited, access to mobile telephony was only possible on contract, that is postpaid to which only a very few people were eligible. The strength of mobile telephony, prepaid pricing schemes, that the ITU labelled "a marriage seemingly made in heaven for Africa's citizens: mobile cellular and prepaid" was not yet developed (ITU 2001a:2). But beyond some similar pattern of cellular introduction, difficulties were also home made. Firstly, the country's division into service areas hampered the network roll-out and this did not change until the licenses were granted countrywide, plus an individually attached service obligation to be negotiated with the regulator. Secondly, the old joint venture between Millicom and TTCL increasingly hampered mobile cellular developments, because both partners regarded the status quo as an interim solution. The Mobitel joint venture with TTCL as evident junior partners, increasingly got in the way of TTCL's strategy to install its own mobile operation. On the other hand, Mobitel's conflict with the regulator on their unlicensed operations in Arusha and Mwanza got more stressful, so that Mobitel was handicapped, too.

The modest growth rate led to proactive sector promoting government interventions and TCC invited more telecom companies to apply for a mobile license.[146] In April 1999 the regulator received six more applications, among

[143] Following Mobitel's unlicensed business roll-out, which was objected by TCC in the first range, but later de facto overruled by the ministry that later legalised Mobitel's step and triggered a major regulatory crisis (see chapter 4.2.6.2.).

[144] C.f. Trasa (1999). Africanconnection (1999) estimated similar numbers, for Mobitel a user basis of 14,600 and for Tritel of 5,000. Figures for Zantel were hardly available, but a rough orientation would be the above given 2,000 subscribers.

[145] In 1995, half of the African countries had introduced mobile telephony, but only 7%, among them Tanzania, also allowed competition in the mobile market (c.f. ITU 2001a).

[146] The decision to open the market for more competitors was heavily contested by the operators Mobitel and Tritel. They strongly lobbied their interests to hold the current number of operators (Daily Mail [Dar es Salaam] 3.5.1999) and also threaten TCC to pull out of the country, if new mobile telephone licenses would be approved (The Express [Dar es Salaam] 22.7.1999).

them the most important cell phone providers on the continent: Mobile Telephone Networks (MTN), Vodacom and Mobile Systems International (MSI) (Daily Mail [Dar es Salaam] 5.4.1999). In the end, Vodacom South Africa together with its local partner Planetel[147] won the tender and their new license was approved in July 1999. Therewith, from the mid 1999 the mobile telephone sector was in a state of reorganisation. The division in service zones was transformed into countrywide operations to which individual service obligations were attached. TTCL sold its stake in MIC and the then planned TTCL subsidiary Cellnet was approved to roll out mobile services. Mobitel's and Tritel's operating license was upgraded for countrywide services, Vodacom received a new license, and finally Zantel's license was also expanded to the mainland (The Guardian [Dar es Salaam] 23.11.1999). Despite the fact that only three operators actually offered services and infrastructure rollout was rather slow, in the end of 1999, TCC laid the basis for the most competitive mobile phone market in Africa. This decision paid off as Tanzania entered a phase of accelerated growth in mobile telecommunication from 2000 onwards.

4.2.6.3. Data Communication Services

In 1995 TCC called for tenders for international data carriers. The data communication operators were licensed as infrastructure providers to deploy a national Internet backbone and provide for international connectivity. Three licenses were issued: for Datel and Wilken Afsat in November 1995 and for Sita in April 1996. Licenses were granted for a 10-year period and cost US$ 100,000 plus a royalty fee of 5% on annual turnover. All three data providers obtained the right to install national communication networks for data traffic[148] and the right to install their own international gateway in order to route their international traffic. The regulator indicated that no more licenses for international data communications would be issued during the ten-year license duration (c.f. Jensen 1999).

The data operators started business September 1996, and in accordance to their corporate profile targeted distinctive customer segments. Datel was a joint venture between TTCL and France Telecom's subsidiary, Nexus International as

[147] Planetel was not unknown in the Tanzanian telephone market. Until the joint venture with Vodacom Planetel was service provider for Iridium's global satellite network in Tanzania (East African [Nairobi] 11.8.1999).

[148] As opposed to voice traffic that fell under the TTCL monopoly.

majority shareholder (36 percent).[149] As part of the TTCL group Datel targeted a broad customer segment and the operator became the key player in the emerging ISP market. SITA[150] has been the world's leading provider of global information and telecommunication solutions to the air transport and aviation industry. It almost exclusively served the airline industry and all announcements to enter a broader commercial market never materialised;[151] with the notable exception of Cyber Twiga, Tanzanias first ISP, which was connected through SITA (see below). Willken Afsat, a Nairobi based company, also originated from aviation, but since the 1990s the company focused on the East African telecommunication market. Core business was advanced satellite communication solutions for East Africa, especially VSAT[152] provision making use of the company's own VSAT hub in Kampala. Though the company advertised being an ISP, it mainly targeted a premium business segment as service and connectivity provider.[153]

The operators had a distinctive profile. This also allows deducing TCC's basic considerations with regard to the intended market structure. SITA's pronounced focus on aviation contravened the claim to develop into a public data communication provider and it was more likely, that the TCC license was meant to ensure that the Tanzanian airline industry and the airlines serving Tanzania could be equipped with state of the art data communication facilities.[154] Wilken

---

[149] Further minority shares were hold by the International Finance Corporation (IFC – World Bank Group) and Proparco, a subsidiary of the Caisse Française de Développement (c.f. The Express [Dar es Salaam] 25.11.99)

[150] Originally "Société Internationale de Télécommunications Aéronautiques", SITA today uses only the shortened version of its name (c.f. Sita 2004).

[151] Announcement quoted in TRASA (1999).

[152] VSAT (Very Small Aperture Terminal) is a satellite communications system. A VSAT end user needs a box that interfaces between the user's computer and an outside antenna with a transceiver. The transceiver receives or sends a signal to a satellite transponder in the sky. The satellite sends and receives signals from an earth station computer that acts as a hub for the system. Each end user is interconnected with the hub station via the satellite. For one end user to communicate with another, each transmission has to first go to the hub station which retransmits it via the satellite to the other end user's VSAT. VSAT handles data, voice, and video signals (Tech Target n.d.).

[153] Early Customers were e.g. Tanzania Breweries, Barclays Bank, Stanbic Bank or Unicef Tanzania (c.f. Wilken Afsat n.d)

[154] The company's history underlines SITA's strategic approach to data communication, which has been aviation only, see SITA (2004). However, to understand the SITA Cyber Twiga agreement, it might be helpful to emphasise, that in 1991 SITA created Scitor to look at the general data network side of its business. Scitor was later renamed in Equant (see Financial Times [London] 2.5.2001). Later, the new business approach by the SITA group also was mirrored in the new licensing setup in Tanzania, in which SITA gained a license for

Afsat has focused on corporate data communication and the company merely served a premium business segment. Finally Datel remained to serve the broader market, and therewith, the TTCL subsidiary was installed in the market segment with the broadest customer basis. In fact, with one exception, all commercial ISPs that subsequently came into being were connected through Datel.[155]

Data communication started at 9.6k[156] for international connectivity in 1996, but subsequently the operators managed to upgrade their data transfer capacity and already in 1997 all operators run at rates between 19.2k and 128k (NSRC 1999) – though still very low in comparison to today's rates, but for that time a noticeable improvement. Prices for connectivity developed in the opposite direction. According to AISI (1999), for the first two years of operation leased lines to Datel and SITA were 17,000 US$ a month for 128k and 12,000 US$ for 64k. These were hardly affordable. According to the same source, only in 1997 Wilken Afsat introduced a more competitive price charging 5,000 US$ for a 64k line per month, a level that became a rough reference price for the other operators, too. A further decrease occurred in November 1999, when Datel introduced wireless local loop (WLL)[157] technology for its Dar es Salaam customers to reduce the price for communications as well as maintaining high quality speed link between points (c.f. The Express [Dar es Salaam] 25.11.99). For a 19.2K wireless link Datel charged 350 US$ per month for each end, i.e. a total of 700 US$ (AISI 1999).

In addition to the three commercial operators, the University of Dar es Salaam also received an international data license, at first only to serve university community. With assistance from the Government of the Netherlands, the University Computer Centre (UCC) was rebuilt in 1995 and was assigned the task to provide computing and Internet services to the university. Originally

---

closed user groups, and Equant took over the general data communication business. However, in 1996 SITA's restructuring was not yet finalised, so that Cyber Twiga could enter an agreement with SITA, but actually meant to be an agreement with Scitor (=later Equant).

155 Whereas SITA and Wilken Afsat operated their own data hub, Datel got its international connectivity from the Norwegian operator Telenor.

156 Unit in which data transfer capacities are measured, 9.6k, or kbps, is a data capacity of 9600 bit [binary digit] or 9.6 kilobytes per second.

157 A wireless local loop is the use of wireless connection as the last mile for delivering telephone services. WLL is a generic term for a method of connection which can be implemented using a number of different technologies. With Integrated Services Digital Network (ISDN) or Digital Subscriber Line (DSL), the local loop can carry digital signals directly and at a much higher bandwidth than they do for voice only (Tech Target Network n.d.).

licensed as a closed user group, and therefore not subject to the data communication license fee, UCC, nevertheless, started a commercial ISP scheme. In the time of inception, the university operated a 64k VSAT link with a downlink in South Africa.

Beneficiaries of early Internet communication were the aviation industry, connected to SITA's network, the high end business customers taking advantage of the services offered by Wilken Afsat and a number of corporate and public customers, among them, most of the emerging ISPs who channelled their data traffic through Datel. Thus, benefits were concentrated in a few customer segments. The distinctive business strategies of SITA and Wilken Afsat left only Datel to serve a broader customer segment. However, Datel was clearly overstrained to deliver appropriate services and to reach the intended scope of countrywide data provision. All three license agreements with the commercial providers came with various obligations, especially with regard to upcountry infrastructure rollout (Int. Mbita 4.6.2002). But it became increasingly evident, that the operators fell short to meet their obligations, either due to limited capacities or due to the general business approach. As a consequence, TCC started to rethink the license setting. This was further supported as the ISPs started to lobby against the then licensing setup. From the beginning of 1999 the regulator was in principle ready to change the general setting, i.e. to introduce extra ISP licenses and to open the market for data communication services to boost the upcountry Internet connectivity (Int. Medeye 10.6.2002, Int. Mruma 10.6.2002, c.f. TCC 2002: 8)

4.2.6.4. Internet Service Provision

The emergence of ISPs was a reaction to the increasing popularity of email usage and the recognition that Internet services could be a viable business. As shown earlier (chapter 4.1.2.3.), FidoNet systems were the first facilitators of electronic communication. Since 1996 the Fido nodes lost their significance, when commercial providers started their operation. Especially COSTECH was as the largest Fido provider was affected and could only later regain some significance in Tanzania's ICT development, when it started to engage in upcountry and rural Internet connectivity. In fact, a growing academic email network and increasing numbers of NGOs connected to global communication networks indicated that commercial email services were just a little step ahead. In February 1994 the Arusha Node Marie started operation as the first commercial FidoNet host in Tanzania. Created as a private cooperative from Arusha based business people and NGOs, the node was established as user-owned society (c.f. AAAS 1996). Within the first years of operation, Marie reached a customer basis of about 120 and proofed that email services could be conducted as a small but viable business.[158] Other commercial FidoNet hosts followed, for example Dar es Salaam based Tanzania Online, including their Zanzibarian partner Zaccoserv, in early 1996 (AAAS 1996). But in contrast to the Arusha based node Marie, which remained for the time being unaffected by commencing full ISP services in Dar es Salaam, Tanzania Online felt the growing competition in Dar es Salaam and was taken over by African Online in May 1997. Therewith Tanzania became the fourth operation within the Africa Online group and their pan African ISP rollout strategy (c.f. African Lakes 2001).

Commercial ISP started soon, after the regulator had introduced the new rules for data communication. Shortly after SITA and Datel started their operations in April 1996, Cyber Twiga commenced as the first Tanzanian ISP on the 25 July 1996, utilising a 9.6k[159] Internet connectivity provided by SITA (c.f. NSRC 1999). Soon, the ISP market crowded up as new ISPs offered their services: Heartbeat Online (later Raha.com) in November 1996, Internet Africa (November 1996), Cats-Net (February 1997), Internet Zanzibar (July 1997), and AfricaOnline (May 1997). The University Computer Centre started a commercial ISP scheme in August 1996 (c.f. Hassan 1997, NSRC 1999, AISI 1999). Connectivity charges were high. Cyber Twiga charged 90 US$ for email only per month and 150 US$ for full Internet access per month, telephone costs excluded. Internet Africa as the second provider introduced a price of 100 US$ one-time connection fee per user and a monthly fee of 50 US$ per user, again, telephone costs not included (Hassan 1997; AISI 1999; NSRC1999).

As the market started to consolidate, the main ISPs took shape – Cyber Twiga, Internet Africa, Raha (former Heartbeat), Cats-Net and AfricaOnline, and as a special case the University of Dar es Salaam. In due course, connectivity prices levelled off at roughly 50 US$ per month connectivity fees plus telephone costs. Beyond the approximation of monthly connectivity fees, differences appeared with regard to pricing schemes, included online time or nonrecurring hookup costs, so that the operators developed a distinctive business profile. Cyber Twiga, for example, targeted particularly corporate customers and expatriates and focused on service rather than on price, so that costs were slightly higher

[158] Usage fees for the service were 50 US$ for registration and onsite installation of software and hardware, monthly usage fees of 30 US$ (NGOs and educational institutions had to pay reduced fees of 15 US$ / month), 0.20 US$ per kilobyte data traffic (incoming and outgoing) (AAAS 1996).

[159] The low data transfer rate was further exacerbated by the fact, that Cyber Twiga started its operation with only two telephone lines for Internet connectivity. No more TTCL lines were available and customer inconveniences inevitable (Int. Messer 20.5.2002).

(60 US$/month). The University Computer Centre offered the cheapest solution (50 US$/month, but only 30 US$ connection fee), and Raha and Cats-Net appeared as the ones targeting the broadest customer segments (Hassan 1997, NSRC 1999).

Service provision was largely concentrated in Dar es Salaam, plus customers from outside who need to call a Dar es Salaam based dial-in number to access the Internet. But despite a considerable number of ISPs, the entrance costs remain high, so that the number of overall customers remained low, too. In 1996, the ITU (2001: 158) estimated 500 Internet users in the country, slowly increasing to 2.500 in 1997 and 3,000 in 1998.[160] The first significant spread of Internet usage could not be witnessed before 1999, in which the estimated user number increased to 25,000. However, if the figures are translated into macro statistical figures, Internet usage figures remained extremely low, accumulating at a user density of only 0.12 Internet users per 10,000 inhabitants (c.f. ITU 2000: A-13).

Internet usage was in its absolutely infancy and both, costs and access were recurring issues of concern.[161] On the other hand, ISP business undeniably gained ground, particularly in Dar es Salaam. Here, services were available, and consequently costs became the key issue. The ISPs continued to complain about high connectivity fees,[162] and inefficiencies on side of the data operators. Though relations among the ISPs were not without friction – competitions was intense and the margins small – the ISPs started to lobby their case jointly and brought forward their central claim to give them the freedom to choose data providers anywhere in the world to connect them to the Internet, and therewith circumventing the disadvantageous local conditions (c.f. Daily Mail [Dar es Salaam] 24.1.00). In return, the ISPs would accept to become subject of a TCC license themselves, also because the providers recognised that they could

[160] Similar figures were provided by the BMI Technology Group, who put Internet subscribers for 1997 at 2.000 and 3.000 for 1998 (quoted in: Sunday Observer [Dar es Salaam] 14.11.99).

[161] Telling in this regard is a rather polemic text highlighting the severance of the Internet situation in Tanzania in the end of 1997. The author severely criticised the regulator as being blissfully ignorant and the operators as being utterly under equipped. Though the article was rather naive and took up a not transferable position of a technical state of the art perspective, it, nonetheless, very well summarised the hardships and shortcomings encountered whilst attempting to establish Internet connectivity at the end of 1997 (c.f. Salim 1997).

[162] In the Daily Mail (24.1.00) an ISP manager stated that average connectivity costs would account for 20,000 US$ per year for an ISP. If compared to the 700 US$ costs per month for a 19.2k line as provided by Datel as given above, the 20,000 US$ would roughly coincide with a line capacity of 64k, which would be a sound assumption of average connectivity costs at the end of 1999.

increase their negotiation power, if they could manage to take up a joint position and to counter TCC's position to discuss licensing issues with each operator individually.[163] At that time, TCC was already unhappy with the performance of the data operators. TCC proposed a new ISP license in April 1999, but to the ISP's dismay TCC demanded high license fees and the ISPs still would have to get bandwidth from one of the public data provider licensees. Thus, none of the original demands were taken into account.[164] ISPs became not only subject to an expensive license fees, they also continued to rely on the data operators licensed by TCC.

4.2.6.5. Payphone Services

Three operators TTCL, Jupiter, and Adesemi provided payphone service. Due to the granted monopoly in fixed line services, only TTCL was licensed, the other two operators had to enter resale agreements with the incumbent. Payphone services started in 1996 and the operators installed card phone boxes providing services for local, national and East African calls. Though TTCL operated telephone boxes before 1996, these run under the old tariff structure and had allowed calls of quasi unlimited duration and caused extremely long waiting lines at public phones booths (c.f. UNECA 1996). Though tariff structure changed and 3 minutes local call units had been introduced, from 1997 the availability of payphone services increased and became very popular.[165] In 1999 716 payphones had been installed in the country (ITU 2001a: 158) and payphone services were regarded as promising market. Nevertheless, the then market leader Adesemi comprising a market share of roughly 55 percent, was

[163] One result of the joint efforts was the creation of COAT (Communication Operators Association of Tanzania) in January 1999, bringing together representatives from the mobile telephone, data communication and ISP business. COAT was the first lobby group in the telecommunication, but in retrospect, COAT failed to meet the expectations, because of too broad issues to be covered in largely distinctive market segments and different willingness to cooperate. The first stronger sectoral lobby group first materialised with the creation of COAT's successor TISPA (Tanzania Internet Service Providers Association) – only comprising the ISPs.

[164] Despite a generally poor record, COAT successfully renegotiated the fees for obtaining an ISP license. This was originally set by TCC at 150,000 US$ and brought down to a nonrecurring fee of 1,000 US$ and a license fee of 1,000 US$ per year (Int. Messer 20.5.2002; see also Daily Mail [Dar es Salaam] 5.1.2000).

[165] Despite the popularity of payphone services, the overall number of phone booths decreased from 715 installed boxes in 1990 to 494 boxes in 1997 (figures taken from ITU 2001: 158). This affected TTCL boxes and mirror the process of TTCL's commencing commercialisation, i.e. to de-install defunct and uneconomical phone boxes. Growth on an economically sound basis could be observed from 1997 onwards.

put under liquidation in 2000, mainly due to poor management, inadequate financing and unfavourable resale agreements with TTCL (c.f. World Bank 2001: 6). As a result, TTCL remained the only significant market participant, the more that Jupiter never gained any significant market shares.

4.2.6.6. Public Radio-Paging Services

National public radio paging licenses were granted to four companies in December 1995. TTCL, Beep-Me Limited, Call Systems Limited, and African Communications Group Telesystems Limited (ACG Adesemi). However, radio paging never became an accepted technology, neither in Tanzania nor in any other African country. Thus, none of the operators ever started operations.

4.2.6.7. Telecommunication Contractors

Roughly 80 private companies have been licensed to import, distribute, install and maintain terminal or network telecommunication and radiocommunication equipment (TCC 2001: 7, see also Daily Mail [Dar es Salaam] 1.5.99). The number of business creations in the period can illustrate the significant impact of telecommunication sector reform on the contractor's segment. With regard to IT vendors offering hard- and software equipment and respective technical support, in fact, most of the businesses came into being between 1994 and 1999. A vendor survey conducted within this research project (see chapter 5.1) revealed that 22.7 percent of IT vendors already were established before the reform in 1993 – also reaffirming the finding that the ban on computers had never been strictly executed. However, the commencement of reform and the arrival of new technologies resulted in a major improvements for the segment and in sum, roughly 63 percent of the IT vendors came into being in this period presenting a strong indicator that reform was set on track.

4.2.7. Human Resource Development: The State of ICT Education

As the market slowly matured the demand for an educated IT workforce also increased, not only from the new IT companies, but also from the private and public sector that needed to develop their IT departments in the enterprises and in public administration. The local education system was offering a bachelor programme at the Computing Centre at UDSM, but student numbers were well below the expected demand of skilled IT professionals. In the mid 1990s a lack of professionals became a pressing issue. Though the demand still was relatively low, the first leading signs of a changing political and sectoral landscape loomed and it became evident that the country needed to adjust its higher IT education.

For this reason the University of Dar es Salaam initiated an Institution Transformation Programme (ITP) in 1993 to strengthen the University's educational profile and make UDSM a national hub for all ICT activities. In 1995 a rough schedule for the transformation was finalised and on the basis of the approved Information Policy Plan (IPP) and the Information Master Plan (IMP)[166] the old outmoded Computing Centre was renamed in University Computing Centre (UCC) and mandated to oversee the implementation of the plans (c.f. Mutagahywa 2003; Int. Nfuka 10.6.2002). The new strategic approach of 1995 aimed at increased performance, efficiency and production of quality products and services. To achieve this goals, UDSM's ICT strategy based on two corner pillars: firstly to develop UCC as ICT service provider for the university including the long term strategic goal to incorporate UCC as a fully UDSM-owned company. This was achieved in 2001. Secondly to create a Department of Computer Science as the main institution of academic teaching at UDSM. The Department finally came into being in 1999. Both achievements were major milestones in the computerisation history of the University. All project components significantly relied on donor funding. The main donors were the government of the Netherlands, the Delft University of Technology and the Humanistic Institute for Developing Countries (HIVOS) as well as NORAD and SIDA (AAU 2000).

For the interim period until the creation of the Department of Computer Science, UCC had to fulfil both mandates: firstly to act as an academic teaching institution, and secondly as infrastructure provider to develop and to maintain the ICT infrastructure for the university (c.f. UCC 2001).[167] But whereas UCC successfully developed and implemented ICT projects to enhance the University's capacity to make use of ICT starting with the provision of Internet services and the deployment of a data communication infrastructure, the interim solution was noticeable at the expense of academic ICT education.

Therewith, in the second phase of the Tanzanian ICT history – after all a phase of some noticeable growth and Internet diffusion – there were only random improvements in higher education. A start was only made, when academic teaching had been transferred from UCC to the newly formed Computer Science Department, which commenced in July 1999. This terminated the interim solution, in which UCC gained substantial profile in its originally assigned task,

[166] While the Information Policy Plan provided the vision, the Master Plan provided the course of action. This general approach was not only developed at UDSM, but was embedded in joint efforts of the Association of African Universities (AAU) to promote ICT in higher education. An overview on the approach developed by AAU can be found in AAU (2000).

[167] See also UCC's website at http://www.ucc.co.tz

but failed to meet the requirements of academic teaching. However, after some build-up work, the Department of Computer Science started to offer a bachelor programme in computer science and a bachelor with computer science (to allow some combination with other subjects). The department targeted an annual output of 100 graduates and the first BSc started with roughly 25 students (Int. Fournadjiev 23.7.2002). The new facility was well equipped and the demand high, but a shortage of academic teachers and reliance on part time teachers considerably slowed down the capacity building process of the department.

Apart from UDSM, the African Virtual University (www.avu.org), a World Bank initiative, established learning centres at UDSM and the Open University of Tanzania (OUT) to offer higher IT education. But despite high claims, the AVU did not meet the expectations in the first years and only offered short-term courses and diploma programmes.[168]

Outside university education the IT training scenery continued to diversify. Apart from various public training institutions that rely on regular short courses – like the College of Business Education and the Institute of Finance Management – a handful of private training facilities emerged, often in affiliation with companies with an existing IT profile. Among them LearnIT of the SoftTech Group (www.learnit.co.tz), the Institute of Information Technology (IIT) (www.iit.co.tz), the NIIT Group (www.niit.com) or the Institute of Management and Information Technology (IMIT) of the Cats Group (www.catsgroup.com). Usually they started with certificated courses – from self-developed certificates to e.g. Microsoft or Oracle certificates – and subsequently prepared for upgrading their portfolio to diploma and degree courses. Thus, until 1999 it was particularly short course end user training that was mushrooming.

The overall situation of computer training improved between 1994 and 1999, but especially targeting IT professional was neglected and only had a slow start. This became increasingly subject of concern, the more that the demand for a professional ICT workforce not only grew globally, but also increased on a national level with increased ICT penetration in the country as major secondary towns like Arusha or Mwanza became IT company locations. Thus, the country was only insufficiently prepared to respond to a growing demand of IT professionals and especially since 1999 a lack of IT professionals was regarded as a key constraint on a ranking list of constraints for growth and development

[168] Among the early problems of AVU were the creation of partnerships among the member universities with very different needs and inappropriate generally US made courses not adapted to local needs (Int. Senkondo 30.7.2002).

in the ICT sector (c.f. Lamtrac 2001: 2). This was also observed by the Ministry of Communications and Transport and they recognised that a shortage of IT professionals might turn into an obstacle of the intended ICT developments. To tackle the issue, and to identify solutions that might also be implemented in the short run, the ministry commissioned a study to identify the countries training needs and determine the prerequisites for quick implementation of ICT training programmes. The output was the Lamtrac report. They conducted a "survey of the need for a vocational training programme for ICT professionals in Tanzania" and compiled a list of recommendations meant to be an emergency response to a flawed status quo. Jointly with other initiatives, the new programme started in 2002, and induced a phase of renewed efforts to improve the unsatisfying situation of professional ICT education in the country.

4.2.8. Concluding the Second Phase

The second phase of ICT history in Tanzania was characterised by a controlled introduction of sector reform as part of an overall reform process. Reform steps were carefully phased in and the legal and regulatory reform preceded private sector participation. Legal and institutional frameworks were put in place and subsequently complemented by the reduction of the TTCL monopoly and introduction of partial liberalisation and competition in selected areas of the market. The new liberalised environment increased market efficiency and attracted a number of new market participants who started to provide telecommunication and ICT related services, among others, mobile service operations and Internet services had been introduced. However, growth figures remained below expectations and service provision was for the most part concentrated in Dar es Salaam. Thus, legal and regulatory reforms indicated a new direction and set the course for new developments, but the expected growth figures failed to appear. This ambivalent record was due to three main factors: (i) firstly markets and targeted customer segments were simply too small to result in any broader impact, (ii) market regulation was often weak and inconsistent and could not provide a level playing field and (iii) the reform introduced private participation as a state-centred top down approach and failed to respond to the private sector needs. Reforms were initiated as a government controlled process and comprised multilevel policy and legal actions aiming at the establishment of strong market mechanisms. Complementary reform measures and the comprehensive approach of the TRP provided general reference points and sectoral legislation laid the ground for the regulative setup and the market structure. However, signals sent from the political arena and received from the market players were often ambiguous or openly contradicted the idea of a fair and transparent competitive environment, such as regulative inconsistencies, preferential treatment of TTCL or a de facto ministerial prerogative on rule making. Furthermore, reform adjustments were insufficiently communicated and failed to provide the necessary responsiveness to the operator's needs. This pattern reappeared as rules and guidelines repeatedly encased the market without any prior consultations or other feedback mechanisms. Despite the undoubted opening of the market, the reforms were state-biased and, in essence, only allowed directed, state controlled market participation, so that confidence in the Tanzanian market environment remained low.

4.3. The Third Phase since 2000: Departure

The efforts and commitment to build a framework for ICT started to pay off since 2000. In politics, efforts to draft and approve an explicit ICT policy took centre stage and ICT gained a much higher recognition within the political decision-making process. The regulator TCC, which merged with the Tanzania Broadcasting Commission in May 2003 to form the new regulating agency Tanzania Communications Regulatory Authority (TCRA), increased its profile, when it amended the regulative framework for increased private sector participation. The market diversified with regard to both international operating companies and small local enterprises, so that the supply of ICT improved noticeably. Furthermore, the ICT diffusion rates deepened and increasing knowledge and capacities of all stakeholders contributed to a more output oriented and demand driven service delivery. On the other hand, developments were not linear and progress was recurrently retarded due to weak political backing, implementation deficits or market failures. Particularly the TTCL privatisation developed into a stumbling block of the sector, as the dispute between the government and the investor paralysed planned activities not only of the incumbent, but also of the other players in the sector.

Nevertheless, the positive developments prevailed. Tanzania caught considerable international attention and the country was increasingly commended as a positive example for African ICT developments.[169] Progress was unspectacular but steady, so that Tanzania underlined her role as a donor's favourite and also became an important recipient country in ICT development cooperation.[170]

---

[169] For example, ITWeb [Johannesburg] 14.3.2001, 10.4.2001, 13.7.2001; Financial Times [London] 4.8.2001; Business Day [Johannesburg] 18.9.2001.

[170] For example, one of the most visible appreciations of Tanzania's ICT development efforts have been the invitation to participate in the G-8 DotForce. The gains of participation in the DotForce were numerous. According to the Tanzanian representatives in the DotForce, David Sawe and Simbo Ntiro, participation significantly contributed to a better understanding of

4.3.1. Keeping the Profile: Continued Efforts of Reform

Since 1998 the Tanzanian Development Vision 2025 has laid the foundation for the country's development paradigm, and it has become a key reference for all policy activities. From 2000, the Poverty Reduction Strategy Papers (PRSP) – as one major input to receive debt relief within the Heavily Indebted Poor Countries (HIPC) Initiative – supplemented the Vision 2025 as a second major policy document.[171] Both documents particularly emphasise the role of good governance and the rule of law, appropriate sector policies and the need for a strong competitive economy as corner pillars of all future Tanzanian development efforts.[172] Therewith Tanzania was in line with the international development paradigm and notably one of those countries in which the overall strategy led to considerable success. The Economic Intelligence Unit has described the Mkapa era as "a new period concord between the government and the Bretton Wood institutions" (EIU 2003: 30) and acknowledged that Tanzania has undergone a prolonged period of macroeconomic stability (cf. EIU 2003; URT 2003, IMF 2004).[173] Over the past years, the overall GDP growth has steadily risen from 3.5 percent in 1999 to 6.2 percent in 2001 and 5.7 percent in 2003. Having achieved single-digits by January 1999, the rate of inflation has

---

ICT, inspired the work of the eThinktank, a think tank that paved the way for the approval of a national ICT strategy, induced a new South-South dialogue on ICT and increased the reputation and credibility of the DotForce members vis-à-vis high level government representatives to push the agenda (Int. Sawe 23.4.2002, Int. Ntiro 22.7.2002).

[171] To be more precise, poverty reduction strategy papers were compiled in series of documents, because PRS was set up as a process and papers were updated on a regular basis. Since 2000, the PRSPs replaced the old Structural Adjustment Programmes, thus also acknowledging the deficiencies of the old programmes to bring sustainable development. The international development community has mandated that all low-income countries, receiving debt relief under the HIPC initiative or concessional lending from the World Bank or the IMF, should develop country-owned Poverty Reduction Strategies. These aimed to benefit the poor, recognised the multidimensional nature of the causes of poverty and the long-term perspective for poverty reduction. Typical priority areas of PRSP were sound macroeconomic and structural policies, good governance and appropriate sectoral policies to be developed in a country-owned participatory process. An in-depth evaluation of PRS process in Tanzania can be found in World Bank/IMF (2004), a detailed background of PRS can be found in Klugman (2002).

[172] The official Tanzanian perspective includes two more papers as basic policy documents: the National Poverty Eradication Strategy and the Tanzania Assistance Strategy (c.f. URT 2004). Though each document has its own key features and characteristics, there is (and must be) a considerable amount of redundancy.

[173] However, despite all efforts of diversification, the economy is still dependent on agriculture which in 2001 accounted for 48% of the GDP and employed roughly two-thirds of the labour force (EIU 2003: 29).

declined consistently, registering 4.6 percent in 2002, and according to official estimates, 4.4 percent in 2003 – achievements that were also reflected by increased official and private capital inflows (c.f. EIU 2004: 5; IMF 2004: 19; URT 2004). In November 2001, Tanzania reached the completion point under the HIPC-debt relief initiative and freed resources were diverted into poverty reduction programmes being implemented under the Poverty Reduction Strategy. In 2003, Tanzania completed the first Poverty Reduction and Growth Facility (PRGF) arrangement. In early 2004, the successful completion of the review process enabled the release of further IMF assistance – the most visible seal of approval awarded by the IMF.

A critical appraisal of Poverty Reduction Strategies and their origins in controversial structural adjustment programmes can be found elsewhere.[174] For this purpose it is sufficient to note that PRS were not a rejection of the old structural adjustment programmes but rather an extension, combining the old features of macroeconomic stability and structural reform with a new focus on poverty reduction. Good policy and a market led development path that explicitly targeted pro poor growth potentials became the fundament of Tanzanian development. However, adopting a PRSP and implementing respective measures have been two very different things. The second and the third PRSP progress report gave a condensed overview of Tanzania's most recent socio-economic performance and outlined the way ahead.[175] Accordingly, poverty reduction will continue to occupy centre stage in the macro and sectoral policies. The reform approaches and new focal areas in the country's poverty reduction strategy line up with other major reform sites under construction, such as public service reforms, parastatal sector reforms, legal and financial sector reforms and a local government reform programme. Taken together these form a very ambitious reform agenda. Inevitable progress has not always been satisfactory. Both, inconsistent strategy formulation and weak implementation capacities recurrently reappeared in all reform areas. But despite deficiencies in the details, reforms resulted in notable improvements in the country's performance record over the past years. Consequently, Tanzania has been

[174] See e.g. UN Department of Economic and Social Affairs (1998); Walther/Hentschel (2002); IMF (2003) Cling, et al. (2003).

[175] The second report highlights what the government did during the past years of policy development and defined focal areas of intervention, such as, among others, to strengthen "ties between the rural and urban economy, improve rural infrastructure, encourage development of SMEs, facilitate provision of micro credits, strengthen local capacity and strive to achieve a more broad based economic growth" (URT 2003: vi). The third report, that also closed the three-year cycle of the implementation of the first PRSP formulated in 2000, drew attention to achievements with regard to the Rural Development Strategy (RDS) and the mainstreaming of gender, environment and governance issues (URT 2004: 1).

viewed as one of the top most improved countries in Sub-Saharan Africa.[176] Measures that targeted an improved business climate included:[177]

- A range of legal, regulatory and administrative reforms that have been under implementation, with a view to strengthening integrity of the government's public financial management;
- New progress in the reform of the tax administration in 2003, including the abolishment of a number of local government taxes that had been seen as nuisance taxes by the business community;
- Land Acts that have been amended to remove impediments to bank lending;
- Approval of new simplified guidelines for business licensing in early 2004;
- Approval of a policy for small and medium enterprises in 2003;
- Establishment of the Tanzania National Business Council has enabled increased dialogue between the government and the private sector regarding development issues or the creation of a regular International Investors Round Table "to be part of the advocacy in creating a positive image of Tanzania" (URT 2003: 24).
- The government has strengthened the effectiveness of the Tanzania Investment Centre (TIC) in promoting and facilitating private sector investment. The TIC has been restructured and now grants investment certificates and tax incentives;
- A new private sector development strategy has been drafted;
- Implementation of a national anti-corruption strategy and action plan;
- Start of licensing for operators for export processing zones (EPZ);
- Approval of the Business Environment Strengthening in Tanzania (BEST) aiming to facilitate and improve government-private sector partnerships.

These measures added to the overall reform record as already introduced in section 4.2.3.1., so that the business community already benefited from new policies. In addition to measures targeting the overall environment, individual

176 This is reflected in the self image of the country and presents the reform status, see URT 2003, but also in various third party assessments from World Bank (2002), IMF (2004) or EIU (2003) to complaisant newspaper coverage in African Business [London] December 2002 and May 2003, or Financial Times [London] 4.8.2001.

177 Based on URT (2002, 2003), PWC (2003) and IMF (2004)

sectors have received special assistance, too. For example, the telecommunication sector, among others, was identified as a so-called "lead sector" eligible for special assistance by the Tanzania Investment Centre (TIC)[178] and thus became an attractive investment destination for foreign investors.[179]

An improving general framework, respective political guidelines and first efforts of implementation of various measures in support of a private sector led development strategy started to take effect. But beyond the general scenario, how have these changes affected the telecommunication sector? What measures have been implemented, how and for what purpose? The new prominence of ICT, discussions about the digital divide and an increasing number of ICT related projects in the country called for urgent policy action and thus the formulation of an ICT policy – in line with the overall development approach – came to the forefront and largely determined the time between 2000 and 2003.

4.3.2. A Catalyst for ICT: the eThinktank

Major ICT policy input was prepared by the eThinktank. The eThinktank was created in February 2000, initiated by a handful IT protagonists and later included decision makers from politics, private sector and civil society who aimed a substantial Tanzanian contribution to the ongoing debate on the global digital divide.[180] At the same time the number of ICT related projects in the country grew considerably, the group felt that there was a growing need for coordination and harmonisation of efforts, all the more, that government responses to ICT were still weak. Officially, the Ministry of Finance was still responsible for all matters of computers and computerisation, but, in practice,

[178] Through the TIC, the government offered Certificates of Incentives for foreign and local investors with a minimum investment capital of US$ 300,000 or US$ 100,000 respectively. The certificate for lead sector investments guarantees, for example, a tax holiday, exemption of import duties on capital goods or 10% on semi-processed goods, investment guarantees and others (c.f. TIC 2002; Int. Rutihinda 28.4.2003, Int. Kuwite 28.4.2003).

[179] Between 1990 and 2001 37 percent of all projects approved by TIC were in the telecommunication sector, followed by manufacturing (23%) and natural resources (9 percent) (Madete/Aiko 2003:6).

[180] The idea of forming a group dealing with ICT was launched by a senior KPMG consultant in 1999 (who soon afterwards retired and left the country) and almost immediately taken up by three Tanzanian ICT pioneers David Sawe of the Civil Service Department (CSD), Simbo Ntiro (then KPMG consultant) and Ali Mufuruki (Infotech Group), who have been key facilitators of the group (Int. Mufuruki 22.4.2002, Int. Asikainen 24.4.2002). A good summary of the eThinktank's background is also given in the Business Times [Dar es Salaam] 1.9.2000.

the Ministry remained quiet about all policy issues. The eThinktank attempted to fill the policy vacuum. Its mission statement reads as follows:

> "Offer ICT leadership by catalysing policy changes and by supporting related developments aimed at enabling Tanzanians to participate effectively in the modern Internet-based global economy, benefiting their Nation and partners" (eSecretariat 2000).

Two major goals were derived from the statement: (i) the development of a national ICT policy and (ii) the creation of a national ICT organisation. To that end the group sought to compile the terms of references for a policy framework and approached UNDP for assistance. UNDP was ready to support the group, but made the involvement of a government agency as an official government counterpart a prerequisite for funding.[181] For this reason the Civil Service Department (CSD) became the official UNDP partner, partly because the department had already gained some experience dealing with ICT and ICT implementation measures in the civil service, partly because the director of the Information Management Department of the CSD, David Sawe, was a key activist in the eThinktank, too.

With UNDP assistance, an eSecretariat as implementing arm of the eThinktank came into being. However, beyond the formal structure, the necessary momentum to get a policy process on track foremost were due to commitment and high quality interaction of particularly David Sawe, Simbo Ntiro, Ali Mufuruki and the UNDP representative Riku Asikainen,[182] who pushed the subject of ICT and put the need of an ICT policy on the agenda. Thus, it was only a handful of people who accelerated the policy developments. The eThinktank's core team managed to pool a policy team in February 2001 and delivered a model ICT policy in December 2001. This was a major achievement; the model policy comprised of a comprehensive strategic approach, and concrete policy objectives to be implemented in a given time frame and in well-defined responsibilities. It especially highlighted areas of:

---

[181] At that time, there was no government involvement in the process, but any effort to support the process from UNDP must include approval by the government, because, according to UN byelaws, UNDP would not be allowed to bypass the government (Int. Asikainen 24.4.2004).

[182] When Sawe, Ntiro and Mufuruki approached UNDP for the support of a national ICT forum, the UNDP country officials chose Riku Asikainen to handle the matter. Asikainen was no regular employee and only UN volunteer at the Tanzania country office. However, these persons pushed the subject of ICT in Tanzania and, as an assorted grouping, they highlight the role of individual commitment and competence in the more recent ICT history of Tanzania.

- Universal access and service,
- Improving regulatory capacity,
- National, regional and international Internet connectivity,
- Establishment of (Cisco) networking academies,
- Internet connectivity for higher learning institutions, research centres, and secondary schools,
- Support of telecentres,
- Promotion of ICT utilisation in the health sector,
- Promotion of ICT in governance,
- Conducting an e-readiness assessment,
- Creation of a national ICT organisation (eSecretariat 2001).

The policy proposal was important for two reasons. Firstly it was the first comprehensive policy document addressing ICT in Tanzania. To a large extent, it anticipated the official policy approved in 2003.[183] Secondly, it delivered input that could not be ignored and forced politics, especially the MoCT out of its idleness in order to reclaim control of the process.[184] In fact, after the ICT policy responsibility was officially assigned to the MoCT (see below), the ministry responded promptly and initiated a national ICT task force to draft a national

[183] The document had its limitations, too. It was too ambitious and, to a large extent, carried the notion of a catch-all document, e.g. with regard to the proposed guidelines for ICT for development in Tanzania that dealt, among others, with infrastructure, education research and innovation or international Internet governance (eSecretariat 2001: 7-8) or the role of ICT in governance that comprised far-reaching measures of internal administration and the establishment of government databases (c.f. eSecretariat 2001: 11). Furthermore, sectoral policy responsibilities had been poorly assigned and the tasks and mandate of the proposed national ICT organisation remained vague (c.f. eSecretariat 2001: 17, 30). Despite the shortcomings, the policy provided a good orientation for the road ahead and made available distinct positions of most issues that would be part of an official policy.

[184] In retrospect the relation between the ministry and the eThinktank was crucial for the policy process. The eThinktank stimulated a competition of ideas, filled unoccupied topics and made proposals on numerous sub-themes of the ICT debate. The eThinktank doubtlessly forced the ministry's hand and it should not be overlooked that the debate was often intense and conflict ridden. Many of those involved, questioned the ministry's competence to handle the matter and many wanted to see it under the responsibility of the CSD, especially after the UNDP support commitment in favour of the CSD. Inevitably, the MoCT opposed the CSD-UNDP cooperation on policy issues. However, with some distance to the then discussions, the debate's sharpness must be put into perspective, because it was never intended to seize any competencies from the MoCT, but merely to catalyse the process (Int. Sawe 23.4.2002, Int. Ntiro 22.7.2002, Int. Kowero 2.5.2002, Int. Mufuruki 22.4.2002, Int. Asikainen 24.4.2002).

ICT policy. These new developments were a major success for the group and the eThinktank can be labelled as a prime catalyst of the ICT policy process.

Within the two years 2001 and 2002, which marked the preparatory policy work, the eThinktank had its maximum influence. It filled the gap left by a lack of political guidance and derived its strength from the manpower, the capacity and the commitment of its members. For most of the early members, this applies until today, but competencies and responsibilities have shifted away from the eThinktank in other political decision making fora. The eThinktank started to change its face, particularly as it successively developed into an online discussion group.[185] Therewith, it lost its previous role and policy issues migrated to other fora. Though the eThinktank remained a lively discussion group on ICT and also acted as an important intermediary in the ICT for development debate. At the time the government officially took over the policy process, the eThinktank had successfully accomplished its tasks and lost relevance.

4.3.2.1. The MoCT and the ICT Policy

Between 2000 and 2003, policy issues were at the forefront of the debate. Already in 1999, first ideas of a national ICT policy to harmonise and streamline multiplying ICT activities in the country materialised. Although the eThinktank's activities were the most visible efforts to put ICT policy issues on the agenda, the government was also pressurised by the ongoing international debates on ICT and the digital divide that called for some official government responses. Secondly, the government was alarmed by increased but fragmented ICT activities by various public and private institutions that lacked coordination and started to evoke reminiscences of failed computerisation efforts some 20 years ago (see chapter 4.1.2.1.). But still, responsibilities were not clearly assigned. Officially, the Ministry of Finance was still responsible for all computer issues, but had already withdrawn from any attempt to frame computerisation politically. The treasury solely maintained control of importation issues and taxation. In contrast, the Ministry of Science and Higher

[185] In June 2004, the eThinktank online community comprised 439 members from the public and private sector and from NGOs in Tanzania, as well as expatriates, members from the donor and the international research community. The online group started in May 2001 and has by June 2004, posted over 4,000 messages, which equals a considerable output of roughly 130 messages per month. High pitch months were July and August 2003 with 336 and 298 messages covering issues such as a Tanzanian Internet Exchange Point, mobile tariffs, mobile vs. fixed line operations, copyright, domain names, Linux and open source, virus alerts, ICT policy and TTCL (c.f. eThinktank 2004).

Education raised some claims to control the policy process,[186] but in the end, it was almost inevitable that the Ministry of Transport and Communication became responsible for the coordination of the political aspects of ICT. The MoCT was innately involved in telecommunication, the sector with most interlinkages to the cross cutting features of ICT, it was one key player in the Telecommunications Restructuring Programme (TRP) – the onset of modern telecommunication in Tanzania – it already supervised the regulating agency TCC and it was already in charge of the National Telecommunication Policy in 1997 – a first attempt to translate technological developments into new political goals. Thus, the Ministry was fairly well experienced and consequently the cabinet officially appointed the MoCT as a national coordinator and a focal point on all ICT related activities in April 2001 (MoCT 2001a).[187] But despite the ostensible agreement on the MoCT's future role, responsibilities were also tied to one condition: the call for a joint taskforce on the issue of a national ICT policy. This strategic move also meant to get the critics on board, both the ones who were merely interested in a sound policy document and those who wanted to ensure their legitimate institutional interests and their adequate consideration in the projected process. Thus, there were many good reasons to attach the policy issue to the MoCT, but the ministry continued to be subject to severe criticism (c.f. GTZ/ZEF 2002: 47). In fact, sometimes the policy process was unbearably slow, but bearing in mind the infancy of the subject and the fragmentation of ICT within the government, as well as strategic delays before and after the general elections 2000, the MoCT started to cope with the issue and improved gradually. It deliberately acted on different levels: (i) it introduced a working structure in the ministry,[188] and (ii) it introduced a first concept of the

[186] Claims to control the policy process were not openly disputed, but it was not without purpose that the Ministry for Science, Technology and Higher Education attempted to strengthen its ICT profile, e.g. as facilitator of large, international ICT related conferences on intellectual property rights (see The Guardian [Dar es Salaam] 20.6.2000) or ICT in higher education (see (see The Guardian [Dar es Salaam] 22.6.2000). Therewith, the ministry underlined that it was not willing to hand over core competencies offhand that were regarded to be well stored in their own house.

[187] It is noteworthy that the appointment related to ICT policy issues and the coordination of the ICT development process. It was not intended to stop any ongoing ICT activities, the line ministries stayed in charge of their sectors, such as education, health, agriculture or e-government – strategy development, and implementation of the latter remained the responsibility of the CSD.

[188] August Kowero was appointed as national ICT coordinator working directly under the aegis of the permanent secretary Salim Msoma, who had already established a good reputation in dealing with telecommunications. From 1999 to 2000, they served under the late minister Ernest Nyanda, in whose incumbency ICT was a subordinate issue only. His successor, Mark Mwandosya, consecutively acquired a sound profile in ICT matters.

role of ICT in the country, which served as a first orientation for the future strategy. It reads:

> "The importance of Information and Communications Technologies (ICTs) services to national socio-economic development does not need to be overemphasized. ICTs impacts cut across all the sectors (productive and services ones) of the entire economy. ICTs services, therefore, improve information exchange among the people, fight poverty and thus enhance economic growth, factors which are in line with the goals and objectives outlined in the National Development Vision 2025. In order to realize the above benefits, there is a need for having proper coordination in the development and implementation of ICTs related activities" (MoCT 2001).

(iii) The ministry compiled a preliminary ICT development schedule that introduced three phases of the ICT development process:

1. Development of an ICT National Policy,
2. Development and preparation of an ICT National Implementation Strategy, and
3. Implementation of an ICT National Programme.

Furthermore, it initiated the process by (iv) setting up a national ICT taskforce to draft a national ICT policy and to advise the government on all issues of ICT. The taskforce was not only an appropriate forum to address the issues, it was also an instrument to minimise institutional rivalry, as all important actors were all on board of the process – branches of government, parliament, private sector and eThinktank.[189] The 15 members making up the taskforce, chaired by the vice-chancellor of the University of Dar es Salaam, Matthew Luhanga, started to execute its work in August 2001.[190] Finally, (v) the ministry secured the necessary funds to finance the taskforce. After various donor agencies had been

[189] This is why the national ICT coordinator August Kowero insistently highlighted the 'joint' element of the endeavour, emphasising that the taskforce was a national rather than a ministerial task. He even went as far as to claim that the different approaches by the ministry and the eThinktank were indistinctive due to overlapping in both personnel and strategic orientation (Int. Kowero 2.5.2002), a statement, however, that would fail to bear against a closer examination.

[190] Other members were Batilda Burian (MP), Henry Mgombelo (MP), Theophilius Mlaki (Tanzania Commission for Science & Technology), Zaipuna Yonah (TTCL), George Nangale (East Africa MP), Beda Mutagahwya (University of Dar es Salaam Computing Centre), Ali Mufuruki (Infotech Group), Anita Ngowi (Vice President's Office), Ami Mpungwe (Intrinsic Technologies Tz Ltd), Suleiman Tawakal (Revolutionary Government of Zanzibar), Ahmed Ali (Revolutionary Government of Zanzibar), Emmanuel Mange (TCC), David Sawe (CSD) and Simbo Ntiro (Consultant), see Daily News [Dar es Salaam] 12.01.2002.

approached, SIDA committed to contribute to the support and allocated roughly one million US$ to support the policy process (Int. Karhammar 23.7.2002).

With this setup, the taskforce assumed the core activities and initiated the drafting process. However, first results were hardly satisfactory because different working groups worked independently and elaborated on various sub-issues, such as infrastructure, e-government, education and private sector development. Inevitably, these were poorly interlinked and had only little coherence.[191] Only gradually working procedures improved and the draft policy gained direction and substance, though still not convincing in each aspect. The Minister of Communication and Transport, Mark Mwandosya, approved the final draft in November 2001. In May 2002 the draft was introduced to the public on the occasion of a national ICT policy workshop held in Dar es Salaam. This was meant to initiate a broader public discussion on ICT, which was still an issue mainly discussed among experts and technocrats. Of course, the workshop did not turn ICT into a mass friendly subject, all the more because the organiser withdrew from their original plan to hold a series of similar workshops in different regions of the country (Int. Ntiro 22.7.2002). Nevertheless, the workshop led to a tangible appreciation of ICT, mirrored in the broad spectrum of attendees[192] and the media coverage of the workshop (c.f. Daily News [Dar es Salaam] 27.5.2002, The Guardian [Dar es Salaam] 27.5.2002). The many responses during the workshop, comments and criticism collected by a drafting committee of the taskforce during and after the event brought new input for a revised policy. The new text was then subject to a parliament hearing and subsequently submitted to cabinet in December 2002, who approved it in early 2003. The valid version of the Tanzanian ICT policy was finally made public in March 2003.

The policy itself firstly introduced ICT and put it in a broader development context; on the one hand, with regard to the prerequisites to participate in a global information society and, on the other hand, to the national development agenda of the Vision 2025. Following the introductory remarks, it gave an overview on the current ICT status in the country that – though soon outdated – shall allow an easy comparison of target and actual figures (c.f. MoCT 2003: 3-8). The main part consisted of ten focal points that were subdivided again into the four subheadings of "issues", "policy objectives", "policy challenges" and "policy statements". The ten focal points were:

---

[191] The first draft version had more than 200 pages, the final document was reduced to 23 pages. The policy is quoted as MoCT (2003).

[192] The author himself attended the workshop and got a picture of the broad interest in the event and the lively discussions on the various topics.

1. Strategic ICT Leadership
2. ICT Infrastructure
3. ICT Industry
4. Human Capital
5. Legal and Regulatory Framework
6. Productive Sectors
7. Service Sectors
8. Public Service
9. Local Content
10. Universal Access

Therewith, the document covered the important areas in the field and, accordingly, introduced some guidelines for implementation, too. A closer consideration, particularly with regard to the applicability of the policy, however, also revealed some weaknesses. The large number of different recommendations, further exacerbated by the side-by-side subheadings, has been clearly at the expense of a more stringent and coherent document. The intractable structure became perceivable firstly in the ten focal points themselves, which were basically an enlargement of just four key issues (i) legal and regulative framework, (ii) access, (iii) capacity building and (iv) content. This inflated the policy, introduced too many wishful ICT target areas that hid trade-offs and did not provide a clear prioritisation of policy interventions. These stipulated, for example, allocation of funds "equivalent to a reasonable proportion of GDP for ICT deployment, diffusion, and universal access" (MoCT 2003: 10), to ensure that a "reliable state-of-the-art infrastructure, of adequate capacity, high-speed and countrywide coverage is developed" (MoCT 2003: 11), or to support "nationwide e-health, e-tourism, e-education and e-commerce transactions" (MoCT 2003: 18). This was a mere listing and hardly a policy road map meant to explain the ratio, the inner coherence and the schedule of the policy. Furthermore, the recommendations made in the ten focal points were clearly government biased and overemphasised direct government activities and, thus, contradicted the spirit of the mission statement that claimed to encourage "beneficial ICT activities in all sectors through providing a conducive framework for investments in capacity building and in promoting multi-layered co-operation and knowledge sharing" (MoCT 2003: 2), and thus call for a strong notion of partnership between the different stakeholders. Secondly, and beyond the general structure, the elaboration of the individual focal points lacked coherence, too. This can be exemplified with reference to the section "legal and regulatory framework" (Moct 2003: 15). The section did not make any reference to the role of the regulating agency, dealt somewhat arbitrarily with the issues of secure electronic transactions and cyber crime instead of highlighting the actual scope of regulation: the implementation and compliance of fair, transparent and, wherever possible, incentive-setting rules for the sector and thereby deliberately balancing technological, political, economic and legal prerequisites of a sound regulatory regime. Similar applies to the section „universal access", which remained nebulous and gave only vague hints on a "reasonable proportion of funds for universal access" (MoCT 2003: 11), claimed to establish a "rural telecommunication development funds" or the shape of "special incentives to investors in rural ICT" (MoCT 2003: 21/22). Again, the policy covered a number of central and indispensable ICT issues, however, the weak interlinkages and the merely additive character of issues diluted the quality of the endeavour.

Further noteworthy, one issue fell off the original agenda almost completely: the creation of an authoritive national ICT organisation. This was a central demand of the eThinktank's policy proposal and was also incorporated in older draft versions of the policy as proposed by the taskforce.[193] Evidently, the organisation, often referred to as NICTCO (National ICT Council / Commission), originally planned as implementation and monitoring body of all ICT related activities, failed to make the cabinet hurdle. Though the idea of a central ICT organ would have been a consequent continuation of efforts to streamline ICT, the plan was callow, too. Apart from the open question of finance, NICTCO would not only have been a self-monitoring authority, it was also ill defined in contrast to other organisations with a similar task, namely TCC und COSTECH.[194] With this decision, the government also avoided potential threats of a new organisation – either due to conflicting mandates or insufficient funding. Instead the government created a national steering committee on ICT that was assigned the mandate to prepare the Tanzanian contribution to the World Summit on the Information Society (see below).

---

[193] Some draft versions of the policy are available on the ministry's ICT website at http://www.moct.go.tz/ict

[194] One of TCC's tasks, in fact, is to advise the government on ICT and to monitor ICT developments in the country (see section 4.2.4). Similar applies to COSTECH, who, according to the Tanzania Commission for Science and Technology Act (of 1986) Part II Section 5.1., "shall be the *principal advisory organ* of Government on *all matters relating to* scientific research and *technology development*." URT (2001a: 6, Emphasis O.N.). Thus, this put a preliminary end to all NICTCO plans, though the approved policy left a little loophole; section 3.1.1.c. proposes to „create an authoritive national organisation to effect, coordinate and review the ICT policy" (MoCT 2003: 9).

A thorough review of the ICT policy and the policy process reveals strengths and weaknesses. On the positive side, the policy was a result of a multiple stakeholder process. Under the auspices of the Ministry of Communication and Transport (MoCT) the ICT taskforce drafted a policy that comprises major aspects of ICT and their highly complicated and interrelated features – social, political, economic and cultural. Secondly, the process itself contributed significantly to an increasing awareness of the potential and significance of ICT, so that knowledge of ICT and their interlinkages to the broader development agenda grew considerably. Finally, the policy set a clear signal to the outside world that Tanzania was willing to take advantage of a changing environment and that good policy can make a difference in harnessing ICT for development. On the other hand, it also shows that the discussion about ICT and an enabling framework for ICT is not yet finalised. This particularly became evident in the form and the structure of the policy that only partly met the expectations and suffered from a lack of prioritisation, underestimation of trade-offs and a notion of arbitrariness through an overloaded listing of wishful targets. Furthermore, the policy did not provide any well-defined targets or indicators to assess future progress. Unlike the National Telecommunication Policy of 1997 that set inter alia the target to achieve a telephone density of six telephones per 100 population by the year 2020, the National ICT policy did not incorporate any concrete benchmark figures but only a declaration of intent.[195] On the other hand, one might very well argue that the quality of a policy is only secondary, if the basic message is convincing. The overall goal is to provide the general direction and not a legal document that defines responsibilities and procedures. Thus, it is not so much a question, whether one could have done better, but whether the goal to introduce sufficient guidance to the sector has been achieved. From this perspective, the policy must be seen in the context of

[195] It is noteworthy that Tanzania agreed on a set of benchmarks and to monitor the progress made to develop ICT. Almost parallel to the approval of the policy in April 2003, representatives from the African states met on Mauritius for a high level conference “Access to ICT for All” as one preparatory meeting for the December 2003 UN Summit on the Information Society in Geneva. The Mauritius conference adopted a declaration and an action plan as well as a set of benchmarks for actions to be taken. Fully aware of the draft texts of the action plan to be adopted by the World Summit on the Information Society (WSIS), the ministers borrowed from the WSIS texts (c.f. ITU 2004a: 29). These claimed: (i) all villages to be connected by 2010, with a community access point by 2015, (ii) all universities to be connected by 2005, also the schools by 2010 and all primary schools by 2015, (iii) all hospitals to be connected by 2005 and health centres by 2010, (iv) 90 percent of the world’s population to be within wireless connection coverage by 2010 and 100 percent by 2015, (v) all central government departments to have a website and email address by 2005 and all local government departments by 2010 and (vi) creation of the African backbone by 2010 (Regional Ministerial Conference 2003), for a public perception see also Business Times [Dar es Salaam 26.9.2003).

legislation, institutions and other sectoral developments, so that timing and sequencing became a factor, too. The then Director General of the regulator TCC, Asupya Nalingigwa, accentuated this thought in a presentation to the National ICT workshop in May 2002. He called it a major shortfall that the legislation that restructured the sector in 1994 was enacted before the policy was put in place, because this meant a lack of policy guidance on how to develop the sector (Nalingigwa 2002). In addition, the policy presentation coincided with the tapering and disturbing details of the TTCL privatisation dispute that adversely affected the positive perception of the Tanzanian ICT record so far.[196] In this respect, the policy was not in sync with major developments and the incongruence of the calendar of the policy and the sequencing of reform, as well as individual events, disturbed the introduction of a credible policy. Thus, a fair judgment of the policy must go beyond mere words, but needs to consider the context. In this regard, the approval of a national ICT policy has been only one – though despite the above-mentioned reservations – a crucial element of the Tanzanian ICT strategy.

4.3.2.2. Contextualising the Policy: Government Footprints in Support of ICT

Beyond ICT policy measures in the narrower sense, the government sought to promote ICT by a number of other supportive measures. Particularly a waiver on taxes and import duties on computers and peripherals and a growing number of ICT model applications implemented by the government strengthened the credibility of the adopted course. In addition, efforts to participate in the global discussion on ICT, especially the preparatory work to the World Summit on the Information Society (WSIS) further strengthened the government's ICT profile.

With effect from July 2001, the government removed Value Added Tax (VAT), which had been introduced in 1998, from computers and accessories. In his budget speech 2001, the Minister of Finance, Basil Mramba, explained the measure was aimed to enable more people to afford buying and using ICT facilities and, in his words, to "hastening the pace of Tanzania entering the 21st century in the era of globalisation, communication and growing dependence on the Internet." He continued: "This is also intended to motivate the youth to engage in computer related activities in the same way as their counterparts worldwide do. It is my expectation that schools, colleges, and other institutions of learning will take advantage of this move to intensify computer teaching." (Mramba 2001). Already in 1999 the government reduced the custom duty rate to five percent instead of the old 20 percent rate and it was further lowered to

[196] For details see chapter 4.3.4.

zero percent in 2001 (see Table 4.2). The removal of taxes and customs duties on computers and peripherals, which was also accompanied with some relief in the importation regime,[197] had a very direct impact, because the prices for computer equipment came down considerably and made the technology affordable for many more users.[198] For example, the elimination of all VAT and income taxes resulted in unusually high growth in computer imports since 2001. 19,500 units were shipped in 2001, and the year-on-year growth for 2001 reached 36 percent and for 2002 28 percent (BMI-Techknowledge Group 2002: 6)

Table 4.2: VAT and Import Duty on Computers and accessories 1998-2004

	VAT	Import Duty
1998	**20%**	**20%**
1999	**20%**	**5%**
2000	**20%**	**5%**
2001	**0%**	**0%**
2002	**0%**	**0%**
2003	**0%**	**0%**
2004	**0%**	**0%**

Source: Budget speeches 1999, 2001 (quoted as Yona (1999) and Mramba (2001)

Secondly, the government developed into a model user of ICT and thus became a strong protagonist of its own agenda. Particularly the implementation of effective e-government provisions turned into a government focal point. In 2004 the government adopted a national e-government strategy and the different programmes under implementation range from a national website to the installation of an Integrated Financial Management System (IFMS), the computerisation of payroll systems or tax administration and to plans to build a country-wide government Virtual Private Network (VPN) providing broadband Internet connectivity to 22 of 24 regions and 86 of 117 districts (Int. Sawe 11.4.2003, Int. Brems 17.7.2002). The e-government strategy will be discussed in some detail in chapter 5.3., here, it is sufficient to point towards the strong signals coming from the government that through its own project activity underlined the credibility of its own policy direction.

[197] According to the Importation and Distribution of Telecommunication and Radiocommunication Equipment Regulations, 2001, wholesale importers of computers and accessories are still subject to an import licence to be granted by the regulator, whereas importation for private use only was exempted from import regulations, if in compliance with the regulator's type approval (URT 2001b).

[198] The government also introduced additional incentives for the sector, such as the deduction of income tax for telecommunication equipment and telecommunication towers owned by telecommunication companies (Mramba 2002), but it also introduced an excise duty on mobile phone airtime by 5 percent (Mramba 2002), which realised an amount of 2,2 billion TSh. in the financial year 2002/2003. In an attempt to harmonise the tax structure in the East African Community, the excise tax was increased to seven percent in 2003 (Mramba 2003).

Finally, the preparatory work for the World Summit of the Information Society (WSIS) reinforced the Tanzanian ICT profile, because the country sought input to the preparatory process in various fora, and WSIS thus contributed to a growing understanding and public awareness of ICT in the country. The summit process comprised a number of preparatory conferences as well as a regional African preparatory conferences held in May 2002 in Bamako, which was again accompanied by a number of side events that led the way to the first summit segment in December 2003 in Geneva. A second follow-up summit will be held in November 2005 in Tunis and will also be accompanied by a second regional African conference to be held in early 2005 in Accra.[199] Whereas in the beginning of the Bamako preparations, Tanzania participated only rudimentary and could not even contribute to all Bamako preparatory conferences – due to financial constraints and limited human capacity (Int. Rweyemamu 16.5.2002), this changed as the first summit segment was approaching. In February 2003, the Government of Tanzania submitted a draft declaration of principles and of the plan of action to the WSIS secretariat as the national input to the global agenda. The papers elaborated on declarations and action plans on five issues, thus more condensed than the ten focal areas of the National Policy and therewith an indication of a maturing debate. In accordance with the WSIS agenda five prime targets of the national strategy were introduced: (i) infrastructure development, (ii) capacity building, (iii) ICT for education, (iv) ICT for health and (v) cultural and linguistic diversity and local content development (c.f. URT 2003a). More explicitly as the National ICT Policy, the country paper made a stronger and more explicit reference to the international agenda and therewith also introduced the terms of reference for the Tanzanian strategy – the WSIS principles,[200] the Istanbul Action Plan approved at the 2002 World Telecommunication Development Conference[201] and the priority actions

[199] For detailed background on WSIS, see ITU (2004d). Apart from the Bamako conference in 2002, the African WSIS calendar also comprised other high-level conferences, among others in Port Louis (April 2003) and Arusha (April 2003) or regional informal conferences in Johannesburg (June 2003), Cairo (July 2003) and Abidjan (July 2003), c.f. MoCT (2003a).

[200] The declaration of principles approved as a result of the first summit in Geneva in December 2003 lists the following key principles of a global information society for all: (i) The role of governments and all stakeholders in the promotion of ICTs for development, (ii) information and communication infrastructure, (iii) access to information and knowledge, (vi) capacity building, (v) building confidence and security in the use of ICTs, (vi) enabling environment, (vii) ICT applications: benefits in all aspects of life (viii) cultural diversity and identity, linguistic diversity and local content, (ix) Media, (x) ethical dimensions of the information society, (xi) international and regional cooperation (WSIS 2003)

[201] The core of the Istanbul Action Plan is a series of six programmes to be implemented by the Telecommunication Development Bureau (BDT) of the International Telecommunication Union (ITU), focusing on regulatory reform, new technologies, e-strategies and applications,

under Nepad's ICT programs to be implemented by the e-Africa Commission of the African Union.[202] Though most of the points covered were also part of the National ICT Policy, the country paper waived some redundancies of the policy and introduced the Tanzanian standpoints to the global arena.

Though the national ICT policy papers and the WSIS input papers were based on the same approach, there was a striking difference in the tone of both documents, highlighting either a strong national commitment to build confidence in domestic politics and markets or the vulnerability of least developed countries to stress the need for continued external assistance. For example, the national policy introduced its overall vision as "Tanzania to become a hub of ICT Infrastructure and ICT solutions that enhance sustainable socio-economic development and accelerated poverty reduction both nationally and globally." (MoCT 2003: 2), the input to WSIS was built upon the same fundament but additionally underlined the disadvantaged developing country situation and the limited scope of market based ICT development strategies. It states that "experience has shown that the private sector is not able to take the responsibility of owning and, thus, carrying out all the rehabilitation [...] of the existing network and expansion of the new one that reaches all people in the rural and under-served areas for creating open access to all" and consequently demanding "local and international development partners should extend their support to infrastructure programmes that set top priority to reaching rural people" (URT 2003a). Such dual, context-dependent approach called for both strong market elements and increased international assistance. Above all, the ability to shift the focus of texts dependent on audience, also indicated that some ease of handling ICT matured.

At that time, the ICT taskforce had already accomplished its tasks, but the preparatory WSIS activities and the request for country specific input to the summit demanded a new coordinated approach. A national ICT commission, as proposed from the eThinktank and discussed in the policy process, could not be enforced. To harmonise the cross cutting issues of ICT, the MoCT resorted to the idea of the national taskforce bringing together the major players in the field. To this end, the ministry formed a national steering committee of 13 members

economics and finance, human capacity building and special assistance to least developed countries (ITU 2002c).

[202] The NEPAD has outlined top priority ICT programs for building the African Information Society; namely (i) the NEPAD e-school programs, (ii) the east coast of Africa sub-marine optical fiber cable and land locked country fiber cable programs, (iii) the NEPAD e-readiness programs, (iv) the NEPAD capacity building programs from e-learning. In March 2001, the African Union established the e-Africa Commission to implement the NEPAD ICT Programs (e-Africa Commission n.d.).

for the preparation for building an information society in Tanzania in July 2003, which was chaired by the Permanent Secretary in the MoCT, Salim Msoma.[203] Together the MoCT and the regulator TCRA have formed the secretariat to the steering committee (MoCT 2003a). In contrast to the national ICT taskforce, which was recruited as an expert body, the steering committee has been built as a political body representing high-level ministerial decision makers.[204] The new institutional setting of ICT in Tanzania had several implications. Firstly, the ICT for development agenda has arrived in the high-rankings of Tanzanian politics with likely consequences on the assertiveness of proposed implementation measures, secondly, MoCT and TCRA have enforced their claim to be the key players dealing with ICT, thirdly the MoCT successfully accomplished the first of the three phases of their strategy and was now ready to enter the second phase to develop a national implementation strategy, as proposed in the ministry's ICT development schedule (see above). Finally, the creation of the steering committee and the relating secretariat answered the claim to create an "authoritative national organisation to effect, coordinate and review the ICT policy" as required in the national policy (MoCT 2003: 9). Although the committee's mandate was closely related to the WSIS-process and it did not have the broad responsibilities as the ones proposed for the NICTCO, together the committee and the secretariat worked de facto as a NICTCO, yet without having solved the pending issues of funding and a determined scope of mandate.

4.3.3. Readjusting Regulation

From 2000, the regulator entered a phase of accelerated regulative adjustments and has continued to improve its working profile consecutively. The most perceptible changes took place when TCC merged with the Tanzania Broadcasting Commission (TBC) to form a new regulatory body – the Tanzania Communications Regulatory Authority (TCRA) to respond to increased

[203] The other members were Mr. P. Kallaghe (State House), Mr. M. S. Mwaffisi (Director Television Tanzania -TVT), Mr. Kassim Mpenda (Director, Tanzania Information Services MAELEZO), Col. A.N. Nalingigwa (Director General TCRA), Prof. M. L. Luhanga (Vice Chancellor University of Dar es Salaam), Mr. Abdi Khamis Abdallah (Permanent Secretary Zanzibar Ministry of Communication and Transport), Ms. A. M. Ngowi (Director, Vice President's Office), Hon. Dr. B. S. Burian (Member of Parliament), Ms. Eda Sanga (Director, Radio Tanzania), Mr. E. K. Kamando (Head, Info. Dev. Cooperation, Ministry of Foreign Affairs and International Cooperation), Dr. B. B. Rufunjo (Director of Transport & Communications, MoCT) and Mr. A. N. Temba (Director of Policy & Planning, MoCT & Secretary of the Steering Committee)

[204] Only Batilda Burian, MP and also member of the TCRA board, and Matthew Luhanga, vice chancellor of UDSM and then chairman of the national ICT taskforce, were members of both bodies.

overlapping of functions and recognising that technological convergence became a major challenge to the existing regulatory authorities. The new regulator started operation in November 2003 and took over the functions of the Tanzania Communications Commission. Apart from the merger, the old TCC had already introduced some sizeable changes since 2000 and, particularly in 2001, a number of amendments to the Communication Act of 1993 were approved. These covered almost all areas of regulative functions. However, the basic approach as discussed in chapter 4.2.4. remained the same and the issues of licensing, interconnection, frequency management and universal access persisted as the focus of TCC and later TCRA.

4.3.3.1. Merging Regulatory Competencies: the Creation of the Tanzania Communication Regulatory Authority – TCRA

TCRA was established under the Tanzania Communications Regulatory Authority Act No. 12 of 2003 for the purpose "of regulation of telecommunications, broadcasting, postal services; to provide for allocation and management of radio spectrum, covering electronic technologies and other Information and Communication Technologies (ICT) applications and to provide for its operation in place of former authorities and for related matters" (URT 2003b). The idea to adjust the regulative setting was not new. Already in the years 1996 and 1997, TCC, jointly with the ITU, conducted a study to determine best ways of managing services under the existing regulatory regimes. The report acknowledged technological convergence as the major challenge to existing regulatory authorities and advised to establish one commission to regulate telecoms and broadcasting in order to ensure economies of scale, seek the uniformity of decision-making and coordinate developments and save resources (c.f. Nalingigwa 2002). Thus, the merger was a reaction to increased overlapping of functions. On the other hand, the implications of the merger have gone much further, because it also changed the internal structure of the regulator and to some extent, weakened regulatory independence vis-à-vis the government. Officially, the agenda responded to technological changes, beyond government control also increased. In accordance with this dual agenda, TCRA firstly sought a smooth transformation process following up the existing regulatory regimes, so that all licences and permits granted by the TCC or TBC remained in force and memberships of the two bodies of international and regional organizations continued.

The organisational structure, however, changed. Whereas the old TCC had a dual structure comprising the commission overseeing the general direction on the one hand, and the directorate, which was in charge of the daily business procedures on the other hand (see chapter 4.2.4.), this was abandoned under the

Figure 4.11: TCRA Organigram

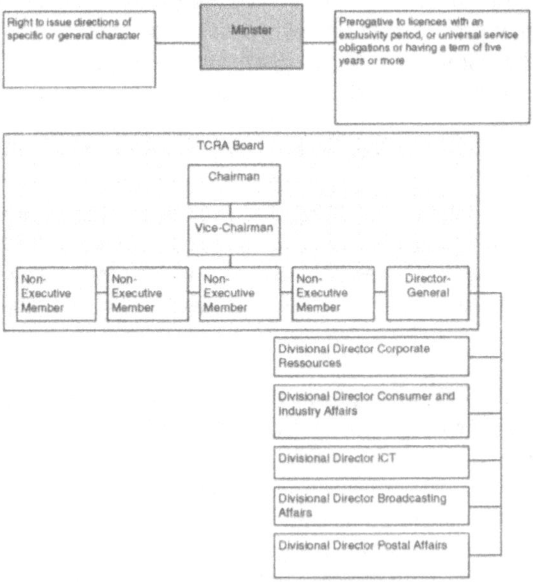

new regime, mainly because of recurrent frictions between commission and directorate as well as between directorate and ministry. Instead, TCRA established a board as the governing body of the authority. The new board was completed in April 2004 and it comprised a chairman, a vice-chairman, four non-executive members and a director general (see figure 4.11).[205] The new setup particularly affected the role of the director-general, who under the old TCC rule, possessed considerable freedom in relation to the commission. He became an integral part of TCRA's central governing body, so that the new management structure contained the powers of the director general and

[205] In accordance with the Tanzania Communications Regulatory Authority Act, the chairman and vice-chairman were appointed by the president, the four non-executive members and the director general were appointed by the Minister of Communication and Transport (URT 2003b section 7/13).

In April 2004 President Mkapa appointed Ambassador Richard Elisante Mariki as chairman and Ms Amina Mrisho Said as vice chairman of the board. In the same month, Minister Mwandosya appointed John Nkoma, the former the Dean of Faculty of Science of Botswana University and former Head of Physics Department of University of Dar es Salaam, as successor of the outgoing Asupya Nalingigwa as Director-General of the TCRA, as well as the four non-executive members Dr. Suleiman Juma Omar, Dr. Vuai Iddi Lila, Dr. Batilda Burian and Mr. Baruany Elijah Luhanga. (TCRA 2004, see also Daily News [Dar es Salaam] 18.4.2004).

strengthened the management control of the former commission. Furthermore, the Tanzania Communications Regulatory Authority Act retained provisions that the minister may give "directions of a specific or general character on specific issues, other than in relation to the discharge of the regulatory function, arising in relation to any sector, for the purpose of securing the effective performance by the Authority of its policy, functions and compliance with the code of conduct" (URT 2003b: section 6.4). This equivalent to a ministerial veto has even been made explicitly with regard to "licences with an exclusivity period or universal service obligations or having a term of five or more years" (URT 2003b: section 6.3). All decisions in this regard have to be submitted to the minister for approval. Therewith, virtually all regulative key issues,[206] except some rather technical regulations, became subject to prior consultations with the minister.

The creation of TCRA was only in parts motivated by the merger of TCC and TBC, though this was proposed since the mid 1990s and was a deliberate decision in face of technological convergence. Containing the power of the director general and strengthening the ministerial prerogative were also strong reform forces. Of course, looking back on the record of TCC the overwhelming majority of disputes as described in chapter 4.2.4. were not between the TCC's commission and the government, but between the executive powers of the director-general and the MoCT. This might explain why a redefined role of the director general has been the major consequence of the merger. It is still too early to assess TCRA and its new leadership, because the recruitment of new divisional directors responsible for the management of different subdivisions was not finalised and a new communications bill was in the decision-making process in the end of 2004. However, a preliminary estimation could conclude that the restructuring degraded TCRA politically, but on the technical level the regulator continued to upgrade its competencies and streamlined formerly overlapping functions. These improvements were also related to a number of regulatory telecommunication programmes. With assistance from the ITU or SIDA, capacity-building measures targeted a broad range of regulatory functions like universal access strategies, spectrum management or interconnection (c.f. ITU-BDT 2001, ÅF International 2002). In the first instance, however, the regulator adjusted the licensing procedures.

[206] The reference to an exclusivity period openly covers all decisions with regard to TTCL, however more strikingly is the reference to a five-year licensing term, because all important licences have a minimum duration of five years (see table 4.3).

4.3.3.2. Licensing

Already with effect from 2000, the regulator had changed the licensing set-up. It introduced ISP licences and abandoned the exclusivity period of the then data communication providers Datel, Wilken Afsat and Sita. It was mainly the poor performance of the operators that forced the regulator to act. All three had agreed on roll-out obligations as a term of their licence, but failed to deliver (Int. Mruma 10.6.2002, Int Medeye 10.6.2002). In addition, TCC was under pressure from the ISPs as well. They complained over exorbitant connectivity costs by the data communication providers[207] and called for the introduction of an independent ISP licence, which would allow to choose other data communication providers than those licensed by TCC (Int. Messer 20.5.2002). Lastly, the looming privatisation of TTCL pushed TCC as well, because the incumbent had already started busineses in each licensing segment and it became evident that the TTCL subsidiary Datel needed to be divested due to the share of France Télécom subsidiary Nexus and other contractual burdens, like with the Norwegian Telenor, who provided international data connectivity for TTCL (AISI 1999; Datel n.d.). But with the divestiture of Datel, new licence opportunities needed to be introduced to get TTCL back in data communications.

These reasons made it easy for the regulator to liberalise public data communication services. The licence holders, Sita and Wilken Afsat successor Afsat Communication, hardly objected the introduction of new licensees, because they were only insignificantly affected due to their business focus. Only Datel objected the new licences (Int. Mbita 4.6.2002). However, at that time, Datel was already in liquidation; in June 2000 final negotiations about the TTCL privatisation commenced and it appeared that the new TTCL investor MSI / Detecon decided to separate Datel from the mother company (see chapter 4.3.5.3.). Without protection of the TTCL family, Datel was simply unable to enforce any claims against TCC's licensing plans. With effect from September 2000, the regulator approved new data communication licences and introduced full competition to that market segment.

In contrast, the introduction of ISP licences was a more controversial point at hand. As stated above, the reason for changes in the licensing set-up had been the high costs of connectivity or, more precisely, the poor performance of the data communication operators, which had been increasingly identified as main

[207] The Daily Mail quoted an ISP executive who specified annual connectivity costs charged by the data communication providers at 20,000 US$ per year (Daily Mail [Dar es Salaam] 24.1.2000).

obstacles to further sectoral growth in the Internet business. In response, the ISPs had started to lobby for an individual ISP licence on the condition that they would get the freedom to choose their global connectivity provider from anywhere in the world. The ISP expected savings of at least 30 percent (The Guardian [Dar es Salaam] 20.8.2003). But to the ISPs dismay, the regulator's response went in a different direction. It liberalised the data communication market, but, nevertheless, incorporated the ISP licence into the licensing regime (TCC 2002: 8-9). Since April 1999, it had invited applications for an ISP licence. Therewith, ISPs were the notable losers of the new licence structure. Their demands were not only completely disregarded, TCC also imposed new financial burdens on them. Only narrowly and due to collective bargaining, the ISPs had averted unreasonable high charges of 20,000 US$ and more (c.f. Daily Mail [Dar es Salaam] 5.1.200, 10.1.2000 and 24.1.2000). Fees were fixed at 75,000 TSh. application fee, 1,000 US$ initial licence fee and an annual fee of 5,000 US$ (see table 4.3). With effect from May 2000, TCC had approved nine ISP licences in the first instance, until early 2004 the number grew to 23. However, TCC's decision was highly controversial, because TCC de facto double charged Internet services by fees imposed on the data operator and the ISPs (for the same data throughput). Furthermore, ISPs had to apply for frequencies in the frequency band if they wished to deploy their own wireless communication devices, so that there were actually no technical reasons to introduce the ISP licence, but only such to diversify the regulator's income basis and keep the room to control Internet services in the country.[208] Whatever motivated the introduction of ISP licences, the rigid way how TCC enforced the new ISP rules damaged the regulator's reputation, which was tarnished anyway, and raised new doubts about TCC's capacities to level the playing field for Internet services.

With regard to the licences in basic telecommunication, TTCL and Zantel continued to be the only licence holders, TTCL for country-wide operations, Zantel for Zanzibar only. With the finalised privatisation of TTCL in February 2001, TTCL's operating licence was renewed under the conditions given in chapter 4.3.4. and accorded by a four-year exclusivity period for fixed line telephony which will end in February 2005.[209]

[208] Particularly the aspect of control must be critically appraised. Only an ISP licence would provide the regulator with direct access to the ISPs and a way to influence questions of content and possible censorships. E.g. TCC has asked the ISPs to block specific websites, particularly pornography, and thus took direct influence on content delivery by ISPs.

[209] The exclusivity period not only affected the fixed line market, it also had implications for the Internet business, because it prohibited Internet telephony (VoIP – Voice over the Internet Protocol) and hence inhibited developments in a promising market segment, for a general

The preparation for the post TTCL exclusivity period, however, only commenced in late 2004. In November 2004, the MoCT introduced a new Electronic and Postal Communication Act, which proposed a complete new licensing setup (URT 2004).[210] At the same, TCRA has introduced a new licensing framework, based on the proposed communication act, that will enter into force in February 2005 and all operators are encouraged to migrate to the new setup as soon as possible. The proposal provided for the following new licence types to replace the existing licence structure:

1. Network facilities licence, entitling the licence holder to construct, maintain, and make available network facilities, and allows the licensees to own and control telecommunication infrastructure and related installations,
2. Network services licence; that is the equivalent to the data communications licence entitling to provide network services, including international voice and data transmission,
3. Applications services license; that is approval for application services in electronic communications and would cover all voice and data services like PSTN, mobile telephony, Internet access and the like,
4. Content applications licence; content applications would include all kind of content services like broadcasting or online publishing.

However, the new guidelines are not yet finalised. Particulalry the proposed bill is a largely unfinished document. Many provisions were incomplete and refer to ministerial declarations to be submitted later. Missing regulative details and the absence of a concrete implementation schedule underline the preliminary character of the bill. However, the basic intention of the new licence setup can be deduced: Firstly, the document complies with rules of technological convergence that is, it abolished a voice/data distinction of electronic communications. Thus, it made provisions for voice over the Internet applications and, in theory, offset all safeguard measures in favour of TTCL. However, if VoIP will actually be introduced, or, if so, under what conditions, is not covered in the document and will, finally, depend upon regulative practices pursuit by the regulator. Secondly, the new licenses foster competition, because the existing monopoly markets of TTCL would be dissolved and merged in distinctive licences for facility based, operation based and application based licenses (c.f. URT 2004). Both, the introduction of competition in former

---

outlook on VoIP see ITU (2001c), the Tanzanian perspective is summarised in TCC (2001a: 3).

[210] The new act will enter into force when approved by parliament.

monopoly segments and the recognition of voice/data convergence would be major achievements in the regulatory setting. On the other hand, the large number of unresolved issues could undermine the document's credibility, particularly because many of the potentially contentious issues need further clarification. For example, the draft bill proposes a complete new cost and fee structure. It proposes that all licence and frequency fees will be limited to covering the administrative costs, and additional resources, such as needed for issuing a universal access fund, would be passed on to the operators by a new concessional levy. This would be a full reversal of the existing structure, which covers most costs by licence and spectrum fees, but does not pose any additional cost burdens on the operators. Detailed cost-models are not yet on the table, but operators have already announced their intent to object to the new cost- and fee structure (The Guardian [Dar es Salaam] 10.11.2004). Other problems might occur with regard to the treatment of licence exempt services, an imprecise borderline between traditional and electronic media – particularly in view of the critical issue of content regulation. Thus, so far, the draft bill reveals little more than the general direction of future regulation. Some aspects, such as voice-data convergence or the break-up of monopoly structures, point in a promising direction, but for the time being, a considerable number of blank positions to be filed later, undermine the credibility of the document. Only after completion of the legal text and, above all, after adopted implementation procedures will assume a definite form, the value of the new bill will become visible.

Beyond the provisional legislative groundwork, concrete effects from the new licensing can, so far, only be detected with regard to a second national operator. In October 2004, Minister Mark Mwandosya announced that Zantel will be approved as the second national telephone operator (SNO).[211] On the mainland, the SNO, most likely, will provide services under the brand of Etisalat, Zantel's main foreign investor. Considering the difficult process to establish an SNO in other African countries, Tanzania took advantage of the mainland–Zanzibar duopoly structure, which allows the installation of an already locally rooted and experienced operator as SNO.

Until the new bill is formally adopted, the existing licensing setup will remain in force. This prescribed the costs of a licence and the respective duration of the operating permit. The structure, as is it given in figure table 4.3, reveals a ranking of licences. The heavy-weights fixed and mobile telephony and public data communication licences were subject to more austere conditions, thus demanding a higher market entrance commitment, but also allowed additional room for manoeuvre in individual licence negotiations.

[211] Quoted in: The East African [Nairobi] 13.10.2004, see also BA No. 210.

Table 4.3: Licence Fee Structure					
Type of Licence	Application Fee*	Initial Licence Fee*	Annual Fee* (Royalty)	Duration (years)	Renewal (years)
Basic Telecommunication Service	**10,000 US$.**	**Competitive**	**0.8% of the annual gross turnover**	**25**	**10**
Land Mobile Cellular Service	**5,000 US$**	**Competitive**	**4% of the annual gross turnover**	**15**	**5**
Data Communications Service				**10**	**5**
(a) Public	**1,000 US$**	**US$ 100,000**	**3% of the annual gross turnover or 30,000 US$ whichever is higher**		
(b) Dedicated / Closed User Group	**500 US$**	**5,000 US$**	**500 US$**		
Internet Service [without VoIP]				**5**	**3**
(a) Commercial	**75,000/= TShs**	**1,000 US$**	**5,000 US$**		
(b) non-commercial	**$50 or its equivalent in TShs.**	**300 US$**	**200 US$**		
Radio Paging Licence	**500 US$**	**2,000 US$**	**3% of the annual gross turnover**	**10**	**5**
Installation and Maintenance Licence	**50 US$**		**on average 400 US$.**	**1**	**1**
Importation and Distribution Licence	**50 US$**			**1**	**1**

* fee to be paid in US$ or its equivalent in TSh.
Source: TCC (2001d)

In addition, all operations that involved frequency requirements were required to make a separate application for radio frequency user licence, which put another substantial financial burden on the operators (MoCT 2001b: 54) (see table 4.4). Therewith, Tanzania has fixed relatively high market entrance costs and the collected fees formed the basis of a self-financed regulator. Though costs were high, the number of operators approved in each segment indicated that these did not hamper the emergence of a lively market. In the beginning of 2004,[212] the regulator approved two basic telecommunication operators, TTCL for the whole union, Zantel for Zanzibar only, and from February 2005 onwards as SNO for the whole union. In mobile telephony four operators were licensed, while it had been five until Tritel ceased its operation in January 2003 (Business Times [Dar

[212] For an overview of all licensed operators see TCC (2003, 2003a), TRASA (2003).

es Salaam] 24.1.2003). Attempts to invite mobile operator MTN, a South African mobile telephony company with a notable expansion record in six other African markets, to take over the Tritel licence failed due to the competitive market structure in Tanzania (Business Times [Johannesburg] 12.2.2003) (Financial Times [Dar es Salaam] 22.1.2003).[213] The greatest impact since 2000, however, has followed the full liberalisation of the public data communication segment, which increased the number of operators from three to eleven and resulted in increased availability of data communication services. At the same time, the number of closed user data communication licences or VPNs increased to nine operators, which mainly came from the banking sector, large companies and institutions of higher education like UDSM.[214] The newly introduced ISP licence particularly hampered the market entrance of small operators, especially those who targeted upcountry customers where business volumes were low anyway. The number of approved licences reached 23 by early 2004. Licences for contractors for installation and maintenance for the year 2003/2004 had risen to 29 and for importation and distribution 26 contractors. Thus, with the exception of basic telecommunication (until February 2005), all segments in the telecommunication market have been liberalised with positive effects on the availability and affordability of telecommunication services. A closer consideration of the different market players is given in chapter 4.3.5.

4.3.3.3. Radio Frequency Spectrum Management

The management and allocation of frequencies was foremost a technical issue meant to ensure the effective use of the frequency band, assign frequencies to operators and to ensure compliance with the given rules avoiding electromagnetic interferences for failure-free wireless radiocommunications. Though each country may allocate spectrum for various services within its own borders, the ITU has been the international body in which the various standards and frequencies are assigned. Default frequencies need to be in compliance with ITU's Table of Frequency.[215] Hence the national regulator has not been

---

[213] The fifth mobile telephony licence, which would also be the maximum number of operators due to limits in the frequency spectrum, could not be reallocated. However the satellite telephone provider, Thuraya, covering Africa, the Middle East and Europe, was granted a licence for satellite telephony, a very small niche market in Tanzania that is only mentioned for the sake of completeness (see East African [Nairobi] 3.3.2003).

[214] Notable exceptions to licensing included the large parastatals Tanzania-Zambia Railways Authority (TAZARA), Tanzania Electric Supply Company (TANESCO) and the Tanzania Railways Corporation (TRC) (see TRASA 2003) and a number of VPNs operated by donors.

[215] The ITU's Radiocommunication Sector (ITU-R) maintains a Table of Frequency Allocations which identifies spectrum bands for about 40 categories of wireless services with

confronted with the task to assign frequencies for given operations, but to decide who is getting how many capacities for which services and in order to ensure smooth operations.

Accordingly, TCC's working focus has been on frequency allocation and monitoring. First systematic recommendations in this regard were elaborated with assistance from the ITU which proposed a strategy aiming to build a computerised spectrum management system providing computerised tools for frequency planning, calculating interference levels, the allocation and assignment of frequencies, and for comprehensive and accurate record keeping (c.f. ITU-BDT 2000). Based on the regulations already in place, the ITU proposal concluded that additional technical and non-technical capacities needed to be installed and a five-phase restructuring programme. This commenced in 2001 and comprised the establishment of a computerized record-keeping system, a computerized frequency assignment system and a frequency planning system, improvement in the monitoring and enforcement capability, and a revision of spectrum pricing (ITU-BDT 2001). To that end, TCC acquired a first mobile radio monitoring station in 1999, installed a fixed radio monitoring station in 2003 and set funds for the financial year 2003-2004 to procure a second mobile radio monitoring station. Together with three stand-alone fixed radio monitoring stations which will be located along the southern, western and northern boundaries and the acquisition of the ITU-R spectrum management software, the regulator will have in place the technical capabilities of effective spectrum management (c.f. TRASA 2003).

This technical upgrading, which was constantly accompanied by respective training programmes, was complemented by legal amendments to the Tanzania Communications Act in 2001 (MoCT 2001b). Licensing conditions, licence classes, quality of networks, inspection rights and rule compliance were put on a new legal basis and made radiocommunications subject to a well defined set of rules and procedures. In effect, all operations ranging from amateur radio stations to wireless communication devices deployed in mobile and fixed telephony or satellite communications were tied to radio frequency user licence and respective fee charges as given in table 4.4.

---

the goal of avoiding interference among different services, c.f. Nunno (2002), a good background on ITU-R can be found at ITU-R (2004).

Table 4.4: Spectrum Charges. Annual Radio Communication Licence Fees

	Type of licence	Licence fee/ Station/Bandwidth
1	**Amateur Radio Station**	**$50.0**
2	**HF Radio Station**	**$80.0**
3	**i) VHF/UHF Radio Station**	**$ 60.0**
	ii) VHF/UHF Radio with pair of frequency	**$70.0**
	iii) VHF/UHF Radio Repeater	**$80.0**
4	**Citizen Band Radio Station**	**$50.0**
5	**HF Cross Border Radio Station**	**$100.0**
6	**Transportable Satellite Transceiver**	**$3,000.0**
	Inmarsat type A	**$2,500.0**
	Inmarsat type B	**$2,000.0**
	Inmarsat type C	**$1,500.0**
	Inmarsat type M	**$1,000.0**
	Inmarsat type Mini M	**$ 500.0**
7	**VSAT Station**	**$1,000**
8	**Public Ground Earth Station**	**$5,000.0**
9	**Public Fixed Radio Station.**	
	First 100 or less Stations	**$70.0**
	Additional 101 or more stations but less than or equal to 300 Stations	**$60.0**
	Additional 301 or more stations but less than or equal to 500 stations	**$50.0**
	Additional 501 or more stations	**$40.0**
10	**Fixed Microwave links base station**	**$70.0**
11	**Wireless Data Radio station**	**$50.0**
12	**Frequency Fee per Cellular zone**	
	First Bandwidth less than 1MHz	**$200.0/25KHz of occupied bandwidth**
	Additional Bandwidth of = 1MHz but = 10 MHz	**$150.0/25Khz of occupied bandwidth**
	Additional Bandwidth of more than 10 MHz	**$100/25KHz of occupied bandwidth**
13	**Frequencies for Wireless Access (WLL)**	
	First Bandwidth of 20MHz or less	**$150/200KHz of occupied bandwidth**
	Additional Bandwidth of ‹20 MHz but = 50 MHz	**$100.0/200KHz of occupied bandwidth**
	Additional Bandwidth of › 50 MHz	**$50.0/200KHz of occupied bandwidth**

Source: (TCC 2001c)

Though technical prerequisites were increasingly in place, the practice of frequency management remained difficult for several reasons. Firstly because operators were reluctant to give back once assigned frequencies.[216] Secondly, illegal frequency use was still common; and, thirdly, enforcement capacities were still under-developed. Further restrictions must be made regarding the scope of spectrum regulation. TCC and later TCRA were occupied with the fundamentals of spectrum management, i.e. assignment and monitoring. Even if results in these fields could be rated satisfactory, the regulator failed to cope with new challenges of wireless broadband technologies and increasing fixed-mobile convergence. New trends have addressed new, more market-oriented approaches to spectrum management, like the use of auctions, spectrum pricing and the introduction of trading rights (c.f. Cave 2002, Nunno 2002, Neto 2004).[217] The complexity of such issues might explain why they have not been addressed sufficiently. However, bearing in mind that the technologies affected by these issues, such Wireless Metropolitan Area Networks (WMAN) or Wireless Local Area Network (WLAN) are not in the distant future, but standards increasingly deployed, the urgent need to take up new concerns of spectrum management became evident.

4.3.3.4. Interconnection

Interconnection has been one of the most contentious issues in Tanzania, like in many African countries and persistent interconnection disputes pointed towards a largely unfinished business of regulation. Interconnection means the linking of public communications networks to allow the users of one network to communicate with users of another network or access services provided by another undertaking.[218] In Tanzania, interconnection agreements on a fair and

[216] Mobile telephony can serve as an illustrative example: Mobile telephony in Tanzania occupied 50 MHz in the 900 MHz spectrum band. Of these, Mobitel occupied 25 MHz, whereas Vodacom or Celtel were assigned five MHz only. The reason was that Mobitel started mobile telephony with an analogous system back in 1994 which needed more frequencies than modern digital (GSM-) systems. However, Mobitel also transferred its analogue to a digital system, but ever since refused to reassign the unused frequencies arguing that they still need them for their analogous system (The Express [Dar es Salaam] 24.8.2000), TCC 2001a: 7, Int. Held 21.5.2002)

[217] At the same time, these new regulative approaches are challenged, too. There is an increasing call for deregulation – e.g. forcefully advocated by the Open Spectrum Movement – which advocates so-called 'smart' and 'polite' technologies that enable ways of controlling interference, even when there are several users of the same frequency in the same place, thus demanding more spectrum for unlicensed use and challenging the core of today's regulative ratio (c.f. Neto 2004: 30, Ikeda 2003, Cave 2002). The proposed communications bill provided for possible licence exemptions, but made these subject to a undefined declaration to be issued by the regulator (URT 2004: 14).

[218] This definition is based on a Directive proposed by the European Commission on access and interconnection in July 2000, quoted in: Intven/McCarthy Tétrault (2000: 3-2).

non-discriminatory basis have been one key obligation for all licensed telecommunications operators. But whereas for end-users these have been safeguards to ensure the interoperability among different networks, for operators the interconnection regime determines the real economic conditions for market entry and operations, because the cost of interconnection has been a major commercial issue (c.f. Bipe 2000: 61). Interconnection agreements provide cost or revenue sharing solutions when a call originating in network A terminates in network B and network B operator can charge network A operator for interconnecting the call on its infrastructure. Network operator A covers all costs by charging the customer, so that, ideally, interconnection agreements allow for transparent and fair cost sharing provisions among telephone operators. Models of interconnection vary – they may be simple, such as revenue sharing[219] or sender keeps all models,[220] or they can be more advanced cost-based agreements that take into account the operator's cost structure or calculations concerning the costs of a network (c.f. Intven / McCarthy Tétrault 2000: 3-23).

The original interconnection regime arranged for in the Tanzania Communications Act and the interconnection regulations 2001 that amended the existing legislation in November 2001[221] provided the general guidelines of the interconnection regime and called for transparent, non-discriminatory and cost-based interconnection charges – which also complied with the WTO Regulation Reference Paper that has defined the central regulatory principles for WTO-members. The individual negotiations of interconnection charges, however, were in the operator's responsibility and the parties to an agreement were "expected to negotiate periodically in good faith (…) to arrive at amicable settlements as to the form and subject of an interconnection agreement" (TCC 2002). The regulator would only intervene, if the parties fail to reach an agreement or disputes arise between parties and they may petition the regulator to arbitrate any open issues (MoCT 2001b: 50).

Evidently, TCC and TCRA interventions in interconnection disputes became more a general rule than an exception. Disputes mainly surfaced in the interconnection of voice networks particularly as the new players, Vodacom in

[219] Typically, new entrants pay the incumbent operator a share of their revenues from interconnection services.

[220] Also labelled as 'bill and keep', which means that no charges are to be paid between interconnecting operators for termination of each other's traffic. Though this is the simplest model, it only works if operators have similar network size, traffic flows and traffic volumes.

[221] The Interconnection Regulations 2001 can be found in MoCT (2001b: 46pp).

August 2000 and Celtel in November 2001, entered the market.[222] In fact, none of four major adjustments made in the interconnection regime were undisputed and the situation became flawed as Celtel started operation and the other mobile operators refused to negotiate interconnection rates with the new operator, because Celtel proposed new 10 US cents interconnection charges in comparison to the existing 25 US cents for mobile to mobile communication (Business Times [Dar es Salaam] 25.1.2002). TCC only overcame the following stalemate by issuing the interconnection determination no. 1 of 16 January 2002 that introduced a 17.5 US cent price cap for mobile-to-mobile interconnection and forced the other operators to enter an agreement with Celtel. Three months later, the regulator introduced price caps for 17.5 US cents for telephone calls originating from TTCL and terminated in mobile networks and a staggered discount solution for mobile calls to the fixed-line network in determination no. 2 of April 2002 (see box 4.5). However, determination no. 2 also had only a short lifetime and was succeeded by determination no. 3,[223] which was disputed as being unrealistic, repeatedly postponed, and, finally, never introduced.

Until that time, all interconnection determinations by the regulator were issued as interim solutions meant to give incentives to the operators to complete bilateral commercial negotiations. Only as these attempts continued to fail and the operators deferred the implementation arguing that the new figures would not allow economically viable operations, the regulator changed its approach and invited tenders for a new cost-based interconnection regime supposed to end the persistent disputes. In the end, Analysys Consulting of the United Kingdom conducted a study on the issue and in September 2004 a new cost-based interconnection regime was introduced accordingly. The new rates are given in box 4.5, but though expectations in the new regime were high, it took only a few days for mobile operators, namely Vodacom, to raise objections and delay the new interconnection regime once more. TCRA was forced to postpone the new

[222] This became also evident by the number of interconnection disputes that appeared in court, thus, that could not be solved by the dispute settlement mechanisms provided by the regulator; e.g. Mobitel against TTCL for outstanding interconnection payments of 5.3 million US$ (Business Times [Dar es Salaam] 4.1.2002), or Vodacom against TTCL for unremitted charges on international calls on its network amounting to 3.2 million US$ (The Guardian [Dar es Salaam] 2.4.2004).

[223] The determination no. 3 proposed an interconnection rate ceilings of 125 TSh. for mobile-to-mobile and fixed-to-mobile as well as 75.00 TSh. for mobile-to-fixed and was actually due to be applicable from October 2002 (see box 4.5). Beyond the pure figures, interconnection determination no. 3 also established two important improvements that became approved, finally: firstly local currency billing was introduced and call durations for charge purposes shall be rounded to the nearest second, replacing per minute charges, to avoid overcharging of calls which do not last to a complete minute (TCC 2002a).

interconnection regime until October 2004 and installed a mediation panel to solve the pending issues (The Guardian [Dar es Salaam] 2.8.2004, Daily News [Dar es Salaam] 9.8.2004, 8.9.2004). Thus, despite undoubted progress in interconnection, that brought down the rates from a level considered as one of the highest in the world to more reasonable rates between 3.9 and 10 US cents (or TSh. equivalent) with considerable effects on end-consumer prices, interconnection remained highly disputed. Box 4.5 gives a selected overview of telephone tariffs (2002 / 2004), successive interconnection rates (2002-2004) (tables at the top of the box) and, in order to understand the economic exchanges, examples of a simplified interconnection-telephone tariff cost model for 2002 and 2004 (at the bottom of the box).

The tables hint towards two main difficulties in establishing a fair interconnection regime. The first derived from strategic grouping in the sector, as TTCL, the TTCL subsidiary Celtel and Zantel managed to enter into mutual agreements and special tariff provisions, whereas Vodacom, Mobitel and, until January 2003, Tritel interconnected through high price agreements with TTCL before the arrival of Celtel and agreements that made use of the 17.5 US cent price caps after the arrival of Celtel. Hence, those with a public stake and some degree of government proximity formed one block and entered into mutual agreements particularly aiming to improve their economic position by lower tariffs[224] and lower interconnection rates. Considering that statistically there are eight times more calls made from the TTCL network to mobile operators than vice versa,[225] the economic relevance of different tariffs and interconnection rates became evident. In the 2002, for example, settings made under interconnection determination no. 2, Celtel customers enjoyed lower tariffs for calling TTCL, and TTCL was disburdened by lower interconnection rates in the high traffic segment of mobile termination calls made to Celtel. Thus, a favourable cost structure of telephone operations gave the TTCL-group a comparative advantage in the market. The new rates of 2004 revised these imbalances, because interconnection rates were no longer dependent on mutual agreements among individual operators, but only on determinations made by the regulator. With new fixed rates the regulator introduced a shared set of binding guidelines, so that the competing mobile operators would solely depend on their

[224] TTCL abandoned operator dependent tariffs in August 2004, when the new interconnection rates were proposed. Before the incumbent charged 295 TSh. only for calls to Celtel and Zantel, calls to Vodacom and Mobitel were charged at 380 TSh., thus considerably higher (Daily News [Dar es Salaam] 3.8.2004).

[225] According to an undisclosed source inside TTCL, quoted in Business Times [Dar es Salaam] 26.4.2002.

services like tariffs, network coverage or quality of the network to attract new customers.

Box 4.5: Tariff and Interconnection Structure 2004 and 2002 (selection)

Telephone Tariffs August 2004 (in TSh.*)

Calls From	TTCL (peak, post paid)	Celtel (peak prepaid)	Vodacom (peak prepaid)
Calls To			
Own mobile network	-/-	265	260
Other mobile network	295	340	340
TTCL local	60	315	360 (0-500km)
TTCL 50 – 500 km	160	350	

Telephone Tariffs March 2002 (in US$*)

Calls From	TTCL (peak, postpaid)	Celtel (peak, prepaid	Vodacom (peak, prepaid)
Own mobile network	-/-	0.33	0.35
Other mobile network	0.33	0.43	0.45
TTCL local	0.04	0.39	0.40
TTCL 50 – 500 km	0.15	0.44	0.45

for rough orientation 1 US$ = 1,000 TSh.
Source: Operators

Interconnection charges 2002-2004			
	Mobile to Mobile	**Fixed to Mobile**	**Mobile to Fixed**
Determination No. 1 of 2004 with effect from October 2004	10 US cents staggered (8.9 US cents /2005, 7.9/ 2006, 6.9 US cents	10 US cents staggered (8.9 US cents /2005, 7.9/ 2006, 6.9 US cents	– 3.9 US cents (single tandem*) / 5.5 US cents (double tandem**) from August 2004 – 3.8 / 5.3 US cents from January 2005 – 3.6 / 5.0 US cents from January 2006 – 3.5 / 4.8 US cents from January 2007
Determination No. 3 of 2002 (suspended)	125.00 TSh. price cap	125.00 TSh. price cap	75.00 TSh for national calls price cap; call volume discount as in Determination No. 2 for international calls
Determination No. 2 of 19 April 2002		17,5 US cents price cap; to Celtel and Zantel: 10 US cents (bilaterally negotiated)	National calls staggered according to distance from 0.064 (local calls), 0.141 US$ (50-500km) to 0.212 US$ (above 500 km) International calls: call volume discount, e.g. 8% for 50,000 – 99,999 call minutes, 11% for 100,000 to 149,999, up to 17% above 250,000 call minutes
Determination No.1 of 16 January 2002	17,5 US cents price cap Celtel – Zantel 10 US cents (bilaterlly negotiated)		
prior	25 US cents (Vodacom, Mobitel, Tritel) 10 US cents (Celtel, Zantel)	TTCL retention rate of TSh. 221 per minute	25 US cents (Vodacom, Mobitel, Tritel)

* Single tandem = data transmissions that make use of one tandem exchange (= switch to control traffic on main trunk lines) before routing the traffic to a local exchange which has direct connections to the end user.
**Double tandem = data transmissions that make use of two tandem exchanges before routing the traffic to a local exchange which has direct connections to the end user.
Source TCRA (2004a), TCC (2002)

Box 4.5. continued

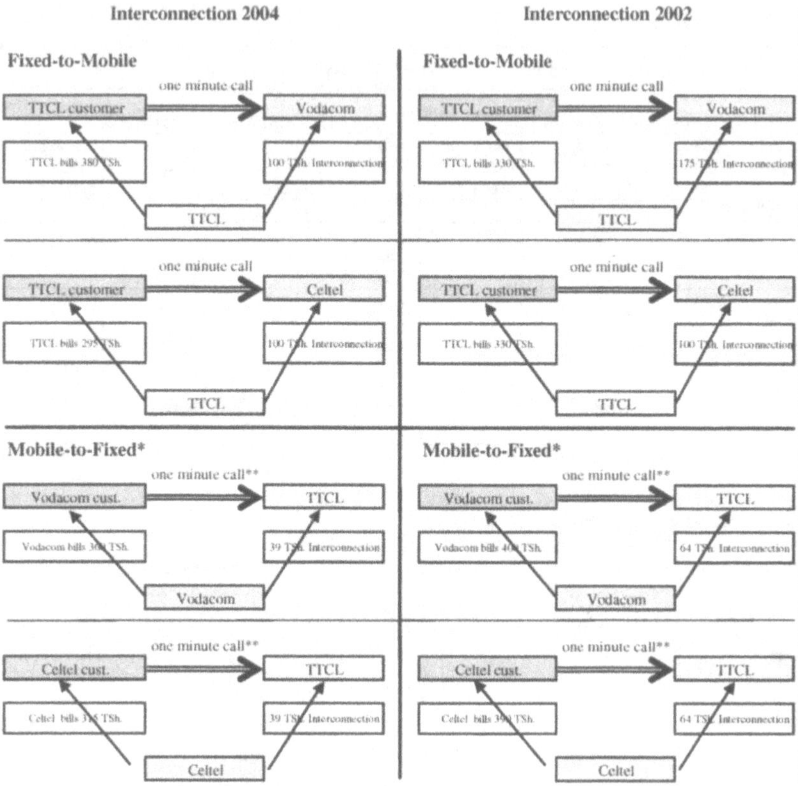

* For simplification figures are given as a 1:1,000 US$ TSh. exchange rate.

** to TTCL within 50 km

The second difficulty, however, to some extent contested the first one of equal treatment. Apparently, the real costs of call termination differ according to the quality and size of the networks deployed. High quality networks with greater coverage need considerably more investment and high maintenance expenses than smaller networks and, thus, calls originating from a smaller and terminating in high quality networks need higher charges to recover the costs. Operators of large scale networks argue that all customers gain from the large networks, simply because the customers of these operators can be called from any other network, thus they add more value to the national network and, consequently, seek higher interconnection rates for calls terminating in their network and lower interconnection for their calls terminating in smaller, less sophisticated networks.[226] Basically, this was the reason why Vodacom, with regard to subscribers the largest operator in the country, has disagreed with the new interconnection regime and provoked an official inquiry on cost-based interconnection rates in September 2004. In the end, Vodacom failed and had to yield to the arguments made by the Analysys study. TCRA declared the determination No. 1 of 2004 as effective from October 2004 (Daily News [Dar es Salaam] 1.10.2004, The Guardian [Dar es Salaam] 1.10.2004).[227]

Apart from the complex technical implementation issues, the interconnection rates of 2004 were an important step in the direction of fairer and more sustainable market structure, in which all operators became subject to the same interconnection rates, with an overall effect of a 40 percent price reduction in mobile telephony. Thus, noticeable progress can be conceded and this applies all the more that all operators except Vodacom – TTCL, Celtel, Mobitel and Zantel –have already announced to pass on the reduced interconnection rates to the customers and, thus, to greater affordability of telecommunication services in the country (The Guardian [Dar es Salaam] 28.7.2004), Daily News [Dar es Salaam] 3.8.2004).

4.3.3.5. Universal Access

Universal service or universal access strategies[228] only came slowly on the agenda and, apart from a limited number of individual projects, rather unspecific

[226] This is the challenge if the regulator targets a cost-based interconnection regime. The calculations get even more difficult, when, for example, policy incentives are to be included like special treatment for rural areas. In such cases, rural interconnection or calls from urban areas terminating in rural areas have higher interconnection rates and, thus, generate a higher cash flow that can be reinvested in rural telecommunications. Hence, interconnection rates are used as a tool to finance universal access policies (c.f. Intelecon 2003).

[227] However, beyond the individual case, the Vodacom objections to the interconnection rates revealed some of the basic difficulties. Cost-based systems need to be built on robust and reliable data and thus a significant amount of detailed and verifiable company information. The readiness to share this information can be put into question. Nevertheless, the consultant claimed to have included the costing of all the individual network elements and then calculated the cost of the entire network based on assumptions and information about the network design based on the firm's financial accounts (Analysys 2004). If this statement and the actual outcome, however, are compared, it indicates that there is still room for improvement in the design of interconnection regimes.

[228] Universal service fosters universal household connections to the public network, thus would not be suitable for most developing countries. In contrast, universal access aims at reasonable individual access to public telephones. This access may be through public payphones, community telecenters, tele-boutiques, community terminals for Internet access and similar means (c.f. Navas-Sabater et al. 2002, Intelecon 2003).

provisions for universal service in various policy documents and mainly undisclosed service obligations for licence holders, Tanzania did not address the issue coherently before 2003, when the Peruvian consultant firm Apoyo was awarded a contract to develop a strategy on universal service and rural telecommunication. The legal basis for this new approach was laid in October 2004, when the government introduced a new draft communications bill that also provided for the creation of a universal access fund (URT 2004: 67).

However, first attempts to address universal service strategies can be traced back to the year 1998 and a partnership between a limited number of least developed countries and the ITU in order to increase the availability of telecommunication services in developing countries.[229] In this context, the ITU had conducted a field study in Tanzania and proposed universal access targets of providing community access points for each community with 500 or more inhabitants by 2020 (ITU-BDT 2001: 7). Furthermore it made a number of rather technical recommendations, among others, with regard to the establishment of interconnection and equipment-sharing agreements.[230] Due to ongoing reforms, at that time particularly under the TRP (see chapter 4.2.2.), the ITU proposals were mainly incorporated in other reform programmes. As a result, ITU suggestions to promote universal access strategies in Tanzania were only taken up in the establishment of a rural telecentre project in Sengerema.[231] First project ideas in this regard already came up in 1996, but the project did not commence before 2000 and the telecentre actually started full operation only in September 2002 (Int. Bugalama 23.4.2003).

In due course, discussions regarding universal access were, to some extent, eclipsed, partly because of only little advancements in Sengerema, mainly, however, due to the ongoing sector reforms that significantly gained momentum, which also led to a redefinition of universal access (see table 4.5). Apparently,

---

[229] The partnership was based on a resolution approved by the World Telecommunication Development 1998. Initially, 48 LDC were chosen to participate in the project, but later it was decided to concentrate the efforts on a smaller number of countries to achieve greater impact with the limited funds available. These were Cambodia, Chad, Comoros, Ethiopia, Haiti (which later dropped out of the programme due to the deteriorating political situation), Sudan and Tanzania (c.f. ITU-BDT 2001).

[230] E.g. a draft agreement on infrastructure sharing among mobile telephone operators was signed in 2001 (c.f. TCC 2001a: 4), but negotiations on individual modalities were never finalised.

[231] The establishment of pilot telecentres as public access points for telecommunication services (Internet, email, telephone and fax as well as training) also went back to a decision made during WTDC 1998 to initiate a programme on "Integrated Rural Development and Universal Access", the so-called Valetta Action Plan (c.f. ITU-BDT 2000a).

the new definitions completely abandoned any public responsibility and, instead, put all weight of telecommunication development on pure market approaches and private sector service delivery. This reflected the optimism in the market, particularly prevalent in 2001 when mobile telephony started to take off, Internet cafes mushroomed and the privatisation of TTCL was finalised and included additional roll-out obligation which required TTCL to expand the network and to install two payphones for every village with 3,000 inhabitants or more (c.f. TRASA 2003; PSRC 2001).

Table 4.5: TCC's understanding of Universal Access

Old Definitions	TCC's redefinition 2001	New proposals 2003
Universal Access A telephone within walking distance Telecentres in every Community	**Universal Access** Mobile coverage of all main population centres Internet Cafes in every community	**Universal Acccess** Reduction of distance to the closest payphone from 26,7km to 5m = installation of 2,046 public payphones Creation of 43 telecentres in district capitals with electricity available
Affordability Subsidised access	**Affordability** Prepaid access	**Affordability** Creation of an Universal Access Fund
Source: TCC 2001b: 11, Apoyo (2003)		

But it became increasingly obvious that a universal access strategy, based solely on market mechanisms, fell short of providing the targeted figures. In addition, TTCL's due diligence revealed that the incumbent's performance was considerably weaker than expected. These shortcomings forced the government to formulate a new start in universal access taking up the old suggestions of roll-out targets and the establishment of a rural telecommunication fund. Universal access became a new priority and this was also evident as the ministry, and not the regulator, was brought in as the principal actor with regard to any universal access decisions.[232] To this end, the ministry invited Apoyo Consultario to formulate a detailed programme for rural telecommunication mainly based on the benchmark figures already proposed by the old ITU recommendations to provide a community access point for each community with 500 or more inhabitants by 2020, introduce telecentres on district level and establish a development fund for rural telecommunication.

The Apoyo proposal, introduced in November 2003, focused on public payphone provision, and thus identified basic telephony and the introduction of respective finance mechanisms as crucial factors in developing rural

[232] This responsibility was approved by law and made explicit in the Tanzania Communications Regulatory Authority Act, section 6.3.2. (see TCRA 2003: 10).

telecommunications. Internet related services in shape of newly created community telecentres were targeted as a secondary complement only. All in all, the proposal called for the installation of 2,046 public payphones and the creation of 43 telecentres in district capitals. The main features can be summarised as follows: Target population: rural villages of up to 3,000 inhabitants without effective access to telephone services (no farther than 1 km from an all-year accessible road).

- Service delivery: Mainly payphones, to a lesser extent telecentres.
- Establishment of a universal access fund under separate management of TCRA, financed in the short term by the diversion of licence fees payments.
- Competitive allocation of funds through minimum subsidy proposals (= reverse auctions), subject to quality requirements and guarantees of fulfilment.
- Incentives: Initial payment before the installation of facilities, the next after the verification of the installation, and the rest of the subsidy distributed in equal installments for a specific period of time (c.f. Apoyo 2003).

The proposal supplemented the TTCL roll-out obligations of serving every village with a minimum of 3,000 population.[233] The second step of the recommended strategy made arrangements for the selection of target villages prioritising on those villages with greater potential impact. These were identified in a two-staged process, firstly taking into account regional varieties including indicators such as infrastructure, economic activity or demography and combined them into regional clusters. Those with lower teledensity, lower income, poor infrastructure and service penetration were considered eligible candidates and eventually two clusters comprising 15 poorer regions of Tanzania became focal zones of the proposed universal access programme.[234] The second step covered the identification of the villages themselves based on a detailed village and household survey to make out the demand for telecommunication and the most promising villages for the pilots.[235] Finally, 550

[233] The proposal very well recognised the fact that TTCL did not comply with its roll-out obligations and recommended to renegotiate the role of TTCL in rural telecommunications and put in place provisions that would allow to bypass the TTCL monopoly in rural areas (Apoyo 2003).

[234] These were: Iringa, Kagera, Mara, Mtwara, Mwanza, Pwani, Ruvuma, Shinyanga, Dodoma, Kigoma, Lindi, Mbeya, Morogoro, Rukwa, Singida, Tabora (Apoyo 2003a)

[235] Apart from introducing benchmark figures on rural telecommunication, the survey revealed a number of interesting facts, among others that 64,5 percent of the surveyed

villages – of an overall 2,046 to be targeted by 2020 – were selected for a pilot programme. These were located in the chosen priority regions Kigoma, Tabora, Rukwa and Singida. Within the poorer regions these are the relatively larger villages with electricity, which have concentrated some trade, transactions and institutions in the districts and were thus regarded as the most promising ones (Apoyo 2003b).

In a complementary attempt, the consultant also developed an Internet diffusion strategy. However, this was made only very carefully and rather generally, taking into account that there has been an increasing displacement dilemma, in which private initiative to start a public Internet access facility has been constricted by donor funded projects in some cases, or as it would be in the given case, of telecentres financed by a TCRA telecommunication development fund. Furthermore, the Apoyo team was well aware of the difficulty in developing a working telecentre model – due to difficulties to provide relevant content for rural users; training of potential users and illiteracy rates in rural communities. The proposed strategy only provided some general recommendations to establish telecentres in 43 district capitals. In accordance to the rural payphone strategy four district capitals in the same regions were proposed in the pilot for the installation of telecentres (Apoyo 2003b).

For financing, the consultant seized the old and recurrently surfaced suggestions of a universal access fund. The working schedule included necessary amendments of the legal framework to build the basis for the fund. Furthermore, a clear mandate and well-defined finance mechanisms were regarded as key prerequisites for successful fund operations. However, the proposal did not further elaborate on the critical questions of fund replenishment and disbursement in the longer term. Only for an initial phase replenishment shall come from earmarked sums from TCC's general reserve account for transfer to the fund by diverting a portion of the telecommunications licence fees.[236] Funds disbursements should be made by competitive allocation process, such as auctions, which would allow a more efficient determination of the subsidy level that the fund must offer to universal access providers[237] (Apoyo 2003). Though

---

households in villages without a payphone make telephone calls on a regular basis, generally by accessing payphones in another community of the district and that the average distance to the next payphone is 26,7km. The average monthly expenditure on telephones is 1.27 US$ (Apoyo 2003).

[236] A direct contribution by the operators was not considered, because levies and fee already in place were regarded to be in an upper limit, so that there would be is no room for a further tax levy on operators (Apoyo 2003).

[237] Determining the least subsidy level has been increasingly done by reverse auction. Drawing on Latin American models, profitable market segments have been disaggregated

the Apoyo proposal was in line with individual measures of successful universal access models, it has not introduced any detailed implementation strategy, but rather an enumeration of different option of how to run a fund solution.

Nevertheless, at the end of 2003 a starting point for a universal access strategy was introduced and awaited further steps of implementation. Possible pitfalls can be identified for numerous topics of the strategy, such as ill-defined responsibilities of the ministry and the regulator or unclear working procedures or a too static approach paying too little attention to local needs. However, finance will be the most critical issue, because relocating from TCRA's general budget can only be a short-term option, if the already tight budget and the multiplicity of tasks are considered. In fact, the draft communications bill diverged from the consultant's recommendations. It proposed a complete new cost structure, targeting a yet to be defined licence and spectrum fee structure to cover the administrative costs of the regulator and an extra levy for the newly created universal access fund. The draft bill specified that contributions to the fund by licensed operators shall not exceed 4.5 percent of the gross revenues generated by the licensee in the preceding year (URT 2004: 68). Though the value is given as a upper limit, a 4.5 percent levy on gross revenues is very high and would mean a considerable cost burden for the operators. Thus, the acceptance of the fund will largely depend on simultaneous relief in the licence and spectrum fee structure, but provisions in this regard are not yet detectable.

The second critical issue relates to the role of TTCL. Since the focus of the fund has been on the provisions of payphones, it directly touches upon TTCL's monopoly including the public payphone market and TTCL's licence obligation. TTCL's exclusivity period did not allow the market entrance of other service providers in rural areas, except TTCL would enter into resale agreements – a solution that had already failed in the past (Adesemi). Also, other options to work around the monopoly, such as a concession to allow Internet telephony (VoIP) would affect TTCL' exclusivity and would mean to end the monopoly ahead of schedule. Furthermore, TTCL's failure to meet its own licence obligations, additionally encumber all universal access efforts, because the proposed programme was designed to complement, not to replace TTCL's service obligation. Apparently TTCL not only failed to provide the new main lines, it also fell short of installing the required payphones. To fulfil their obligations, TTCL would need to install approximately 11,750 rural and 2,750

---

from unprofitable segments and the latter became subject to a reverse auction and licences were awarded on the basis of least subsidy required (Intelecon 2003). Uganda has developed into the African forerunner applying this instrument. Successful bidders for different ICT services received a subsidy amount ranging from 30 to 50 percent of the investment costs to start ICT operations in rural areas (UCC 2003).

urban payphones (Apoyo 2003). The actual number of installations, however, has been only roughly 2,000, of these some 750 in Dar es Salaam, 800 in the regional capitals and only 280 in district capitals (Apoyo 2003). Evidently, these were severe shortcomings in TTCL's service delivery and the failures put considerable pressure on the universal access fund, too. Thus, the success of the fund will not only depend on the availability of resources, but also on ways to clear the pending issues with TTCL and provide new solutions for non-compliance and the treatment of TTCL's licence obligations.

4.3.3.6. Concluding Remarks on TCRA

Since 2000 TCC and, later, TCRA have made considerable progress in improving regulative capacities and, therewith, the regulative environment in the country. TCRA has proved to be a maturing organisation that not only has addressed key regulatory issues such as licensing, frequency management, interconnection and universal access, the performance record also included new provisions for consumer protection, dispute resolution, numbering and tariffs and the approval of a long awaited Tanzania Internet Exchange Point (TIXP) set-up by the Tanzanian ISP Association (TISPA) in July 2003 (c.f. TCC 2001b, 2002; see also chapter 4.3.5.4.). Furthermore, considerable efforts were made to identify a right formula to build internal capacities and implement adequate working structures (c.f. ÅF International 2002). Recalling the main critique, which was raised during the second phase as a lack of credibility, transparency, capacity and inconsistent regulation, not all problems have been solved. However, such general critique would be completely inappropriate to characterise today's TCRA performance. There were ambivalent decisions, weak implementation and enforcement capacities, but all related to individual proceedings, not to the general character of the regulator. Though there was still a gap between the scope of tasks and the overall performance record, the gap has narrowed steadily in the given period.

The challenges ahead will be the introduction of a second national operator (SNO), the elaboration of the new licensing setup as proposed in the draft communications bill, and possible provisions for VoIP, which would largely affect fixed line market developments. The second challenge will be the advancement and implementation of the universal access fund, where the preliminary work has been done and TCRA, in collaboration with the MoCT and possibly PSRC, will enter the more difficult phase to solve the pending issues of finance and the role of TTCL and, finally, to implement and manage the fund. If one concedes that the TCRA's technical capacities grew successively, the outlook for TCRA will lastly depend on TCRA's ability to develop credible and sustainable regulation and to implement them accordingly.

Thereby, the regulator's freedom to act independently and the factual role of the ministerial prerogative, will be critical factors that will determine the regulator's performance.

4.3.4. The Story of TTCL Part II: The Disputed Privatisation

The privatisation of TTCL was the most central episode in the Tanzanian telecommunications sector between 2000 and 2004. As described in chapter 4.2.5., the government launched the tender for a 35 percent stake in TTCL in June 1999, 29 telcos filed a tender and eventually six candidates prequalified. These were:

1. A German Dutch consortium Detecon /MSI;
2. An Indian consortium of Mahanagar Telephone Nigam Ltd (MTNL), Telecommunications Consultants India Ltd (TCIL), and Videsh Sanchar Nigam Ltd (VSNL);
3. Mauritius Telcom
4. MTN of South Africa
5. Sasktel of Canada
6. Vodacom of South Africa with WorldTel

In a multilevel selection process and a subsequent renunciation by Mauritius Telecom, the Indian TCIL-VSNL-MTNL-consortium, and MTN of South Africa to submit bids, the three remaining candidates Detecon/MSI, Vodacom South Africa / WorldTel and Sasktel International were asked to finalise their documents. The selection criteria, as determined by PSRC for identifying the winning bidder, were a combination of price or sale proceeds and the total number of telephone connections to be delivered in the four years of the granted exclusivity period (c.f. PSRC 2000a). PSRC had already determined an internal benchmark for both price and roll-out amount. According to these assumptions the bidders were expected to offer a minimum price of at least US$ 50 million with the proceeds of the bid to be injected into TTCL for future capital expenditures and to install at least 300,000 new telephone lines (World Bank 2000). The bids submitted were as follows (PSRC 2000a):

	Price in Mio. US$	Roll-out amount until 31 December 2003
MSI/ Detecon	**120.00**	**800,100**
Vodacom South Africa / WorldTel	**66.00**	**850,000**
Sasktel International	**60.5**	**575,000**

Apparently, the Detecon / MSI bid significantly exceeded those of the others with regard to price and was also at the upper limit with regard to the roll-out commitment. On 26 June 2000 the government identified the consortium of MSI and Detecon[238] as the preferred strategic investor for TTCL. After further negotiations the basic elements of the deal comprised :

- MSI/Detecon will pay US$ 120 million, US$ 60 million upon signing of the agreement and the balance of US$ 60 million will be paid after finalisation of the auditing of TTCL accounts for year 2000.
- The investor will get full management control
- The purchase price will be paid into TTCL in the form of capital increase
- TTCL and its partners will be issued five new licences (fixed basic telephony, mobile telephony, data services, paging, and ISP services);
- TTCL will have a four year exclusivity period for fixed line voice telephony
- The new TTCL will increase the number of connections from the current 162,000 to 810,000 in the next four years, of these 450,000 should be using fixed or WLL direct exchange line (Fixed DEL) and the remaining 350,100 lines should be using any suitable technology (i.e. mobile technology).
- TTCL will install two public payphones per every 3,000 inhabitants in settlements with more than 3,000 inhabitants.
- For the time being, the government retains 65 percent which will later be sold to private investors in the following arrangement: five percent TTCL workers; ten percent local financial institutions; 14 percent international financial institutions and 36 percent will be retained by the government (c.f. PSRC 2001, Mwandosya 2001).

In addition to the price and roll-out proposal, the consortium also suggested a number of bonus items beyond the requirements of the tender, such as an information campaign to introduce the new TTCL and explain the benefits of privatisation or investing in the company's training facilities. Furthermore, the consortium announced a number of performance expectations based on the investor's assessment of TTCL, which anticipated that:

[238] From the perspective of the consortium, the bid also placed emphasis on the comparative advantage of both partners, that is, MSI with considerable experience in mobile telephony and the Deutsche Telekom subsidiary Detecon as the partner with experience in the fixed-line business (Int. Ickstadt 13.5.2002).

- Investment in infrastructure will be US$ 900 million in the next ten years,
- the customer basis will increase to 1.3 million by 2009, which would mean to reach a teledensity of 5.5 per 100 inhabitants ten years ahead of schedule as proposed in the National Telecommunication Policy,
- 30 percent of the roll-out will be in rural areas,
- TTCL will create 4,200 new jobs over a ten year period, and that
- There will be preferential dividend payments to the government beginning from the first year of operation (World Bank 2000).

Thus, the bid in fact promised a bright future for TTCL. Based on the above figures the transaction was closed and from 23 February 2001 the consortium took over the management of TTCL operations. The first US$ 60 million tranche was disbursed in early March 2001.

However, after this promising start, confusions arose in November 2001, when the audit report for the year 2000 – the key prerequisite for the disbursement of the second US$ 60 million tranche – was rejected by the management due to a number of irregularities. At the core of the issue was, that a large proportion of the company's recorded revenue in the year 2000 – and in fact already in the previous years – was never collected. According to MSI, the total uncollected debt other than from government sources amounted to TSh. 75 billion at the end of 1999 and rose to TSh. 115 billion at the end of 2000.[239] Other contested issues related to interconnection payments and changes in taxation rule.[240] In consequence, the investor refused to disburse the second tranche and pushed for renegotiations on the outstanding sum.

The dispute quickly developed into an astonishing row of resentments, menaces and leaked indiscretions, in which particularly the investor was purposefully discredited, for example, accused of having offset money from TTCL sources, to be insolvent, to liquidate TTCL and put all efforts in the new mobile operation Celtel or to conspire with the TTCL auditor PriceWaterhousCoopers (PWC) against TTCL and the government and the like (c.f. Business Times [Dar es Salaam] 11.12.2001, 21.12.2001, The Guardian [Dar es Salaam] 12.6.2002). In

---

239 Figures were made available by MSI in a press release published on 10 June 2002 in full page adverts in all major national papers. See also the reporting in The East African [Nairobi] 12.11.2004 and in the Business Times [Dar es Salaam] 21.12.2001.

240 Interconnection payments were also due in cases when the respective bills could not be collected, and these burdened the margins in that segment in an unexpected quantity. Changing taxation rules mainly related to omitted tax relief through depreciation allowances (Int. Ickstadt 13.5.2002).

addition, careful threats were lanced, such as to waive the exclusivity period or to regain full management control of the incumbent (The Guardian [Dar es Salaam] 25.4.2002, The East African [Nairobi] 10.6.2002). Alarmed by the mounting bad press, the consortium, namely MSI, and the TTCL auditor PWC, decided to react and published adverts in the national papers attempting to reinstall their integrity and to introduce their thinking on the matter.[241] Of course, the different press statements and campaigns were merely meant to influence the public opinion and to publicly pit one's strength against the other. Behind the scenes the dispute was fierce, too, but tackled on a more technocratic level. Soon after the issue emerged in late 2001, both parties agreed on a joint investigation team with an independent chairman. Only after the joint team failed to reach consensus, the conflict regained momentum. New voices from the parliament and from the workers union (COTWU – Communication and Transport Workers Union) to cancel the whole privatisation became louder (Business Times [Dar es Salaam] 28.6.2002, The Guardian [Dar es Salaam] 25.10.2002, 29.10. 2002), and forced the investor to seek a court injunction in London High Court' Commercial Division to restrain the government from further action and to protect the MSI/Detecon shares in TTCL. The injunction was granted in May 2002[242] and forced both parties into renegotiations as the government mandatorily challenged the case in the same court.[243] Since the ongoing dispute had already damaged TTCL's operation, both sides increasingly recognised the devastating effect of the dispute and finally agreed upon a new arbitration round.[244]

[241] See footnote 238 for the MSI advert, the PWC advert was published on 7.6.2002.

[242] Though invalid until endorsed by a local court, according to Tanzania's Attorney General Andrew Chenge (quoted in The East African [Nairobi] 9.9.2002).

[243] The original contract of February 2001 provided that any dispute, controversy or claim arising out of or relating to the agreements or the breach termination or invalidity which cannot be settled amicably shall be settled by arbitration in accordance with the UNCITRAL Arbitration Rules in force at the date of the dispute. The appointing authority shall be the London Court of International Arbitration (LCIA). The agreement states that the arbitration shall be conducted by three arbitrators in accordance with the UNCITRAL Arbitration Rules and should take place in Geneva, Switzerland, or any other venue as may be agreed by the parties (quoted in: The Guardian [Dar es Salaam] 9.10.2002.

[244] To this end MSI also agreed to suspend their TTCL management control. TTCL Chief Executive Officer Fred van der Voort resigned his post in December 2002 and George Mbowe, long serving senior official in the Tanzanian establishment, among others, former executive chairman of the PSRC, took over the TTCL CEO post immediately (The Guardian [Dar es Salaam] 11.12.2002). Five more MSI/Detecon management staff were replaced by Tanzanian counterparts in March 2003 (The East African [Nairobi] 3.3.2003).

The arbitration led by KPMG of London was a relatively quiet process, particularly because both parties saw the urgent need to come to an agreement and thus have committed themselves to abide by the international expert's determination. The dispute already had a negative effect on TTCL's performance and investment plans had been postponed. However, commencing the mediation alone, brought some relief and TTCL reinitiated a number of modernisation projects mainly targeting greater Dar es Salaam. Nonetheless, it took the arbitrator and both arbitration teams more than a year to come to a decision. On 29 January 2004 the dispute was settled and results were introduced to the public on 18 February 2004. The arbitrator largely followed the arguments made by MSI. The revisited accounts of the financial year 2000 were corrected downwards. The arbitrator determined that TTCL earnings before interest, corporate income tax, depreciation, amortisation and all extraordinary items for the year ending 31 December 2000 was TSh. 9,129 billion and considerably deviated from those 1999 benchmark figures of TSh. 31,216 billion which were taken as the basis of the TTCL divestiture.[245] Thus, the arbitrator confirmed that TTCL performance was much weaker than originally stated and in consequence adjusted the outstanding second tranche from US$ 60 million to US$ 5.28 million – of these US$ 4.96 million as actual price and US$ 321,000 as accrued interest (The Guardian (Dar es Salaam] 18.2.2004, 20.2.2004). Though the price was reduced significantly, but both parties declared the dispute to be settled. In parts this might mirror the determination of both to come to a solution, but, considering the actual abatement of price, it also indicates that the step was justified by the size of the deviation, thus a significant misinterpretation of the company's actual state.

Though various elements of the dispute were made public, it should be noted that the details of the transaction were confidential, so that it is hardly possible to make a fair judgement about the deal. However, considering that the main issues of concern, low collection rates and a high level of account receivables were already known in 1999 (c.f. World Bank 2000), a crucial role must be attributed to the PRSC which did not take into account the financial insecurity of the selling plant and made tentative accounts the basis of the divestiture. The consequences surely had a distressing effect not only on TTCL itself which lost planning reliability, withdrew into low-profile operations and fell short to meet its roll-out obligations, it also affected the whole communications sector that suffered from absent activities from its most important player. Lastly, the dispute also affected privatisations in general. As the biggest divestiture of that

[245] Figures were given by the Tanzanian leader of the arbitration team, Marten Lumbanga at a press conference quoting the figures calculated by the KPMG arbitrators, in: The Guardian [Dar es Salaam] 18.2.2004.

time, it contributed to the bad press of privatisation, but foremost questioned the credibility of privatisation, because the boundaries between the commitment of due diligence and possibilities to sue the public seller on the basis of incorrect accounts became increasingly indistinct. Generally this means that those who privatise have to assume liability for accurate company figures and a due diligence does not stand for an automatic disclaimer, if the company figures are verifiably incorrect.

4.3.5. Market Developments since 2000

Though TTCL's disputed privatisation retarded the sector's development, the overall success of reforms achieving macroeconomic stability, an improved business environment that successfully induced a private sector led growth strategy, new policy guidelines and an improved legal and regulatory framework in the telecommunications sector resulted in notable improvements. A bias towards the fixed-line business still dominant in 1998 gradually levelled off and, instead, mobile telephony became the engine for growth in the telecommunications sector, accounting for a high 45 percent of revenues in the Tanzanian telecommunications market. New potentials of data communications that stood for a quarter of the overall market share in 2003 supported the overall growth trend, too (see figure 4.12). Market size grew from US$ 143 million in 1998 to US$ 389 million in 2003, which equals a compound annual growth rate of a notable 19 percent. At the same time, the TTCL dominance declined from 88 percent in 1998 to 69 percent in 2003; a figure that still proves TTCL's dominant position, but also shows the general tendency of increasing diversification in the market.

Figure 4.12: Telecommunications Market Figures 1998 and 2003

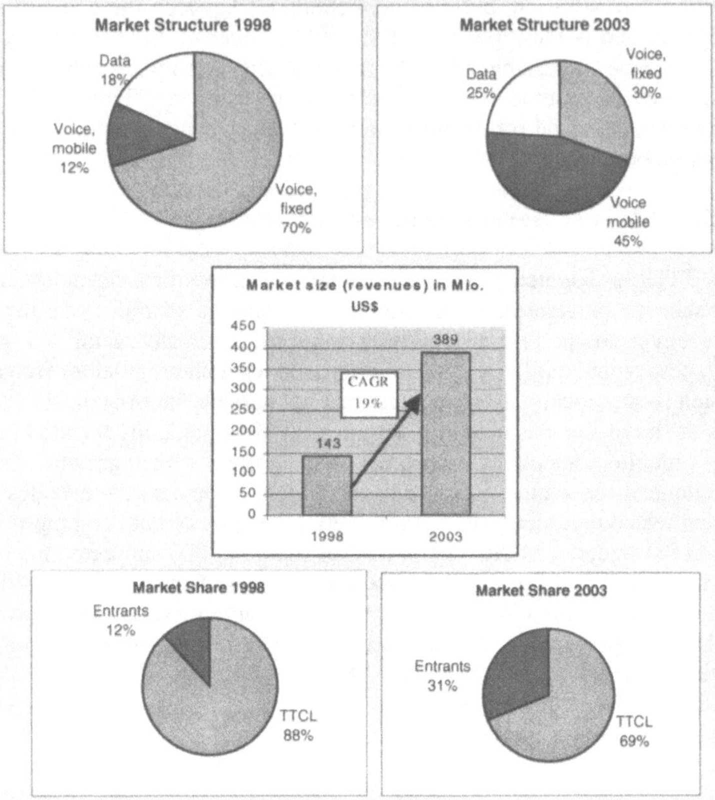

Source: Apoyo 2003

This change towards an open market is also shown in figure 4.13, which gives an overview of the structure of the Tanzanian telecommunications market and the position of different market participants. In contrast to the market schemes introduced in figures 4.2 and 4.9, figure 4.13 proves a further diversification and thus a less volatile basis for various services available. The advancing sectoral developments tied up to the reforms introduced since the end of the 1990s. Particularly liberalisation measures attracted new market entrants who became agents for development. With the commencing competition in basic services and international gateway operations from February 2005 onwards, all market segments have been, at least, partly liberalised. The impact of Zantel/Etisalat must be awaited. In mobile telephony, the newcomers Celtel and Vodacom became the most powerful players and, particularly, the arrival of Vodacom pushed mobile telephony in the country. In the same vein, the data communications business benefited from liberalisation, as the exclusivity for the three licence holders under the old regime was abolished and the number of providers rose to eleven. A final observation can be made with regard to the company's strategic positioning. Apart from TTCL that was deliberately granted a licence in each segment, other market participants also started to enter into strategic partnerships or joint ventures as indicated by the horizontal dashed lines in figure 4.13. Thus, the newly created framework for telecommunications attracted new entrants, resulted in a balanced mix of different companies ranging from international players to home-grown small and medium enterprises and their strategic positioning in the market to make best use of the given environment. The following sections covers a more detailed outlook on the different market segments and how the market participants responded to the new environment.

Figure 4.13: Tanzania Telecommunications Sector 2004

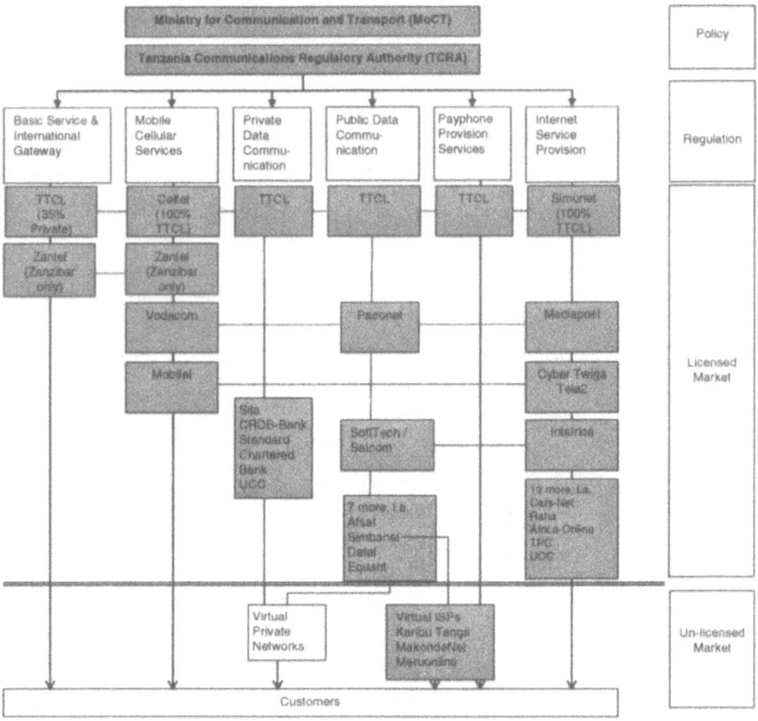

4.3.5.1. Fixed-line Developments

Since fixed-line developments on the mainland solely depend on the performance of TTCL, the disputed privatisation aggravated the deployment of main telephone lines in the country. Figure 4.14 gives an overview of main telephone lines in operation and also includes the country data for Kenya and Uganda to allow some comparison. For Tanzania one can see a steady increase in telephone lines foremost installed under the TRP, until reaching a preliminary maximum of about 170,000 telephone lines in 2000, followed by a decline to 150,000 in 2001 and recurring growth to roughly 160,000 in 2002. This number corresponds to a fixed-line teledensity per 100 population of 0.47, which was well below the Kenyan, but considerably better than the Ugandan figures. The graph mirrors the phases of privatisation and the adjustments made by the investor MSI / Detecon as they disconnected all defaulting debtors with the exception of government agencies from the TTCL network. The result was a decrease of 50,000 main lines[246], a quantity that could not be recaptured in the following years, and right from the start, burdened the task to comply with the agreed licence obligations.

Figure 4.14: Main Telephone lines on Operation

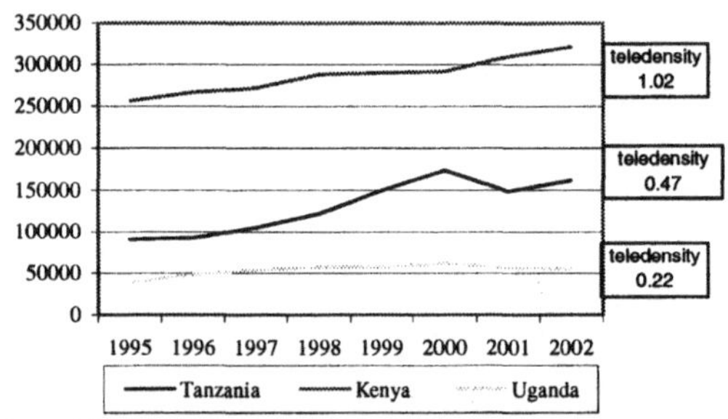

Source: ITU 2004, 2001

TTCL was granted a four year exclusivity period and, in turn, accepted the obligation to expand the telephone network from 162,000 lines to 800,100 lines, of which at least 450,000 should be added to the fixed-line network, the remaining 350,100 lines to be installed as deemed technologically appropriate. Furthermore, TTCL agreed to install public phones in all settlements with more than 3,000 inhabitants (with at least two payphones per each 3,000 inhabitants). However, as the privatisation dispute lingered, and it also it turned out that

[246] A similar reduction also occurred in the privatisation of the Ugandan PTO UTL in 2000, though on a much lower scale.

TTCL was in a poorer state than expected,[247] due investments were recurrently postponed and the company's roll-out targets more and more fell behind schedule. For the year 2002 the rollout-schedule allotted 41,000 new telephone lines, of which TTCL only set up roughly 26,000. On that account, TCC initiated non-compliance proceedings and imposed a penalty of US$ 6.4 million on TTCL, (Business Times [Dar es Salaam] 19.4.2002, The Guardian [Dar es Salaam] 6.8.2002) a clause against which TTCL appealed, but no decision on the matter has been made until now. Yet, TTCL has been far from delivering 800,100 telephone lines. Even if some 350,100 due lines will be shifted to the mobile phone subsidiary Celtel, the targeted number of 450,000 fixed-line lines will be difficult to reach. After efforts to expand the existing network continued on a low level during the dispute, some new actions have been taken after dispute settlement, so that capacities were on the rise, again, and today's network capacity has reached 250,000 lines (TTCL 2004).

More problematic, however, than the gross provision of telephone lines has been the obliged duty to supply public telephones. Earlier attempts to increase the number of public payphones by entering resale agreements with other operators, namely Adesemi and Jupiter, failed because both fell into financial difficulties and had ceased their businesses (see chapter 4.2.6.5.). However, taking into consideration the existence of roughly 4,000 to 5,000 villages with 3,000 inhabitants or more, and the actual provision of only about 2,000 payphones, the targeted figures to be reached at the end of the exclusivity period in early 2005, can hardly be realised. Neither the current distribution of public payphones overwhelmingly being installed in urban areas, nor the overall targeted 14,500 payphones (11,750 rural and 2,750 urban) (Apoyo 2003) make it very likely that TTCL will hit the target. Part of the problem has been TTCL's network, a typical Public Switched Telephone Network (PSTN), composed of 133 switching centres of different hierarchies and hybrid technologies, i.e. digital and analogue, interconnected via transmission (radio links and cable links) and access networks providing last mile connectivity to the end user (c.f. Yonah 2002, 2001). Though there has been a telephone switch in almost every district, considerable quantities of the network technology still are outdated or incompatible. Thus, still after ten years after the commercialisation of TTCL commenced, the working focus on network modernisation prevails as TTCL's

[247] TTCL then CEO, Fred van der Voort, cited technical and logistical problems that hampered the company's efforts, the poor state of the existing network that could not be extended economically without major changes in the network architecture. As a consequence, the management needed to perform groundwork first before connecting new subscribers. Other difficulties related to a polluted customer database, mismatch between outside plant and exchange capacity, lack of market information, and mismatch between available network and demand (Business Times [Dar es Salaam] 19.4.2002, 5.7.2002).

most urging working focus. Current efforts in his respect have been the modernisation of the international gateway,[248] phasing out analogous network systems, replacing small capacity transmission links by high capacity digital systems and installing broadband digital networks capable of delivering quality voice and data services (TTCL 2004).

A further major innovation has been the introduction of fixed prepaid services in April 2004. With the new service TTCL addressed two objectives, firstly the company drew upon the success model of prepaid mobile telephony aiming at benefiting from the experiences of well performing prepaid markets. 80,000 new customers have been targeted in the first six month of operation (East African [Nairobi] 29.3.2004, 3.5.2004). Secondly, TTCL sought to solve the problem of unpaid bills and reduce the number of bad debtors that had already caused concern in the privatisation dispute. The introduction of prepaid services involved considerable network modernisation done by newly installed so-called Intelligent Network platforms, enabling the company to increase the quantity and quality of telephone calls and to introduce the technical prerequisites to offer prepaid fixed-line services. The services will be available in Dar es Salaam, later Arusha and Mwanza will be targeted. Considering the prepaid developments in mobile telephony, the prospects of success seem promising, however it remains to be seen whether the customers will accept fixed-line prepaid services – all the more because the new prepaid system comes with higher costs for monthly rental dues and an extra fee per day (TSh. 180) (East African [Nairobi] 3.5.2004), so that there seem to be only little incentives to migrate to prepaid billing.

---

[248] The story of TTCL's international gateway is as an illustrative example of how rumours aggravate the business climate, and also how the new TTCL management made deliberate use of the Tanzania typical dual mainland-Zanzibar structure, but failed completely to communicate its strategy. TTCL's international traffic has been routed through the company's own satellite earth station in Mwenge (Dar es Salaam). However, in 2002, TTCL transferred all its international routes to the Zanzibar earth station operated by Zantel. At the time when the privatisation dispute had already reached a climax and the investor had been confronted with a bad press anyway, the Mwenge closedown was largely perceived as a further step to misuse valuable TTCL assets (c.f. The Guardian [Dar es Salaam] 15.11.2004). This was not the case, in fact TTCL had cancelled disadvantageous contracts with the satellite operator, temporarily channelled the traffic through Zanzibar and finally entered into new agreements with a new carrier Link Africa, a full subsidiary of Celtel international. The cancellation period was used to modernise Mwenge earth station a process finalised in November 2003 (TTCL 2004). Thus, regarding the international gateway, TTCL is today better positioned than in 2002. The company successfully took advantage of the duopoly structure in Zanzibar, but communicated the actual prudent step so badly, that there still are besetting rumours that the investor plans to liquidate the company in 2006 (The Guardian [Dar es Salaam] 30.1.2004.

Finally, looking at TTCL's overall performance since privatisation, there is no doubt that the failures particularly in the fulfilment of the roll-out obligations can be attributed, to a large extent, to the long running dispute over the investor's payment of the second US$ 60 million tranche. It reduced the company's planning reliability, indulged new bank loans for further investments and made any future developments unpredictable. It is only hypothetical to ask, whether TTCL would have met the obligations without the dispute. 800,000 targeted telephone lines would have equalled a teledensity of three telephones per 100 inhabitants, a figure that could not be regarded as over ambitious. Nevertheless, even during the dispute and under new management, TTCL continued at least some low-profile operations, particularly aiming at building a national backbone necessary to deliver voice and data services. Bearing in mind the huge land areas to be covered, this will continue to be a challenging task, all the more, because TTCL will face a direct competitor for the first time in its company history.

In February 2005, the exclusivity period for TTCL will end and Zantel will become the second national operator in Tanzania, presumably operating as Etisalat. Already in 2002, Zantel attempted to operate a GSM service on the mainland, but failed because TTCL had always insisted on its monopoly on international routing (c.f. Business Times [Dar es Salaam] 25.1.2002, 12.7.2002), so that Zantel was not allowed to use its own gateway for mainland operations and, thus, has given up any expansion plans until 2005. In mid 2004, Zantel operated a GSM network in Zanzibar with approximately 71,000 subscribers and covered approximately 80 percent of the isles. For mainland operation, Zantel Chief Executive Officer, Mohamed Salim, announced that Zantel would "immediately cover six major towns and cover the whole of the country in five to six years" (quoted in: The Guardian [Dar es Salaam] 25.9.2004). However, Zantel has to prove if the company is able to deliver competitive services on the mainland, and so far, it concealed any details on its expansion plans. Furthermore, the regulator has only started provide the regulatory framework for the new SNO, so that there are a number of uncertainties regarding Zantel's mainland strategy. Considering the past performance, however, it might very well turn out that Zantel will not primarily target the telephony end-consumer, but the international voice and data communications business. In this business segment, the company has real assets. It has already operated a well performing international gateway, serving for example, Kenya, Malawi and other neighbouring countries (East African

[Nairobi] 13.10.2003).[249] Zantel has also been signatory to a memory of understanding to realise the Eastern Africa Submarine Cable System (EASSY)[250] that will provide the last link to the existing cable system SAT-3/WASC & Atlantis-2 to completely encircle Africa by high capacity optic fibre. Since EASSY has been the only large scale cable project in the region, comprises a number of follow up regional infrastructure development projects[251] and has also strong backing from Nepad's e-Africa Commission (c.f. Nepad 2004, EASSY 2003), Zantel would develop into an important player in the regional wholesale data communication market.

4.3.5.2. Mobile Telephony

Mobile telephony took off in Tanzania in 2000 and, thus, the growth pattern corresponded to other African countries that had liberalised their mobile telephony market and enabled mobile telephone companies to develop the sector. Since 2001 the number of mobile subscribers already reached the level of fixed-line telephony and, since 2001, mobile telephony surpassed fixed-line telephony (see also figure 4.15). In Tanzania, new dynamisms can be attributed to the arrival of Vodacom in mid-2000. Vodacom Tanzania, a joint venture

---

[249] To this end Zantel also signed an interconnection agreement with the African communications provider Gateway Communications in October 2004. The agreement allowed Gateway, which has already been active in East Africa (Uganda, Kenya, Somalia) to enter the Tanzanian market, and Zantel to further improve its international calling services (BA 231/2004)

[250] The other signatories have been: Djibouti Telecom, Telkom Kenya, Uganda Telecom, MTN Uganda, Rwanda Telecom, Botswana, Telecom, TDM – Mozambique Telecom, Malagasy Telecom, Telkom South Africa and Sentech (c.f.EASSY 2003). It is worth mentioning that TTCL took part in the planning of the process but later withdrew without giving any reasons. Without being too speculative, EASSY is the largest infrastructure project in the region worth some estimated US$ 200-230 million, covering a coastline of 8,840 km, one cannot assume offhand that TTCL ,as one of the biggest carriers in the region, would simply leave the project and yield the data capacities to its expected competitor. Two possible answers can be put forward: either TTCL mutually agreed with Zantel on a specific mode of task sharing and, of course, access to the cable, or TTCL regarded its participation in the project as sufficient, because the TTCL shareholder Detecon is also shareholder in Uganda Telecom, and has strong ties to TDM of Mozambique, both are signatories to EASSY. For the proposed structure of EASSY see also chapter 4.3.5.3.

[251] E.g. a proposed Lake Victoria ICT Backbone Ring linked to EASSY, advocated by the East African Community. Resources for some preparatory work have been made available by SIDA, however the future of the project is unsettled (Int. Yonazi 11.4.2003). Another project would be an 8,840 km Kigali-Kampala-Nairobi-Mombassa cable project planned by Telkom Kenya, Uganda Telecom and RwandaTel, also with a direct connection to EASSY (ARB 2004: 16200).

between Vodacom South Africa and the local telephone company Planetel (see table 4.5), was licensed in December 1999 and started operation in August 2001. It was particularly its aggressive roll-out and marketing strategy that forced the other operators to increase their efforts not to be outpaced by Vodacom's rapid development strategy. Until 2000, Mobitel, Tritel and Zantel only kept a relatively low operation profile with limited network coverage and postpaid services only. The new competitive environment consecutively changed this pattern as Vodacom brought forward a "record roll-out" of its Tanzanian network.[252] Tritel responded by introducing the first prepaid cards[253] to Tanzania in March 2000 and Mobitel launched its GSM network in September 2000 in a move to maintain its market share and to successively replace its existing analogous ETACS system (Tomric [Dar es Salaam] 20.9.2000). Despite these efforts, Vodacom quickly became the market leader followed by Mobitel, who started to invest in its new GSM network and became the only serious competitor to Vodacom for the time being. As a Zanzibar only provider, Zantel did not play a role on the mainland and Tritel did not have the capacities to follow the roll-out pace of Vodacom and Mobitel.

Figure 4.15: Mobile telephone subscribers

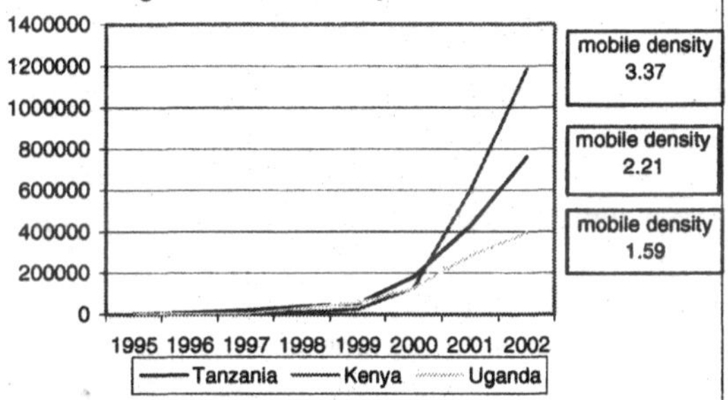

Source: ITU 2001, 2004

A new impetus came at the end of 2001, when the TTCL subsidiary, Celtel, started its operation as the fifth licensed mobile provider, making Tanzania the

---

[252]The first phase of Vodacom's network deployment was done by Siemens who was awarded a turnkey contract for the supply of equipment and infrastructure worth roughly US$ 80 million. On the first day of operation alone, some 3,000 subscribers signed up to the Vodacom network, bringing it up to 15,000 in the first month of operation (IT Web [Johannesburg] 27.9.2004).

[253] However, the price level introduced was too high to establish prepaid mobile telephony for a broader user basis. New Tritel customers had to pay US$ 90, Tritel postpaid customers, who whished to shift to the new service had to pay US$ 45 for a prepaid SIM card. Prepaid cards, branded Jambo, were sold in denominations of 25, 50 and 75 US$ (The Express [Dar es Salaam] 9.3.2000), a price too high to attract new customers.

most competitive market in Africa. As indicated above, TTCL's mobile licence was attached to the privatisation package and the unused licence was the main reason for MSI to invest in the incumbent. Celtel was installed as a 100 percent TTCL subsidiary, but with full MSI management control – a decision that was never challenged in the course of the privatisation dispute. Celtel started operation on 1st Novermber 2001 and soon established itself as a forceful market player, immediately seeking to expand its network and attract new customers. Thus, from the end of 2001, the Tanzanian cell phone market took a new start. With Vodacom as the evident market leader and Celtel, that quickly challenged Mobitel as number two in the market, mobile telephony became accessible and affordable to a growing number of people. Coverage increased, especially when Vodacom and Celtel embarked on a strategy of rapid network expansion as a key of success. The second factor of success has been the introduction of prepaid services and the gradual decrease of minimum recharge card denominations to two US$, later converted into TSh. pricing schemes pioneered by Celtel with lowest availability of 2,000 TSh. Today, prepaid accounts for 98 percent of total mobile subscribers in the Tanzanian market (Vodacom 2004: 32). Furthermore, handset prices have gone down from US$ 100-300 in 2000 to today's roughly US$ 30 for a basic mobile phone, and average telephone tariffs for e.g. calls within one network decreased from 35 US cents in 2000 to 26 US cents in 2004, today to be charged in local currency.[254] Since all operators have introduced similar prepaid schemes, the successive improvements were largely due to aggressive network expansion strategies. For example, Vodacom's first roll-out phase took only eight months and was almost immediately followed by a follow-up contract with Siemens worth US$ 52 million to further increase coverage and capacities of the existing network (IT Web [Johannesburg] 25.5.2001). As competition increased, a first consolidation phase surfaced and those operators that could not follow the pace of Vodacom and Celtel were increasingly pressurised. At the end of 2002, the GSM pioneer in the country, Tritel, declared its insolvency and ceased operations on 31 January 2003[255] with TCC (outstanding licence fees), TTCL (outstanding interconnection fees) and the Tanzania Revenue Authority (TRA) as the largest creditors (The Guardian

[254] Tariffs and the different pricing schemes are available on the companies' websites, www.vodacom.co.tz, or www.celtel.co.tz. The other companies do not provide a website, or just one with some very rudimentary information, but did so in the past. For any information given on websites not longer available one could search the Internet archive (www.archive.org), inserting the desired URL in the search field to get an overview of past website versions. (e.g. www.mobitel.co.tz or www.tritel.co.tz). Of course this instrument is very useful for any other searches, too.

[255] Tritel's network covered Dar es Salaam, Zanzibar, Mwanza, Dodoma, Arusha and Moshi and had roughly 17,000 subscribers at the time of insolvency (Tritel 2002, Business Times [Dar es Salaam] 24.1.2003).

Table 4.6: Key features of mobile telephone operators in Tanzania					
	Shareholders	Profile of main Shareholder	Type / Creation	Subscribers (Aug. 2004)	Coverage
Voda-com	**Vodacom South Africa, Planetel, Caspian Construction**	**Targeting African premium markets with Tanzania as test case for roll-out strategies in large territorial states. Good cooperate citizen strategy. Presence in five African markets.**	**GSM 2000**	**816,169**	**49 cities plus coverage of main roads**
Celtel	**TTCL (100%) with full management control of Celtel International**	**Focusing on African Markets, Expansion seeker Presence in 14 African markets.**	**GSM 2001**	**360,977**	**54 cities plus coverage of main roads**
Mobitel	**Millicom International (MIC)**	**Global operator with a focus on emerging markets in Asia and Latin America. Presence in 4 African markets with Tanzania as the most important one.**	**GSM 2000** **ETACS 1994**	**276,755** **30.000**	**16 cities**
Zantel	**Etisalat Kintbury Meeco International Government of Zanzibar**	**Low operational profile Focus on international connectivity SNO aspirant**	**GSM 1999**	**70,847**	**Zanzibar (80%)**

Source: Interviews, Vodacom (2004), MIC (2004), Etisalat (2004), TCRA (2004b).

[Dar es Salaam] 31.12.2002, Financial Times [Dar es Salaam] 22.1.2003, Business Times [Dar es Salaam] 24.1.2003). Efforts by the regulator to offer the released mobile telephony licence to the South African mobile telephony operator MTN remained unsuccessful and MTN's denial to operate in the country indicated that the market was well supplied with the remaining four operators. Particularly the strong presence of Vodacom and Celtel did not leave any notable room for new competitors. Thus, in 2004 the Tanzanian mobile telephone market consisted of four operators with the following key features:

The mobile telephony market structure underlines the significance of an aggressive roll-out strategy as a prerequisite to gain from economies of scale. A combination of network expansion and the introduction of prepaid options reducing the entry price have been the key to success. Vodacom and Celtel have embarked on such strategies, in both cases supplemented by other measures meant to create a distinctive company profile. Celtel has been the pioneer in Swahili telephone services and local currency charging[256] and thus capitalising on an image of a locally rooted company concerned about the needs of the broader population (Int. Mloka 23.5.2002). Vodacom, on the other hand, successfully attempted to establish itself as a good corporate citizen, e.g. running special employment schemes for handicapped people and other social and

[256] Whereas Celtel introduced its TSh. tariff system from the beginning of its operation, Mobitel converted from US$ to TSh. in July 2003 and Vodacom only in April 2004 (Business Times [Dar es Salaam] 11.7.2003, The Guardian [Dar es Salaam] 2.4.2004).

environmental sponsorships, or has been the facilitator of major sports events like the Vodacom Premier League that started in 2002 as the national football league.[257] In contrast, Mobitel focused on non-core activities and sought a competitive edge by introducing new mobile telephone applications,[258] which turned out to be too early for a young and immature market. As a consequence of non-core business activities, Mobitel's market share declined from 30 percent at the beginning to approximately 19 percent at the end of 2003.[259] However, there have been indications of rethinking in Mobitel' approach and the company announced that they will put new efforts into network extension in the near future. This included a proposed roll-out plan to build a new data backbone from Dar es Salaam to Mwanza and to extend its presence particularly in the lake zone, capitalising on a prosperous mining industry, and the northern territories targeting the main tourist destinations in the country (MIC 2004: 18, The Guardian [Dar es Salaam] 24.9.04).[260] Thus, with two major competitors, renewed efforts by Mobitel, the market entry of Zantel as fixed-line and mobile provider on the mainland and a renewed regulative framework,[261] the market outlook is rather promising, particularly as network extension will be at the fore of at least the three major operators.

---

[257] Vodacom has also introduced a popular magazine called Vodavoice that not only addresses issues of mobile telephony usages, but also the company and its various social and environmental activities.

[258] Mobitel introduced short message services (SMS) to Tanzania that were soon offered by the other operators, too. However, most newly introduced services like downloads of ring tones or news alerts in different areas of interest, online customer accounts or Internet by call or car tracking systems never paid off (c.f. IT Vision [Dar es Salaam] Issue 08, May/June 2002, IT Vision [Dar es Salaam] Issue, 07, April 2002), because they were technically immature or did not fit the needs of the local market. The acquisition of Cyber Twiga motivated by the merging of technologies could not meet the expectations to develop as an Internet platform for mobile telephony.

[259] Estimation according to Vodacom (2004: 32).

[260] The renewed relevance of the Tanzanian market as the biggest Millicom operation in Africa is also underlined by the fact that Millicom has acquired an additional 26 percent stake of Millicom Tanzania Ltd from the Government of Tanzania – shares that were returned to the Treasury when TTCL cancelled its joint venture with Mobitel – bringing its ownership to 84 percent (MIC 2004: 18).

[261] Though the details of post-exclusivity period are not yet announced, one option would also comprise to allow the mobile operators to run their own international gateways, that is to separate the licence for fixed-line services and international voice traffic, so that the future market liberalisation would include the introduction of a new national operator and a liberalised market for international voice traffic in which mobile operators could channel their international traffic through their own gateways (Sunday News [Dar es Salaam] 24.10.2004).

In the first instance, mobile telephony growth had been a result of market liberalisation and respective company performances. The success of mobile telephony, however, must be put into perspective, because mobile telephone coverage is still limited if compared to the actual size of the country. Figure 4.16 shows the coverage of mobile networks and reveals that mobile telephony has mainly covered the larger cities – and some chosen main roads concentrating on greater Dar es Salaam, the northern territories, the lake zone and the transit to Zambia. Hence, the roll-out map equalled the relative economic strength of the regions covered. In 2003, accumulated network coverage was only about six percent of the land surface, reaching roughly 25 percent of the population (Vodacom 2004: 33).

Figure 4.16: Mobile Telephone Coverage 2003

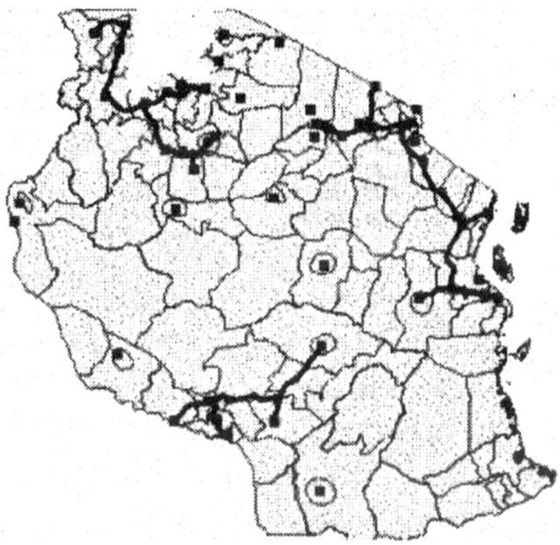

Source: adapted from Apoyo (2003)

Though operators have announced to extend their coverage further and thus indicated that they see further room for viable business operations,[262] the total of unserved areas is still significant, as the figure shows. Especially disadvantaged regions in the east, south east, south and central regions are hardly covered and point towards the limits of pure business models in delivering telephone services to remote, semi-urban or rural areas. Here, the prospects of mobile telephony will depend on the availability of new universal access instruments to provide the necessary incentives for investment. Licence obligations, attached to the operators in the past, were defined too narrowly and either were within the operator's deployment strategy, anyway, or the operators very soon surpassed

[262] According to operator's estimations, the potential addressable telecoms market in Tanzania is three to five million subscribers for the next five years (Int. Sherwood 7.5.2002, Int. Held 21.5.02).

the service delivery obligations made in the licences. For example, Vodacom was obliged to install 50 rural telephone shops within their 15-year licence period, but only after two years of operation Vodacom had detailed plans to deliver 200 community service shops over a two-year period not only in rural areas, but also in densely populated areas to provide services for those who could not afford their own handsets (Int. Sherwood 7.5.2002).[263] Thus, licence obligations have been fulfilled and new instruments are not yet in place. The new universal access fund or other instruments to promote rural telecommunications might provide new opportunities for the operators and, of course, for the users of rural telephones – all the more that the viability of cellular phones installed as village payphones has been clearly demonstrated in Bangladesh, India or Uganda and elsewhere (c.f. Intelecon 2003).

Apart from network coverage, the quality of service and the type of services offered will be critical for future developments. The popular short message services (SMS) have indicated the potential of data transmissions over cellular networks. Though clearly limited in its capacity, installed networks might be upgraded to cost-effectively co-locate wireless Internet technologies, also labelled as 2.5G networks (GRPS),[264] but, so far, operators were reluctant to capitalise on wireless Internet applications. Third generation networks have been deployed for fixed-line telephony, so that the increasing fixed mobile convergence will challenge the telephony and Internet landscape in the near future. But also with limited capacities that prevail today, new applications drawing upon data communications and the development of local content have been on trial,[265] and new services appropriate to local needs might develop into a new promising segment of mobile telephony.

4.3.5.3. Data Communications

In early 2000, the regulator liberalised the market for data communications providers and abandoned the exclusivity status of Sita, Wilken Afsat and Datel, due to disappointing provision of reliable, affordable and sufficient bandwidth

---

263 Vodacom has introduced a franchise scheme to set up the teleshops. They cost approximately 15,000 US$ per unit and are sold at cost. Vodacom provides investment capital assistance and numerous kinds of training from technical installation procedures, administration and accounting to branding and marketing (Int. Sherwood 7.5.2002).

264 For an introduction to different network standards see ITU (2002a:8).

265 In contrast to older attempts made by Mobitel to make use of SMS news alerts and the like, the current trials are centred on local relevant content, e.g. a cooperation between the agricultural NGO Business Information Service, that provides cash crop market information, and Vodacom, that makes the services available through SMS (Int. Mwakalinga 9.4.2003).

for communication services. The licence segment was separated into public data providers and private data providers, recognising different needs of the sector. At the end of 2003, eleven public data operators and nine private operators had been licensed. In addition, the parastatal corporations TAZARA, Tanesco and TRC have operated licensed exempted private telecommunication networks for their own corporate use.[266] Therewith, the market diversified and supply grew considerably. Private networks were particularly deployed in the banking sector (CRDB Bank, Standard Chattered, ULC Tanzania). Sita, the service provider for the aviation industry, became a dedicated private user, too, and continued to focus on its core business. Finally, the University Computing Centre was assigned a closed user licence to develop into a full ICT service provider for the University of Dar es Salaam and its affiliates. Altogether, the number of private dedicated networks remained rather limited. Only those with considerable security concerns, like the banking and aviation sectors, made use of the new licences. In addition, TCC regarded them as an appropriate tool to licence the multiple services of the University Computing Centre that have ranged from commercial to pure university services. The majority of corporate customers, however, channelled their data traffic through the licensed public data communication operators and took advantage of the diversified supply that enabled potential customers to choose from different companies with distinctive business profiles. New ambitious players, often with international financial backing, entered the Tanzanian market and added to the already established market players. Table 4.7 gives an overview of the most important public data communication operators.

With regard to the individual licence holders, particularly Datel has changed. After Datel's divestiture, shares were remitted to the treasury.[267] In 2001, Afripa Telecom, a South African solution provider for data communications[268] was found as a new investor. Since then, Datel attempted to overcome the stagnant phase mainly making use of Datel's old, though modernised infrastructure and

[266] Hereby, they often deployed excess capacity potentially useful for lease or resale arrangements. For example, TCR, with funds from the German Development Bank (Kreditanstalt für Wiederaufbau – KfW), run a fibre optic cable from Dar es Salaam to Morogoro to Dodoma. The project already began in 1994 and the installed overcapacities were meant to be leased to TTCL. Though a Memorandum of Understanding between TRC and TTCL had been signed, the agreement never entered into force. Ever since, installed overcapacities remained unused, because the regulator prohibited any resale agreements other than with TTCL to protect the monopoly. Cable capacities have been greatly underutilised, despite considerable demand expressed by several data operators, ISPs and television stations (Int. von Maltzan 2.5.2002, Int. Habicht 24.5.2002).

[267] Up to present day, the government holds a 35 percent share in Datel (Int. Mbita 4.6.2002).

[268] For details see the Afripa Telecom website at www.afripatelecom.net.

customer basis, but could never retain the old status of number one infrastructure provider in the country, again. Secondly, Sita[269] transferred its public licence to its former subsidiary Equant,[270] which has been one of the major data communication providers worldwide and, for example, has a network reach of 220 countries and territories (c.f. Equant 2004). The Tanzanian licence only has been part of a global strategy to serve customers throughout the globe. Thirdly, Afsat,[271] formerly Wilken Afsat, with headquarters in Nairobi and a focus on the East African markets, has been the only remaining company from the old licence structure and is still pursuing the time-tested strategy targeting the premium business segment, among others, BP, Barclays Bank, Citibank, Stanbic or PriceWaterhouseCoopers (Int. Goyayi 17.5.2002). Therewith, Afsat has been the leading provider in the premium business segment. With ongoing modernisation of the national network, Afsat also planned to go beyond the premium customer segment and target the smaller businesses as ICT trickles down. For this reason, Afsat has offered leased equipment for those who cannot afford to buy it themselves (Int. Goyayi 17.5.2002).

Table 4.6: Main Public Data Communication Operators

	Shareholders / Partners	Est.	Coverage (2003)	Strategy
Simunet	**TTCL** (Simunet is licensed as ISP, but de facto data communication provider)	**1999**	**44 Points Internet Points Presences (PoPs), in 20 regions growing toward district level**	**Building of national Internet backbone**
Afsat	**Formerly Wilken Afsat** (Afsat UK, Afsat South Africa, Wilken (Kenya)	**1994**	**Approx. 100 VSAT stations, approx. 200 area networks ((WAN, LAN), broadband international data gateway**	**Targeting the premium business segment**
Datel	**Afripa Telecom** (since October 2001, until 1999 TTCL)	**1996**	**30-35 VSATs Satellite earth station for international connectivity**	**Targeting government agencies, corporate customers, ISPs**
Simbanet	**Part of the Simba Group** (with operations in Tanzania, Kenya, Nigeria, India and the UK)	**2000**	**VSAT stations in more than 11 regions Concrete expansion plan Virtual ISP scheme**	**Targeting government agencies, corporates customers, ISPs**
SatCom	**SoftTech-Group** Local consortium	**2002**	**Satellite hub inaugurated January 2003 with a planned presence in all regions**	**SatCom is part of the all telecom solution provider strategy by the Soft Tech Group. Targeting corporate customers, government agencies**

269 Company website at www.sita.aero

270 Equant has been outsourced from Sita in 1991 to provide global data services. In 2000, France Télécom acquired the majority share in Equant (Equant 2004). See also their website at www.equant.com.

271 Company Website at www.afsat.com.

Paconet	**Paconet Mauritius** (financed by AIG African Infrastructure Fund)	**2001**	**Earth satellite station in Dar es Salaam**	**Creation of an African network with points of presence in al major African cities**
Equant	**Member of France Télécom Group**	**1995**	**n.a.**	**Global, customized and integrated communications services for multinational corporations**
Source: Interviews, company websites				

Among the major new entrants Simunet, Simbanet, Paconet and SatCom, the TTCL subsidiary, Simunet, has been the most important new market entrant. Simunet was incorporated into TTCL in 1999 after Datel's sell off as the data communication arm of TTCL. Thus, Simunet took over Datel's role as explicit service provider for the broader market, assigned the mandate "to increase Internet usage among all Tanzanians in order to improve their socio-economic well being" (Simunet n.d.). To this end, the company has built the national Internet backbone and provided IP data solutions for Tanzania. In contrast to former Datel operations, Simunet, in fact, developed a fairly good reputation and expanded the availability of Internet access points in the country. The corporate plan has provided to execute the service functions in three phases: the first, to build a strong Internet presence in the regions; secondly, to provide solutions for the last mile connectivity, which also includes service packages for business customers in the form of Virtual Privat Network solutions (VPN); and thirdly, to install points of presences (PoPs) on the district level and in community centers (c.f. Simunet n.d.). In the year ending 2003, the first and second phases were in progress and, for example, 44 functional PoPs covering 20 regions and growing towards the districts have been established (see figure 4.17). Therewith, Simunet operated the largest IP network in the country. Interregional traffic is done by microwave, the local backbone by copper cables, fibre optic and microwave, making use of TTCL / Simunet today's three layer backbone, comprising the public switched telephone network (PSTN), terrestrial links and wireless local loop solutions for public Internet connectivity (Int. Yonah 15.4.2002).[272]

[272] Additional information is taken from messages from Simunet's general manager Zaipuna Yonah to the eThinktank dating from 8.7.2003, 9.7.2003 and 30.7.2003.

Figure 4.17: TTCL IP Points of Presences (POPs)

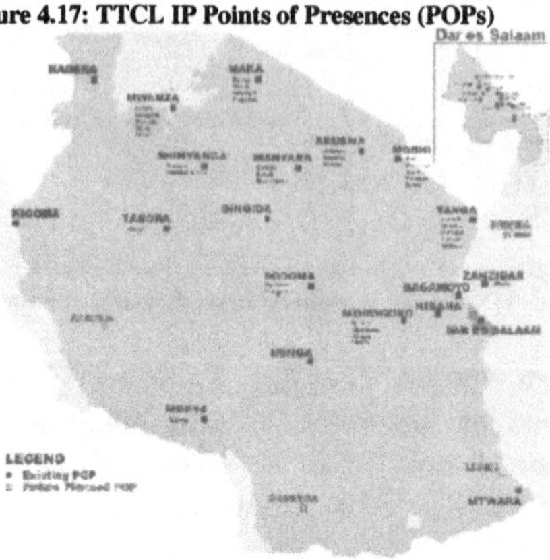

Source: TTCL

In addition to Simunet, Simbanet and SatCom became strong players in the Tanzanian market, because they have not only build distinctive company profiles, they were both also locally rooted and, thus, committed to the development of the national market. Simbanet[273] was founded in 2000 as spin-off of the established Tanzanian IT company ICS (International Communication Systems) in reaction to TCC's new licence set-up in the data communication segment. The business started in Dar es Salaam, later expanded its presence to Nairobi and Lagos, established a software branch in India and opened an office in the UK; the latter became the headquarters of the newly established Simba Group (Int. Horneman 21.5.2002). Though the majority of Simbanet's business has been in Dar es Salaam,[274] they built PoPs in 11 regions with concrete plans for further expansion. Furthermore, Simbanet introduced some new innovative ideas to the data communication market, offering next to a typical service package of Internet access and network solutions, a so called Virtual Internet Service Provider Programme (VISP), a franchise model in which Simbanet provides installation, maintenance and technical support and leaves all non-technical operations like marketing under the responsibility of the franchiser. In fact, this model has lowered the market entrance costs for ISPs and, at present, Simbanet runs 21 franchise operations in Tanzania (Simbanet 2002: 7).[275]

---

[273] Company website at www.simbanet.net.

[274] According to Horneman (Int. 21.5.2002), roughly 80 percent, however, among these some high profile operations like providing connectivity to the Ministry of Finance with about 50 government branches mainly in Dar es Salaam.

[275] Among them TaboraOnline, BagamoyoOnline, MakondeNet, TangaOnline, BukobaOnline or Morogoro.Net.

SatCom Networks Africa[276] came into being in 2002 and launched its services only in January 2003. The company has been member of the Dar es Salaam based SoftTech Group. The new satellite service makes use of a state-of-the-art satellite hub in Dar es Salaam and provides VSAT connectivity through Intelsat satellites. Like Simbanet, SatCom is built upon strong local roots and meant to complement SoftTech's self-image of a total telecom solution provider that already included telecom consultancies (SoftTech Consultants), ISP (Intafrica) and training facilities (LearnIT) (Int. Bhatt 29.5.2002). The SatCom corporate strategy targets PoPs in every region with economically more interesting regions Arusha, Moshi, Mwanza and Zanzibar to be connected first. The corporate strategy particularly dwells upon the groups' overall portfolio of complete turnkey network solutions designed to fit specific business requirements (Int. Bhatt 29.5.2002).

Finally, Pan African Communication Network, Paconet for short,[277] round off the distinctive business profiles in the market segment. Paconet, a subsidiary of Paconet Mauritius,[278] established the Tanzanian branch in 2001. In contrast to other operators, Paconet pursued a pan African strategy that aimed to create a fully meshed African voice and data network with PoPs in all major African cities (Paconet 2003). Paconet does not target national markets, but inter African data traffic. The operation in Tanzania is meant to create one hub for an African network with country-to-country links for direct communication, e.g. between Tanzania and DR Congo, without switching through one of the major US or European hubs. The available bandwidth is sold to other data carriers that want to avoid the expensive satellite transit to Europe or the US for their regional communication. At the same time, Paconet sought to strengthen corporate ties in Tanzania and has committed oneself to a number of partnerships and affiliations, such as with Mediapost for ISP solutions, Macromania for web solutions and Planetel as the local shareholder in Vodacom, to gain a presence in all Tanzanian ICT market segments (Int. Ogembo 24.5.2002). Therewith, Paconet is placeholder for Vodacom's aspiration to operate its own international gateway, if regulations will be adjusted accordingly.

---

[276] Company website at www.satconet.com

[277] Company website at www.paconet.net

[278] Paconet Mauritius is funded by the AIG African Infrastructure Fund established in March 2000 to invest in infrastructure and related industries in Africa. The American International Group (AIG) is the principal sponsor of the Fund. Additional investment comes from development agencies such as the World Bank's International Finance Corporation (IFC), the African Development Bank (ADB) or the European Investment Bank (EIB) and others. The fund focuses on countries that are believed to have a positive investment climate for infrastructure (EMP n.d.).

All in all, the established operators and, particularly, the new entrants significantly expanded the reach of IP data services in Tanzania. Furthermore, they have deliberately positioned themselves in the market to develop a distinctive company profile on which viable business operations can be founded. However, market diversification had its limits and the operators also faced a set of shared problems and challenges. Firstly, all operators had similar roll out strategies, i.e. building a strong customer base in Dar es Salaam and, in a second step, targeting prime regions such as Arusha, Mwanza, Moshi and, to some extent, Zanzibar. Upcountry infrastructure deployment in economically marginalised regions, or to the district level, is only done little by little, so that the operators have a rather identical regional coverage – to some extent except Simunet, that has been the market leader with regard to both, scale of operation and regional coverage. The situation aggravated further, because all operators have predominantly concentrated on the same customer segment, which has been the high potential customers from the business sector and government agencies. Attempts to move beyond these business patterns are only in their infancy. The sector awaits new political or regulatory incentives, foremost from the universal access strategy, which might make available new investment opportunities for rural telecommunications, and, secondly, to allow Voice-over-IP (VoIP), which is currently prohibited in order to protect the TTCL voice monopoly, but will be back on the regulatory agenda at the end of TTCL monopoly in 2005.

Another key characteristic of the market is related to costs. Though the times of exorbitant pricing are over, bandwidth charges are still high – often too high for small scale rural operations. Costs depend on quality, capacity and business orientation and vary accordingly. Connectivity costs for a leased dedicated line with a capacity of 64 kpbs – a conventional size in connection capacity – ranged from roughly 3,000 US$ in the premium business segment covered by Afsat,[279] 2,000 US$ as offered by Simbanet, to the low end offers by Simunet and Datel, where such data lines are available for 400 to 600 US$.[280] Despite a downturn,

[279] Apart from such high-priced packages, it is also noteworthy that Afsat run a general agreement to provide low cost connectivity for selected members of COSTECH's network or COSTECH affiliates. E.g. the Sengerema telecentre, which operates under COSTECH supervision, has paid only the equivalent of 300 US$ in TSh. for its Internet connectivity (Int. Mlaki 23.4.2002, Int Bugalama 23.4.2003). This clearly indicates a way forward that agreements with umbrella organisation can benefit a considerable number of member organisation.

[280] The figures are for orientation only, because they are hardly comparable. The operators do not only operate different technologies, they also differ in technical support, after sales services and their agreements with international satellite operators who connect them to the Internet via satellite and a downlink in Europe and the US.

the price-levels still do not offer appropriate connectivity solutions for rural Internet connectivity, where only one access site is connected to the line and, as a result, making connectivity the most burdened individual cost factor for rural data services. Without further decline in prices additional regional coverage will be difficult to attain.

A last common feature of data operators in Tanzania has been their sole dependency on satellite connectivity for their international gateway and the deployment of VSAT stations for national connectivity, usually built as relatively small point-to-point deployments for trunked Internet access. Both international gateway and national VSAT deployment rely on the availability of satellites,[281] and, although African satellite coverage has increased in recent years and new satellites are likely to come in the next years (c.f. Imfundo 2003, ARB 2004: 16009), the operators also count on the realisation of the Eastern African Submarine Cable System (EASSY) and related projects to close the last remaining link from South Africa to Djibouti of a complete cable ring around the continent (see figure 4.18). Zantel has been the Tanzanian partner of the EASSY project (see above). The availability of a fibre optic cable would not only bring new opportunities of connectivity, advantageous cost and capacity structures, but would also assure new competitive elements in the international connectivity market – with likely effects on the price structure.

[281] An additional problem of satellite access, especially in rural areas, is the considerable under-utilisation of capacities, that is relatively costly connectivity solutions offer 24 hours operations, but, due to limited resources, e.g. electricity, the Internet access sites only run for some hours and stand idle for the rest of the day. Thus, most of the time, rural operators are charged for capacities de facto not absorbed. However, if operators are not willing to adapt their pricing schemes to actual usage patterns (e.g. log into a satellite for some hours only or charging per traffic volume), satellite connectivity will remain rather expensive, including the peculiar side effect, that low usage rural customers, in fact, subsidise the traffic-intensive business users who have to pay the same money for significantly more data traffic.

Figure 4.18: Structure of the proposed EASSY Project

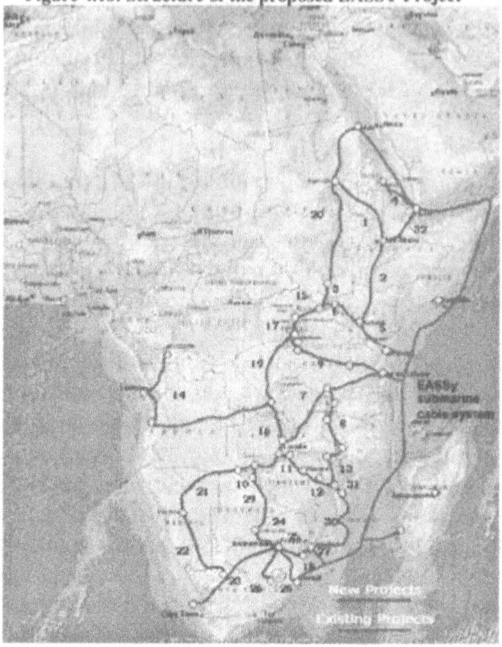

4.3.5.4. *Internet Service Provision*

Due to the positive changes in data communications, the market for ISPs also improved and the expansion of Internet services, in turn, had direct effects on the number of Internet users in the country. Figure 4.19 shows the gradual increase of Internet users starting in 1999 with steady increase ever since. However, the chart also reveals that growth has been considerably more dynamic in Kenya, where, despite a difficult political and economical environment technology, diffusion rates were supported by other benefiting factors such as higher literacy rates, advantages in the educational system, a higher rate of urban population,[282] or the role of Nairobi as an economic hub for the region and a longer history of an open economy that made it less difficult to adapt to new technologies. However, the figures should not be overvalued – they are not more than a rough approximation. The majority of the population has made use of the technology in public access facilities like in Internet cafes or telecentres, and also stand-alone computers with access to the Internet are generally used by more than one user. Thus, de facto user numbers are much higher than given in the official statistics as provided by the ITU, and a current

[282] Data taken from UNDP's Human Development Report 2004, see UNDP (2004).

estimation for Tanzania would be a number somewhere between 300,000 and 500,000 more or less regular Internet users.

Figure 4.19: Internet Users

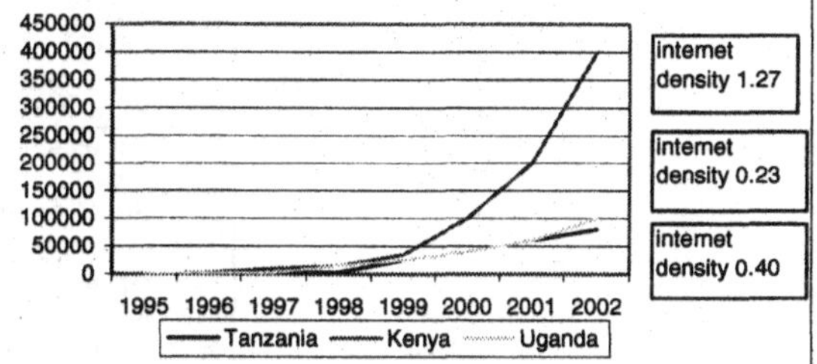

Source: ITU 2001, 2004

Within the new licensing setup, 19 licensed ISPs successfully had applied for a new licence – not taking into account those ISPs covered by Simbanet's Virtual ISP Programme. First licences were approved in early 2000, and, within the same year, a group of ISPs that already controlled the ISP market before the new licences, successfully applied for an ISP licence. Beyond the core group of ISPs, a mix of various businesses have successfully applied for a licence, mainly either aiming to serve some market niches or as place holders for future business operations. Thus, the basic composition of the market consolidated by 2002 and has been characterised by a stable number of prime ISPs – Africa Online, Cats-Net, Raha, Internet Africa, Cyber Twiga, UCC and the Arusha Node Marie, a number of newcomers – Mediapost, Newafrica, the Tanzania Posts Corporation (TPC) or Acexnet, and finally some smaller providers with undetermined ISP business strategies, like JR Electronics, ET Communications or Sawan IT.com. An overview of the most important players is given in table 4.8.

Table 4.8: Overview of main ISPs

	Est.	Dial up subscribers (2002/2003)	Coverage	Business Strategy
Africa Online	**1997**	**1800**	**Dar es Salaam, Arusha, Mwanza** **Advocates national ISP dial-up access**	**Target premium customer segment – individuals and business.** **Entrance into data communications failed due to high costs (licence fees)**
Arusha Node Marie	**1993** (as Fido Provider)	**2500**	**Arusha**	**Cooperative non-profit organisation.** **ISP solutions for Arusha**
Cats-Net	**1997**	**4000**	**Dar es Salaam, Arusha, Mwanza, Sansibar**	**Target broad customer segment** **Special scheme for Internet cafes**
Cyber Twiga	**1996**	**2000**	**Dar es Salaam, Arusha,**	**Target premium customer**

	Est.	Dial up subscribers (2002/2003)	Coverage	Business Strategy
(Tele2)			**Moshi, Mwanza, Sansibar, Tanga, Singida, Mbeya Advocates national ISP dial-up access**	**segment – individuals and businesses. Acquisition by Mobitel to develop mobile Internet solutions has not paid off so far**
Internet Africa	**1996**	**2500**	**Dar es Salaam Gradual roll-out strategy with the start of SatCom**	**Provision of business solutions**
Newafrica	**1998**	**50**	**DSM**	**Emphasis on local content provision. Difficulties led to a partial retreat in offline publishing**
Raha	**1996**	**6,000 – 7,000**	**DSM, Sansibar, Arusha, Moshi, Mwanza Further coverage in preparation**	**Target broad customer segment. Outside Dar es Salaam service provision is done with locally rooted partners like Zanlink, Cybernet or Kilinet**
TPC	**1999** (IT Depart-ment)	**-/-**	**Dar es Salaam, Tanga, Moshi, Arusha**	**A national TRP network is under preparation aiming to establish public Internet services in the 14 head post offices in the postal zones (Dar es Salaam, North, South, Lake, Coast, Zanzibar), with possible project expansion to all post offices. Innovative products like 'email to box number' under preparation**
UCC	**1996**	**300 – 400**	**DSM, Dodoma, Mwanza**	**Target the broad customer segment as part of the commercial expansion of UCC**
Mediapost	**2001**	**2,000**	**Dar es Salaam, Arusha, Mwanza**	**Target corporate customers**
Acexnet	**2001**	**n.a**	**Dar es Salaam**	**Specialised in wireless solutions for Internet access in Dar es Salaam. Low cost offers**
Source: Interviews				

Some characteristics of the ISP market are noticeable. First, the introduction of the ISP licence did little more than expand the income of the regulator, so that the quantity of main providers has remained the same, but the business environment for ISPs became more difficult, because they were additionally burdened by the newly imposed licence fee. Only one new entrant, Mediapost, started operation in 2001. Mediapost Tanzania is a subsidiary of Mediapost South Africa, who formed a joint venture with Planetel and Paconet meant to strengthen the business profile of its Tanzanian operations. The group built a presence in all licensed segments and Mediapost became the ISP to all group members, so that it started service from a stable and strong customer basis (Int. Ogembo 24.5.2002). Mediapost could also draw upon its African experiences. Since the end of the 1990s, it has been the only company after Africa Online and

M-Web which attempted to build on an African business strategy with branches already established in South Africa, Rwanda and Congo Brazaville, and expansion plans to Kenya, Uganda, Senegal, Nigeria, Cameroon, Malawi and Zambia (Business Times [Dar es Salaam] 3.5.2002, IT Vision [Dar es Salaam] Issue 07, April 2002). The second African player in the Tanzanian ISP market, Africa Online,[283] belongs to the leadoff ISPs in the country and has entered the market by taking over the local ISP Tanzania Online in 1997. Africa Online particularly targeted corporate customers and higher income individuals, and, like others, especially benefited from the growth in the ICT market in 2000-2001, where the company grew threefold with regard to its subscriber number (Int. Habicht 24.5.2002). Furthermore, the company has established a number of telecentres, branded e-touch, a public access service, which offers e-mail and Internet access through public access points (c.f. Africa Online 2001). However, due to AfricaOnline's limited coverage, e-touch centres never got a significant market share in the fast growing Internet café and telecentre market. All other main ISPs in the country have been Tanzanian start-ups highlighting the role of local entrepreneurship for Tanzanian Internet development. This pattern applies for other African countries as well. Due to small business margins and cumbersome to develop end-consumer markets, there are hardly pan African ISP growth strategies. Instead, most ISPs are founded on local experience and entrepreneurship as main facilitators for end-consumer Internet connectivity.

This also becomes evident if one looks at the ISPs business strategies. All ISPs rely on customers in Dar es Salaam and, to a considerably lesser extent, in secondary cities, particularly Arusha and Mwanza. No ISP had expansion plans beyond secondary city access. Thus, the main business is in Dar es Salaam. Here, the main focus is on corporate customers including government agencies, international NGOs and donors. Internet Africa, as part of the SoftTech Group, has been able to offer the most comprehensive service packages, also including advanced software solutions, for example, for audit and financial management solutions. Together with Africa Online, Cyber Twiga and Mediapost, these ISPs have pursued the most explicit prime customer segment only strategy. In contrast, Raha and Cats-Net have moved beyond the corporate customer strategy and supplemented their business portfolio by targeting broader customer segments, too, by aggressively promoting new services and offer solutions for

[283] Africa Online had a moving history from a Kenyan start up company, African expansion into five African markets, acquisition by African Lakes, a joint venture with WorldCom subsidiary Uunet Africa, expansion to four new markets, difficulties due to overexpansion of the business profile, rehabilitation and refocusing on the core business of Internet service provision in sub-Saharan Africa (c.f. Bittner (2001), BA No 209 (May 2004). Today, Africa Online operates branches in Côte d'Ivoire, Ghana, Kenya, Namibia, Swaziland, Tanzania, Uganda, Zimbabwe.

individual middle class subscribers. Both ISPs became the largest operations with regard to subscribers. Especially Raha set an innovative course for new upcountry ISP development by partnering with locally rooted providers like Zanlink (Zanzibar), Cybernet (Arusha) or Kilinet (Moshi), who can bring in their local expertise and reputation to attract new customers. Consequently, Raha's customer basis has been more evenly distributed, i.e 30 percent are corporate customers, a little less are government bodies and the rest, mainly individuals. A similar pattern can be observed for Cats-Net (Int. Sheriff 5.6.2002, Int. Sachdev 14.4.2003, Int. D'Souza 29.5.2002).

Costs for Internet connectivity show only little variations, also indicating an almost consolidated market segment. Connectivity solutions range from leased lines for high usage customers (1,000 – 1,500 US$ for a dedicated 64 kpbs line), to special service provisions, e.g. for Internet cafe connectivity as provided by Raha and Cats-Net to dial-up accounts, usually supplemented by other services like web hosting and website development. Dial up costs start between 30 (Cats-Net) 35 (Africa Online) and 40 US$ (Internet Africa), of which only the Cats-Net solution offered unlimited Internet access, whereas the average cost for unlimited access is available from 60 US$ upwards.[284] All ISPs have introduced a variety of pricing schemes to fit the individual or corporate needs. This also included, for example, a website development packages that have been available for roughly 125 US$, plus a monthly hosting and maintenance fee of 25 US$ (Raha) (Int. Sheriff 5.6.2002, Int. D'Souza 29.5.2002, Int. Messer 20.5.2002, Int. Bhatt 29.5.2002, Int. Habicht 24.5.2002, Int. Ogembo 24.5.2002).

Next to the main commercial ISPs, niche market ISPs could build a distinctive profile, among them, those that developed as local providers. The Arusha Node Marie (ANM), that started 1993 as the first FidoNet e-mail provider in Arusha, was upgraded to a full ISP in 1997. Due to its history, ANM has worked as a non-profit organisation which provided ISP and web services to its members, some 750 registered clients coming from the business sector (50 percent), NGOs (20 percent) and individuals (30 percent), and ranging from heavy users like the East African Community (EAC) to members disburdened from service charges, such as secondary schools, hospitals museums and the like (Int. Rowberg 14.4.2003). The ISPs regional approach also applied for the Raha partners

[284] The new market entrant Acexnet, a company established in the US to serve the East African market, has partnered with Simbanet to facilitate its Tanzanian operations. Acexnet started business only in 2001, specialised in wireless solutions for Internet access, particularly WiFi solutions and aggressively promoted a 25 US$ fee for unlimited monthly Internet access. Similar very competitive pricing schemes were introduced for other services as well, particularly aiming to attract corporate customers (see company website at www.acexnet.com).

Zanlink (Zanzibar), Cybernet (Arusha) and Kilinet (Moshi), as well as the ISPs operating under Simbanet's Virtual ISP Programme in Tabora, Bagamoyo, Mtwara, Tanga, Bukoba and Morogoro. In addition, the University Computing Centre belongs to this group offering dial-up services in Dar es Salaam, Dodoma, Arusha and Mwanza for favourable 30 US$ or even 10 US$ for UDSM members. Since the incorporation of UCC as a UDSM owned company in 2001, new efforts were made to increase ICT service provision for the general public, and UCC has successfully managed to develop this as a second business, complementing its main task to provide ICT services to the university (Int. Nfuka 10.6.2002, UCC 2001).

A new approach to ICT services has been proposed by the Tanzania Posts Corporations, seeking to capitalise on its strong regional coverage and its high acceptance as communication facilitator. The idea to establish TPC as an ISP came up in 1998 and originated in an EAC initiative in the same year, entitled 'East African Postal Automation Project' with the objective to improve the quality of service and introduce new value added services (Int. Gewe 5.6.2002). Thereby, the 1999 created IT Department had not only the task to develop elctronic postal services like distribution of P.O. Boxes, box tracking and rent or money transfer solutions, it also attempted to work out ideas for tailored postal Internet services. To this end, TPC advocated to link all their postal offices starting with 14 head offices spread over six postal zones – North, South, Central, Lake, Coast and Zanzibar – with Dar es Salaam already computerised and nine branches with some first progress. Targeted Internet service packages would include innovative ideas such as a combination of email and snail mail, within TPC labelled as hybrid mail, a service that would allow to write emails to a P.O. Box and vice versa, or the introduction of digital signatures for confident and protected email correspondents (Int Gewe 5.6.2002).[285] However innovative these services might have been, a lack of funds recurrently delayed their implementation. If the project will be realized, it could particularly take advantage of the high acceptance of postal services in rural regions, so that IT services in post offices could lower entrance barriers to ICT utilisation.

However, not all business strategies turned out to be successful. Newafrica which came into being in 1998 as a subsidiary of the local investment company Green & Growth sought to establish the company as content provider by introducing a comprehensive African web portal offering in depth information on African countries and on individual issues like travel, economy, news or art and culture and online directories of businesses, public sector agencies and professionals. Conventional ISP services supplemented this portfolio (Int.

[285] For available TPC services, see also their website at www.tanpost.com.

Lyimo 5.6.2002, Int. Allcard 5.6.2002, Int. Pastory 5.6.2002). Though the web portal became increasingly successful with more than 100,000 visitors per month, the service failed to generate the income needed to finance the staff intensive content provision, so that the company was forced to return to traditional offline publishing and gradually also lost its weak role as ISP that never connected more than 50 subscribers.

Considering the last years of ISP development in Tanzania, the sector grew and comprised a mix of companies able to deliver a broad range of services. Thus, the sector advanced considerably in the given period. However, beyond the individual business profiles there are also a number of problems faced by all providers. All ISPs had difficulties to extend their coverage beyond Dar es Salaam and some primary regions, predominantly targeting a premium business /government / NGO segment, and only a few have successfully attempted to address the market beyond these segments. However, it must be emphasised that this service bias has also been a result of increasing business activity and economic growth in Dar es Salaam and new business opportunities outside Tanzania's economic hub. Thus, ISPs benefited from increasing business activity in the country and, naturally, concentrated their business in the prospering regions. Furthermore, the scope of expansion opportunities remained limited. The main impediment to expansions has been a lack of finance. For end-consumer services, ISPs rely on the infrastructure provided by the data carriers, and, even if data carriers took the risk to serve low margin areas, they forward the inevitable higher costs of bandwidth provision to their customers, among them the ISPs. Yet, connectivity costs account for 20-40 percent of all incurred costs for ISP operations,[286] so that increased upcountry service delivery will further exacerbate the ISPs financial room for manoeuvre.

To bring down the connectivity costs the ISPs strongly advocated a Tanzanian Internet exchange point (TIXP). To this end, the ISPs revived the Tanzania ISP Association (TISPA) that was originally formed as a representation of interests, particularly against the regulator's licensing regime. The ratio of the Internet exchange point was to interlink all local ISPs to reduce international data traffic, because without IXP, traffic from, for example, Raha to Cyber Twiga was routed through the expensive international circuits to be delivered only next door. Consequently, the motto has been "to keep local traffic local", aiming to reduce costs of bandwidth for local traffic for ISPs and end-users, and provide greater efficiency and convenience for local content accessibility (Sheriff 2003).

[286] The ISP with the lowest connectivity charge quota has been Africa Online with 20 percent; the one with the highest quota Mediapost with 40 percent (Int. Habicht 24.5.2002, Int Ogembo 24.5.2002)

After considerable difficulties,[287] the TIXP started trial run in July 2003, with the following as local ISP peers: Africa Online, Cats-Net, COSTECH, Raha, SatCom (incl. Intafrica), Simbanet (incl. Accexnet) and UCC (Sheriff 2003).[288] In the end, TIXP brought not only some relief with regard to connectivity costs, it also represented the first attempt in which ISPs successfully and jointly lobbied their case.

Though TIXP can be labelled a success, it could not change the main problem of weak financial resources. The newly introduced ISP licence fee has intensified the difficult financial situation and significantly narrowed the financial scope of the ISPs. The thin capital resources left little room for investment and access to bank capital has been restricted, because commercial banks have been reluctant to grant credits without bank securities other than land titles or other tangible assets, however, in any case, hardly responsive to ISPs and other ICT companies that cannot offer more than little equipment and a number of virtual customers. The current banking practices have been hardly in favour of supporting ISPs. A lack of opportunity and a lack of incentives hampered ISP's business activities. The existence of smaller ISPs, e.g. running under Simbanet's virtual ISP scheme, or rural Internet cafes proved the viability of rural services, though they might be difficult and cumbersome. The interviews within this research and many discussions on the topic have shown that there is a surprisingly high number of well qualified rural entrepreneurs willing to take the risk of rural or semi-rural Internet operations. Their biggest impediment is the restrictive banking environment and a clumsy bureaucracy – well known arguments in the debate on private sector development or SME promotion. Thus, promoting a sound business environment in general, and elaborating on a comprehensive universal access strategy with fair and transparent mechanisms for filling and

---

[287] The length of the proceedings might mirror the difficulties encountered. First talks started in 1997, and it took six years to install a little equipment and link the ISPs. The main difficulties were foremost to build confidence among the operators, that the IXP will benefit all of them, and to overcome the mutual mistrust that paralysed any joint action. Secondly, some operators misconceived that the regulator (or TTCL) might be responsible for installing an IXP and underestimated their role as market participants. Finally, some difficulties arose when TISPA applied for a licence to operate the IXP, because, according to the regulations, a licence holder must be a registered company, not an association. Currently, TISPA is underway to reregister, and TCC turned in and approved the trial run. Funding for the TIXP came from DFID. (Int. Medeye 10.6.2002, Int. Mruma 10.6.2002, Int. Sangiwa 22.7.2002, email from David Sawe to the eThinktank dating from 2.7.2003, Sheriff 2003).

[288] As non-ISP peer, the Bank of Tanzania is also linked to the IXP (Sheriff 2003). Apparently missing is TTCL, that refused to link to the IXP, because it feared a loss in revenue due to less international data traffic.

allocation of a rural telecommunication development fund in particular, might bring a new dynamism to upcountry ISP expansion and ICT utilisation.

4.3.5.5. Payphones Services

With the commencing boom in mobile telephony, the rapid mobile penetration pace eclipsed the payphone market, so that it became less important between 2000 and 2003. The main competitor to TTCL, Adesemi, got into difficulties, though not only due to losses in the operation of payphone services, but also due to unfavourable resale agreements entered with TTCL as the sole provider of fixed-line services. In 2000, Adesemi was put under liquidation, and Mobitel took over the insolvent company and added Adesemi's roughly 500 payphones installed mainly in urban areas to the Mobitel network (The Express [Dar es Salaam] 9.11.2000). During the same time, the privatised TTCL introduced a new payphone brand, Rafiki,[289] and gradually increased the number of payphones from 780 in 2000 to roughly 2,000 in 2002 (ITU 2004: A-74). In comparison to other growth figures in the sector, that was only a modest increase, all the more that the majority of payphones was installed in urban areas. However, as the basic pattern of mobile telephony penetration as a mainly urban or semi-urban phenomenon became evident, payphone services slowly regained attention and even mobile operators started to reconsider payphone services as a business solution that will bring telephone services to those who could not afford a mobile handset. Finally, the universal access strategy, proposed in November 2003, predominantly focused on payphone services. The consultant calculated an economically viable payphone centred universal access model, which would comprise the installation of 2,046 payphones within four years and a pilot component in which 550 payphones are to be installed within a six month period (Apoyo 2003b). These are ambitious targets and hardly achievable by TTCL alone. Thus, implementation will depend on TTCLs willingness to untighten their monopoly, enter new and fair resale agreements or the like. Considering TTCL's past performance, TTCL concessions seem unlikely and the universal access project would have to wait until the monopoly will end in early 2005. However, both ways would bring new opportunities to payphone providers, so that new market stimuli can be expected in the near future.

4.3.5.6. Telecommunication Contractors

Licences for importation, distribution, installation and maintenance of telecommunication and radiocommunication equipment are approved or

[289] Swahili for ‚friend'.

renewed on a yearly basis. As shown in chapter 4.2.6.7., this segment has been widely consolidated and, thus, the number of approximately 80 private companies in this field of operation remained very stable and unchanged in the considered period.

4.3.6. Human Resource Development

A lack of well educated IT professionals and shortcomings in IT education characterised the period from 1994 to 1999. However, as weaknesses in the human resource situation were increasingly regarded as critical impediments for future ICT utilisation and government officials sought new ways to meet the anticipated demand of IT professionals, subsequently new programmes have been implemented, particularly targeting the tertiary education at the University of Dar es Salaam (UDSM) and the Dar es Salaam Institute of Technology (DIT), to a lesser extent also attempting to introduce computer courses in secondary schools. In addition, the private learning institutions that came into being due to weak training facility provision by public institutions also managed to strengthen their profile. Thus, gradual changes in the education system that could be observed since the mid 1990s continued, and although on a very low level and not without difficulties, human capacity building steadily improved since 2000.

The most profound changes were related to strategic adjustments at the University of Dar es Salaam. Following up the Information Policy Plan (IPP) and the Information Master Plan (IMP), both approved in 1995, providing for the establishment of the University Computer Centre (UCC) as ICT service provider for the university and the creation of the Department of Computer Science for tertiary IT education, and a short interim period in which UCC also took over tasks originally assigned to the Department of Computer Science, both institutions have only fully evolved since 2000. In January 2001, UCC was incorporated as a full UDSM owned company and became the central hub for ICT development in the country. Its mandate covered Internet service provision, software development and system integration, network design, installation and management, provision of training, including common business applications and also some technical software packages for professional users, and finally UCC has been involved in a number of consultancy services (c.f. UCC 2001).

With the creation of the computer science department in 1999 and after roughly a year to set up the structures of the academic department, the explicit task sharing between both institutions commenced in 2000. Though this meant a retreat from academic teaching, UCC was now in the position to focus on ICT infrastructure development and service delivery, which significantly strengthened the overall capacities of the university as a whole. Among others, UCC has achieved the following:

- Development of a Financial Information System (FIS), currently being extended to faculties and institutes,
- Development of an Academic Register Information System (ARIS),
- Electronic handling of student administration and applicant admission,
- Development of a Library Information System (LIBIS), which made library holdings accessible on the Internet (at www.libism.ac.tz/adlib/),
- Building of a database of theses and dissertations as part of a Database for African Theses and Dissertations project under the framework of the Association of African Universities,
- Development of an e-learning platform 'Technology Enhanced Independent Learning Environment' (TEIL) for selected UDSM faculties,[290]
- Development and implementation of a University-wide data communication fibre optic network within the main campus (WLAN)),
- Installation of high speed wireless links to off campus colleges (University College of Lands, Architecture and Survey (UCLAS) and Muhimbili University College of Health Sciences MUCHS) (WMAN),
- Installation of departmental networks,
- Development of an Instructional Technology Unit (ITRU) to facilitate the integration of technology in teaching and learning and in the research process (Mutagahywa 2003, Nfuka 2002, Int Nfuka 10.6.2002, The Guardian [Dar es Saalam] 17.5. 2003).

Apart from these broad undertakings, UCC also continued to be responsible for providing training courses for university members and the general public. Courses were mainly regular short courses comprising the most common business applications (e.g. Microsoft Office plus Internet and email) and tailor-made packages (e.g. desktop publishing). Starting from 2001, professional courses granting a certificate in software development have been added to the training portfolio.[291] Training facilities have been located in Dar es Salaam

[290] The Faculty of Law was first selected to implement TEIL and did so in 2000. The Faculty of Commerce and Management and the Faculty of Engineering followed subsequently, see the TEIL website at http://www.teil.udsm.ac.tz.

[291] A typical training package was made up of different modules, e.g. introduction to computers and operating systems plus Microsoft Word, Excel and Access plus an introduction to Internet and email, and was offered either as daily or evening courses and was available

(university campus plus a city centre branch) Arusha, Dodoma and Mwanza; plans are underway to establish further regional training centres in the country (UCC 2001, The Guardian [Dar es Salaam] 24.2.2004). Within only a short period, UCC managed to cope with a multiplicity of different services from Internet provision, networking and training services. Assistance for these ambitious programmes was made available from different sources, usually with one donor for one specific working programme; among others the Dutch NGO Hivos, the Flemish Interuniversity Council (VLIR) of Belgium, Carnegie Foundation, as well as from the Swedish and Norwegian development agencies SIDA and NORAD (Int. Nfuka 10.6.2002).

A further innovation, which caught particular attention, has been the introduction of the Cisco Academy at UCC in 2000. The Cisco Networking Academy Programme for least developed countries has been a partnership training programme between Cisco and UNDP, and had extended the existing Cisco programme to 34 least developed countries, so that today, academies have been set up in 149 countries (UNDP Tanzania 2001, Cisco 2003). The academy is a comprehensive e-learning programme that provides students with Internet technology skills, so that successful students have the knowledge to "design, build, maintain computer networks of local, national and global businesses" (UNDP-Tanzania 2001). The programme has been delivered in modules and students in the first four semesters have been prepared to qualify as a Cisco Certified Network Associate (CCNA), and students who completed semesters five through eight are prepared to take the Cisco Certified Network Professional (CCNP) exam. The programme has employed an e-learning model using a combination of web-based, instructor led learning and hands-on lab teaching methods.

UCC has been selected to serve as a regional academy supervising and coordinating all Cisco academies in the country, and train the trainers for the local academies. In a first phase, five local academies were established at (i) Department of Computer Science (ii) Dar es Salaam Institute of Technology, (iii) UCC Arusha Branch, (iv) Zanzibar University and (v) Kasulu Teachers College.[292] Course fees ranged from 500 US$ in Dar es Salaam to 200 US$ in

from 120,000 TSh. (UCC 2001). To increase the recognition of its courses, UCC introduced the International Computer Driving Licence, developed by the South African based ICDL foundation (The Guardian [Dar es Salaam] 18.6.2002). ICDL has matured in recent years and became one of the most recognised certificates for basic computer knowledge in Anglophone 'Africa, c.f. the ICDL website at www.icdl.org.za.

[292] The original proposal was for seven local academies, but two candidates did not join the academy, because they objected to Cisco's business approach and the exclusive training on Cisco equipment. These were the Vocational Education and Training Authority (VETA) –

Kasulu – costs, with which the Kasulu Teachers College might be labelled "the cheapest Cisco Academy in the world" (Int. Calder 1.4.2003). In due course, nine more academies have been established in Dar es Salaam, Mwanza, Arusha, Zanzibar, Morogoro, Bukoba and Iringa, and the gradual expansion indicates the success of the Cisco programme. In 2002, 89 students graduated from the programme and in 2003 250 students from around the country were expected to pass a Cisco examination.[293] Because of the outstanding performance of the programme, UCC became a promising candidate to turn into a so-called Cisco Super Academy. This is one of two proposed academies to oversee the academy programme for the whole of Africa.

Next to UCC, the Department of Computer Science became a second hub for ICT training in the country. Since its inception in 1999, the department sought to "become a centre of excellency for computer education on university level" (Int. Fournadjev 23.7.2002). The department was set up with assistance from Delft University of Technology which provided methodological assistance for curricula development, textbooks and computer equipment. Further assistance came from the Dutch government and Cisco, the latter supplied equipment in line with the Cisco Academy programme. Until the academic year 2001/2002, the department offered a BSc programme as well as diploma and certificate courses. In 2002 the first master programme was introduced (MSc). Roughly 70 students have been admitted per year, so that the department had to supervise 200-250 students per three-year course duration, and thus, managed to approximate the outlay capacity of 300 students (Int. Fournadjev 23.7.2002).

Apart from education, the department established a good reputation as software solution provider and conducted a number of innovative projects, particularly adjusting existing programmes to the local context and, thus, made software solutions fit local needs. This has included, for example, the Kiswahili Linux Project that translated the open source Linux operating system into Swahili to ease computer utilisation for Swahili speaking users and promoted a wider acceptance of open source software in the country.[294] Furthermore, in December

---

supported by their main donor SIDA, who later initiated their own IT-Fundi project (see below) – and Mzumbe University, which, however, became a Cisco Academy later on (Int. Magoha 11.6.2002).

[293] According to UCC Cisco coordinator, Jabiri Bakari, quoted in The Guardian [Dar es Salaam] 17.1.2003.

[294] For details see the KLP website at http://smartcampus.udsm.ac.tz/KLP/outcomes.htm. The approach to translate popular software into Swahili was, in due course, reflected in other programmes as well. E.g. Microsoft introduced a Swahili version of its operating system in 2004 (The Guardian [Dar es Salaam] 14.6.2004), Google presented a Swahili search engine (http://www.google.com/intl/sw/) and only recently COSTECH, with support from the ITU,

2004 software developers at the UDSM department of computer science have released the first version of OpenOffice.org in Swahili – called Jambo OpenOffice.org. The initial release of the software will only work on the Linux operating system, the final release, however, planned for February 2005, will also work on Windows. The project has been funded by SIDA, and has caught attention in so far, as it targeted the whole Swahili speaking regions in East Africa. Therewith, the new software has a potential scope of bringing affordable and well tailored software solutions to more than 70 million people living in the region (c.f. Marson 2004).

Like UCC with Cisco, the computer science department was credited for its good performance record and became a lead partner for a computer training programme run under an African Virtual University scheme. Though in the past AVU failed to deliver the expected results due to static learning modules, it has regained some significance in distance education delivery in recent years (Int Senkondo 30.7.2002). Since February 2004, UDSM, as the lead partner university jointly with the AVU and the Royal Melbourne Institute of Technology (RMIT) in Australia, started to offer a 4-year computer science degree programme delivered as a distant education programme (Bachelor of Business Administration – BBA) (AVU 2004). In Dar es Salaam some 50 students have enrolled in the first year of the programme (The Guardian [Dar es Salaam] 24.2.2004).

Only recently, the Faculty of Commerce and Management at UDSM launched a postgraduate diploma programme in ICT Regulation and became the first East African country that set up a university course aiming to train students in regulatory matters. The so-called NetTel Africa Programme was established when the Telecommunications Regulators Association of Southern Africa (TRASA) identified the need to build capacity in telecom policy and regulation. Seven African universities and one training organisation in Southern Eastern Africa, joined by several U.S. universities and African regulators, collaborated to develop an online postgraduate and masters degree curriculum in telecommunication policy and regulation. The programme at UDSM had

---

started a project to translate open source software into Swahili. Starting with a focus on Internet browsers, the project aims to bring the Internet within the reach of the Swahili only speaking population in Tanzania and East Africa (Ali Ayub Kalufya in an email to the eThinktank, dating from 13.10.2004, see also the project website at http://www.kilinux.org, and the media coverage in Sunday News [Dar es Salaam] 14.11.2004). Further software developments at the department have been e.g. a hospital management system, a university management system and several tailored solutions for the business sector (The Guardian [Dar es Salaam] 27.7.2002, IT Vision [Dar es Salaam], Issue 03, December 2001).

roughly 20 students in the first year (The Guardian [Dar es Salaam] 18.3.2004, Dot-Com Alliance 2002). Therewith, UDSM has complemented the foremost technological training with regulatory aspects of ICT and thus started to accommodate the importance of regulation in tertiary education.

Finally, the Dar es Salaam Institute of Technology (DIT) made considerable advancements in computer education. In contrast to academic teaching, the DIT approach has been more oriented towards practical needs. Actually, DIT's ICT profile was not very pronounced until 2001, but the Institute took advantage of a SIDA initiative to strengthen the ICT human resource capacities in country. The initiative followed the recommendations of a SIDA commissioned study, conducted by the Swedish Consultant Lamtrac, on the need for vocational training of ICT professionals in Tanzania. The report identified a lack of IT professionals as a critical constraint for growth and development of the ICT sector (Lamtrac 2001: 13, see also Buisness Times [Dar es Salaam] 24.5.2002) and recommended as one key measure the promotion of a practice oriented training scheme aiming to train students to effectively facilitate the implementation of basic information technology (Lamtrc 2001: 15). As a result, SIDA launched the IT-Fundi[295] project and DIT successfully applied to host the project in 2002. DIT designed the project as a three-year course, including obligatory traineeships after the first year of study. Furthermore, the project provided an exit option after each year, to be responsive to market needs and to allow companies to temporarily release staff from work to take a one year course for some computer and network essentials (Int. Masika 24.7.2002). SIDA provided roughly one million US$ to support the computer training programme, equipment and upgrading the DIT infrastructure installing a LAN with highspeed Internet access (c.f. Kondoro 2002).

The IT-Fundi project commenced in April 2002, and initially started with a modest 25 students; the targeted outlay capacity has been 75-100 students per year (Int. Masika 24.7.2002). In addition to the IT-Fundi project, DIT also managed to host other training schemes, as it was awarded to become a Cisco Academy and an ITU approved training centre,[296] so that it subsequently strengthened its ICT profile and developed as a major training centre in the country.

The above thee institutions have been the spearhead institutions of ICT education in Tanzania. They built considerable capacities attempting to thwart the undersupply of IT professionals. Within a relatively short period and with

[295] Fundi, Swahili word for 'expert'.

[296] Financed by ITU resources channelled through TCC/TCRA (Int. Masika 24.7.2002).

considerable donor assistance, the training institutions developed appropriate curricula and installed state of the art equipment for first class computer education. At the same time, however, fundamental problems cannot be ignored.[297] Firstly, the availability of equipment did not guarantee that students or staff took full advantage of the installed equipment and did not prevent an accumulation of idle capacities. An unused computer aided decision-making system in the Faculty of Commerce and Management, already installed in 2000 (Int. Matambalya 17.5.2002), might be just one example of persistently underutilised computer systems and under-worked state of the art computer labs. To some extent, this disaccord has been a typical transitional phenomena in which an overall undersupply of trainers and trainees must be tackled at the same time.[298] However, this might become less a strain after the first generations of IT trainees will have finalised their studies and be available as trainers, themselves. The more alarming observable fact though, has been a shortage of eligible students. As one consequence, all three institutes have been unable to operate at full capacity and even attempts to lower entrance qualification[299] brought only little relief. Beyond a question of pure utilisation, these observations, however, have been distressing for a more fundamental reason: An education system that fails to provide pupils with the basic prerequisites for university entrance qualification will have serious problems to meet the minimal requirements needed to take advantage of information society opportunities.

Still complementing the public education system, the private teaching facilities continued to expand their role as ICT training providers. Particularly those already introduced in chapter 4.2.7. became key players in an emerging market of training providers. LearnIT, the Institute of Information Technology (IIT),[300]

---

[297] Still in 2003, the shortage of qualified personnel and quality ICT training was regarded as a major obstacle to ICT development (c.f. Mendes et al. 2003).

[298] In fact, the recruitment of qualified personnel had been a major problem in all three institutions. Approved posts remained vacant and the training institutions relied on part-time lecturers or borrowed trainers from the private sector (Int. Nfuka 10.6.2002, Int. Fournadjiev 23.7.2003, Int. Masika 24.7.2002). According to Vasco Fournadjiev, lecturer at the computer science department, low university salaries on the one hand, and high educational entrance qualifications on the other (Master, PhD, etc.) aggravate the shortage of qualified teachers and make university careers rather uncompetitive (Int. Fournadjiev 23.7.2003). The introduction of expert assistance schemes has been one solution to ease the problem at the computer science department. Fournadjiev himself has been part of such a scheme, his salaries are mainly paid by the Dutch Ministry of Foreign Affairs.

[299] DIT, for example, lowered admission level from previously required C grades in mathematics, physics and English, dropped them to D and offered the affected students bridging courses to bring them to C-level (Int. Masika 24.7.2002).

[300] An overview of IIT is given in IT Vision [Dar es Salaam] Issue 02, November 2001.

the NIIT Group or the Institute of Management and Information Technology (IMIT) have embarked on a twofold strategy, both attempting to increase the acceptance and the reputation of the courses offered. For short courses, the institutes placed an emphasis on certificated courses, that is, internationally accepted courses certified for specific products like Microsoft, Oracle or Cisco. Increasingly popular has been the International Computer Driving Licence (ICDL) that goes beyond company or product related training approaches (Int. Bhatt, 29.5.2002, Int. D'Souza 29.5.2002, Int. Chagani 29.5.2002). A similar approach applied to advanced diploma courses, usually designed as one-year courses. Here, the training institutions partnered with recognised international training institutions, mainly well-known universities, which provided curricula, materials and conducted the examinations. For example IMIT made use of a training scheme developed an supervised by the Manchester based National Computing Centre (NCC), and LearnIT[301] offered a computer science degree granted by APTECH, a recognised global IT training provider (www.aptech-worldwide.com), the NCC and the London Metropolitan University. Thus, private training institutions advanced considerably. However, the growing number of training institutes must not deceive that the private institutions had similar problems as their university counterparts and, foremost, that the quality of courses was not always assured. Thus, the conclusion of the Lamtrac report of 2001 can still claim validity: "It is very difficult to judge the quality of the existing training programmes, but we doubt that all providers would qualify as career /advanced career education providers. However, some of the providers operate according to high international standards [...]. They also have ambitious curricula and apply applying selection criteria for applicants and provide for job practice" (Lamtrac 2001: 6). The private training institutions profited from a weak education system and successfully focused on potential markets not served by the public education system. Even as the public institutions caught up, the private training facilities successfully supplement public tertiary education and those which have capitalised on internationally recognised certificate courses still rank at the top of broad ICT end user training in the country.

4.3.7. The Role of Donors in ICT Development

The previous chapters have recurrently referred to donor assistance in ICT promotion and have already indicated the many facets ranging from multi million US$ programmes like the the Telecommunications Restructuring

[301] LearnIT, together with DIT and UCC, have also been national training partners of the Institute of Communication and Development (IICD), a Dutch NGO highly active in the field of ICT promotion in the country (see chapter 4.3.7.). The national training partners received input from the IICD to provide institutional management and technical skills training, tailored to the needs of IICD projects in Tanzania (Int. Moens 18.7.2002).

Programme (TRP), to support for ICT in a narrower sense or efforts to integrate ICT support in existing programme portfolios. The high number of projects and the fact that ICT project components often have been hidden in larger approaches have left a patchy donor environment that is difficult to assess. Nevertheless, some basic findings on donors and different donor approaches can be made.[302] First of all, within the complete scope of development cooperation, ICT promotion and related measures only played a minor role. For those with some profile in ICT promotion, two approaches can be distinguished, firstly those donors who made ICT a sectoral focus, and, secondly, those who have integrated ICT into their existing project portfolio.

In recent years, SIDA has developed as the primary donor in the field of ICT and has integrated ICT issues into existing programmes and adopted ICT promotion as a strategic area for Swedish development cooperation. Sweden has supported the telecommunications sector in Tanzania since the 1970s with technical assistance, training and equipment; it was a prime donor in the TRP between 1993 and 2001, and provided support to capacity building of the MoCT, particularly the national ICT taskforce, and to strengthen TCC's regulatory capacities. Other support was given to UDSM to improve the ICT infrastructure and training facilities and the DIT, in support of the IT-Fundi programme. Apart from these strategic ICT targeting measures, Sweden also assisted the Local Government Reform Programme inter alia, with assistance to implement a financial management and accounting system and brought forward some pilot attempts of rural connectivity within its rural development strategy (Int. Karhammar 23.7.2002, 27.3.2003, Int. Jensen 27.3.2003, Rylander (2002)). Sida's ICT activities comprised policy and regulation, human capacity building and – though to a considerably lesser extent - rural ICT development. These activities made SIDA the most advanced ICT donor in Tanzania, also underlined by the fact, that in late 2002 Sida has created the post of an ICT project manager, currently held by Nils Jensen, to ensure SIDA's appropriate responsiveness overseeing an increasingly complex sector.

Secondly, the World Bank has been a prime facilitator of ICT. Especially the Bank's predominant role in the TRP made the Bank an early supporter of telecommunications sector reform in the country. In recent times, however, the Bank changed its approach to ICT, and, to the same extent as the overall responsibility of ICT was shifted to the World Bank's private sector investment

[302] An up to date overview on donor's (bilateral and multilateral) general ICT strategies is given in OECD (2003).

arm, the International Finance Corporation,[303] the Dar es Salaam World Bank office subsequently withdrew form the concrete project level. Exceptions related to global World Bank projects, that remained in the Bank's project portfolio: Firstly, the Tanzania Development Gateway as part of the World Bank's Development Gateway, a comprehensive information portal providing access to knowledge for development.[304] Resources to set up the gateway were made available by the World Bank's InfoDev initiative (Int. Nzuki 20.5.2002).[305] The support covered planning and implementation and also included an e-readiness and e-needs assessment which was carried out by COSTECH. The website (www.tanzaniagateway.org) has been accessible since early 2003 and was among the first country gateways of today's 12 gateways in operation in Africa. The second World Bank project supported the Tanzania Global Development Learning Centre (TGDLC) as part of the Global Development Learning Network (GDLN).[306] The TGDLC has been one of the first ultra modern distance learning centers providing facilities for videoconferencing, Internet and the like in developing countries. The core function has been to provide ICT-based distance learning courses or online conferences for decision-makers and mid-level professionals (Int. Senkondo 30.7.2002). The centre started operation in June 2000.

UNDP Tanzania has also built some distinctive ICT profiles. They have provided support to the eThinktank and the MoCT for preparation of the national ICT policy, to disseminating development information through the Internet,[307] facilitated the coordination of donor activities around ICT,[308] and supported the training of IT professionals through the Cisco Networking Academy Initiative. In addition, UNDP promoted ICT in individual projects

[303] IFC is the largest multilateral source of loan and equity financing for private sector projects in the developing world. With regard to telecommunications, IFC is shareholder e.g. in Celtel International or Paconet. For an overview see www.ifc.org. Within the IFC the Global Information and Communication Technologies Department (GICT) is the main body in charge of the Bank's ICT approach, see http://info.worldbank.org/ict/gict.cfm.

[304] see www.developmentgateway.org.

[305] see www.infodev.org.

[306] see www.gdln.org.

[307] Particularly supporting the development of the information portal TZ Online www.tzonline.org, the eThinktank website www.ethinktanktz.org and the portal www.icttanzania.org.

[308] UNDP hosted an ICT Donor Working Group to improve the coordination of assistance activities and share information about programmes on the sector. Regular participants of the group meetings were Sweden, Finland, United States, UNDP and ILO. The author participated in the monthly group meetings during his field visits in Dar es salaam.

such as the computerisation of Bukoba Hospital (Int. Magoha 11.6.2002, Int. Asikainen 24.4.2002). Furthermore, the Tanzanian USAID office had some advanced plans to extend USAID's broad ICT supporting Leland Initiative[309] to Tanzania. From 2005 onwards, the programme will presumably target rural ICT developments (Int. Moore 18.6.2002, Int. Msaki 18.6.2002). Finally, the Dutch NGO International Institute for Communication and Development (IICD) has focused on ICT. IICT has been particularly active in promoting ICT applications for good governance, education and enhancing livelihood opportunities in rural areas (Int. Moens 18.7.2002).[310]

The majority of donors, however, approached ICT without specific working programmes, but have incorporated the issue into their existing programme portfolio or supported individual ICT projects on case-by-case basis. Particularly Finland and Denmark targeted ICT as a cross-sectoral issue and have been the most advanced donors who have dealt with ICT as a cross cutting issue. For example, within the Finnish regional approach to development cooperation and their special assistance scheme to Lindi and Mtwara region, the large scale Rural Integrated Project Support (RIPS) has included components of ICT infrastructure building, computerisation of local administration, ICT training programmes and the like (Int. Eirola 8.7.2002). Denmark, to some extent, has incorporated ICT into their sectoral programmes, particularly in private sector development approaches (Int. Laudridsen 29.7.2002). Most other donors also had individual ICT projects or even rather hidden ICT components within projects in their working programmes, but usually on a rather small scale. These projects have covered a very broad mix of different projects, like the support for the Tanzania Internet Exchange Point (DFID), ICT in education (DFID, Netherlands), computerization of the Tanzania Revenue Authority or ICT in health projects (GTZ), fibre optic installations (KfW), support for a financial management and accounting system or road planning (NORAD), e-governance programme for Zanzibar Municipal Council (UNESCO), projects for small enterprise promotion (ILO) or just the donation of equipment (Japan) (Int. Mey 18.6.2002, Int von Maltzan 2.5.2002, Int. Sauvik 17.7.2002, Int Abeynayake

[309] For scope and strategy of the Leland Initiative covering 20 African countries, see USAID (n.d.).

[310] Projects included the Kinondoni Municipality Project supporting ICT applications in local administration, development of new tools for ICT in primary and secondary education, and finally, rural ICT development which started with a series of round tables for ICT in Mwanza region in 1998, and resulted in the support of agricultural price and business information services (Int. Moens 18.7.2002, Int. Mwakalinga 9.4.2003, Int. Maselle 24.3.2003).

17.6.2002, Int. Egawa 17.7.2002, Int. van der Zwaan 26.7.2002, Int. Baxter 26.7.2002, Int. Riad 15.5.2002).[311]

These projects should not be taken as an exhaustive list of ICT projects in the country, but the listing shows that most donors have some ICT projects in their overall portfolio, but have been reluctant or had difficulties to translate their projects into a strategic approach, so that a basic distinction can be made with regard to core and occasional ICT donors.[312] Considering those donors with a higher ICT profile, one can also conclude that most project approaches focus on the creation of a good environment for ICT – policy, regulation and human resource development – have increasingly put an emphasis on rural telecommunications development and, finally, targeted applications or programmes in individual sectors, particularly governance and health. This triangular approach also outlines the road ahead for future ICT related donor activities.

As the overall number of projects increases, donor coordination will become more important. Experiences from the UNDP hosted ICT Donor Working Group have shown, that there are limits to donor coordination, because project approaches have been too different or donors simply were not willing to share everything. Nevertheless, some minimum provisions for information and experience sharing will remain critical, particularly because projects and programme objectives will become more ambitious than they were in the past. Secondly, there has been the general tendency to incorporate ICT into existing sectoral programmes. This approach provided the opportunity to merge ICT opportunities with sectoral competencies. However, this would also include a concrete definition of ICT for sectoral developments and this is still an

[311] As a special case, Belgium has become an ICT supporting donor, too. The project built upon an old cooperation agreement between Belgium and the Tanzanian Ministry of Defense, aiming at improving the communication capacities of the Tanzanian armed forces. Considerable capacities for radio and satellite communication had been built, but much of the capacity remained unused. For this reason and due to other general policy considerations, the Belgian government insisted on opening the military communication network for civil use, and this resulted in a proposal, developed by the Tanzanian government, to upgrade the military network to a country-wide government network. This US$ 60 million project has been approved and became the fundament for the Tanzanian e-government strategy (Int. Sawe 11.4.2003, Int. de Schelleskens 17.7.2002, Int. Brehms 17.7.2002, see also chapter 5.3).

[312] Though some donors have seen a relevance to incorporate ICT as a cross cutting issue, like e.g. gender or environmental issues into their overall project approach, this would also set additional hurdles within given project circles. So that it has been foremost a lack of capacity that deterred donors from developing an appropriate ICT strategy in support of a given sectoral focus (Int. Laudridsen 29.7.2002, Int. Sauvik 17.7.2002).

outstanding work step on the part of most donors. Thirdly, project coordination is not only required among donors, but also in the relation between donor projects and market developments. Though the ICT market still is small, there has been vital activity in many areas from access provision and infrastructure building to setting up training facilities. Donor projects did not always meet the concerns of a young and fragile market. Particularly in the area of infrastructure and training, donor projects have competed with local companies. Early market assessments and increased partnerships with local companies might help to avoid such uneven competition and, in addition, might induce new market improvements. Finally, as for the country in general, the biggest challenge has been rural ICT development. Here, it will be crucial that development projects will be in line with and supportive of the official rural development strategy.

4.3.8. Concluding the Third phase

This chapter has dealt with the role of politics and policy and the efforts of the Tanzanian government to create a sound environment for private sector led ICT developments and the responses of the market participants as agents of change. The third phase of Tanzanian ICT development has witnessed an accelerated ICT development and has been characterised by an improving political and economical environment, advanced regulatory capacities and an emerging ICT market on the one hand, and slow policy implementation, setbacks and limited market reach on the other. Though positive developments clearly preponderated, the overall record has been erratic.

Policy efforts to promote ICT comprised different policy levels related to an outer and an inner frame of policy interventions. The outer frame targeted macroeconomic stabilisation measures and strengthening the general policy framework in favour of a market oriented development strategy, particularly targeting a stable macroeconomic framework and a sound business environment conducive for local and foreign investors. Adjustments and reforms focusing on ICT in a narrower sense made the inner frame of ICT promotion. New policies had been introduced and new institutions came into being, both meant to build confidence in the market and level the playing field for private sector participation. In addition, the education system came under scrutiny and first adjustments were made, particularly with regard to tertiary education, to address the shortage of human IT resources in the country. Especially policy and regulative capacities have increased subsequently, provided sufficient sectoral guidance and enhanced investors' confidence in a newly emerging market segment.

This not only resulted in extended market reach of ICT, but also to the formation of a new business sector in the country. Private companies took advantage of political change and a vibrant new ICT industry has been a first visible development impact of a private sector growth strategy. A healthy mix of different companies materialised from large international corporations, often joint venturing with local companies to numerous locally rooted small and medium enterprises, which started operation either as successful start ups or adjusted their previous business profiles. Furthermore, the telecommunication sector significantly contributed to the national income,[313] it created new jobs in the first labour market[314] and, both companies and employers, became regular tax payers. Therewith, the ICT industry became a modern sector in the local economy. A diverse market, adequate policy guidance and promising rudiments of regulatory oversight have made the ICT sector a model for other sectoral development strategies, too.

A more in-depth assessment, however, also revealed limitations. Though not in the general setup, but with regard to political and technical implementation deficits caused by very different reasons. Deficiencies related to the quality of policies, which have tended to be over ambitious and, therefore, lacked a clear focus or valuation of different and sometimes contradictory policy objectives. For example, The ICT policy failed to set well-defined target figures, did not provide clear-cut responsibilities and did not sufficiently distinguished between direct and indirect measures in support of ICT. Secondly, implementation capacities have often been weak, partly because the political backing was insufficient or the implementing agencies lacked the needed resources to comply adequately with their mandates. Evidently, the handling of the TTCL transaction by PRSC, MoCT's faint leadership in the ICT policy process or the regulator's weak rule enforcement with regard to, for example, interconnection, have been indicative examples in this regard. Thirdly, the Tanzanian policy environment

[313] The latest available Economic Survey states that the contribution of the transport and communications sector to GDP increased to 5.5 percent in 2002, compared to 5.4 percent in 2001. Therewith, the sector is well behind agriculture, but, together with mining, tourism and financial services, a stable pillar of the Tanzanian economy (c.f. URT 2003c)

[314] There are no exact figures on employment in the telecommunication sector. However, the private sector companies interviewed for this report stood for 5,000 jobs, most of them, roughly 3,000 from TTCL. This figure stands only for the supply industry – operators, carriers, ISPs, vendors – thus does not account for IT professionals in other industries or in the public sector. According to a Vodacom source, the company had roughly 180 direct employers, but provided more than 5,000 indirect jobs. Though this ratio of direct and indirect jobs is not representative, it indicates that the overall number of jobs in the Tanzanian ICT supply sector might be well above 50,000, if a careful ratio of 1:10 between direct and indirect jobs is applied.

has been largely characterised by a retarding moment. Naturally, this correlates with insufficient political input and weak implementation capacities, but also touches upon an unfinished transformation process, in which the role of the state has been radically redefined and political behaviour and structures of decision-making have not yet fully uncoupled from previous state-centred development strategies. Finally, the above-mentioned deficiencies have been reinforced by some hidden agendas that can be partly deduced from a mismatch between official announcements and political action. Hidden agendas have been motivated differently. Whereas the preferential treatment of TTCL can be seen as a comprehensible safeguard measure for the incumbent's transition, the ministerial prerogative in regulative measures was overtly beyond any safeguard intentions and solely restricted regulative independence. All in all, the complex interlinkages between TTCL, the MoCT and the regulator, best personalised by the TTCL chief executive and long served CCM cadre George Mbowe, contributed to sometimes half-hearted reforms and revealed persistent efforts to hold tight different mechanisms of political control.

These restrictions to the general good performance qualify the Tanzanian ICT policy record and explain why considerable progress has been recurrently disrupted. However, it would be wrong to attribute all shortcomings to the political sphere, there were also failures on side of the business sector, which had to learn to cope with the new environment, too. Weak entrepreneurship, one-sided business strategies, little response to customer's needs, misinterpretation of market potentials or failures to take a common business stance to lobby joint positions, vis-à-vis, the state made business difficult and threw back the coming out of a strong and self-confident business sector. Furthermore, economic relations within the business sector impeded further sectoral growth, particularly with regard to the banking sector and the difficulties of the IT sector to access the capital markets for further investments.

Returning to the general picture, the above-mentioned deficiencies have limited the market reach, but they must be put in perspective of the overall trend of increased market efficiency and improved ICT supply in the country. Beyond the market, however, where the ICT agenda is politically and socially motivated and seek the inclusion of the poor and marginalised, developments have been slow. A long awaited rural communication development strategy, first announced when market reforms commenced in 1994, is not yet in sight. Some preparatory work has been done, particularly since 2003, but the crucial details addressing questions like 'what to finance?' and 'how to finance?' are still missing. A new push towards rural telecommunications might be reinitiated with the coming end of the TTCL monopoly in February 2005, which might not only provide the occasion to introduce a more liberal framework for the remaining monopolies in fixed-line telephony and international voice traffic, but also a juncture to finalise the rural telecommunication policy. The creation of a rural telecom framework providing special incentives, investment schemes or subsidies, would be the missing link between a market oriented and rural development strategy. This might gradually change in the near future. TCRA conducted a first workshop on the new licensing and rural telecommunications policy proposal on 29 October 2004 and provided provided some general strategy outline, reiterating fund like solutions, proposals to encourage small operators, developing the policy context in line with the regional development initiatives of the EAC, SADC and Nepad, and making available the excess transmission capacities of e.g. the Tanzania Railways Corporation or the Tanzania Electric Supply Company (The Guardian [Dar es Salaam] 1.11.2004).

Finally, the focus on the general business environment and the expected and, in fact, realised positive impacts on development have proven to be a necessary, but not sufficient condition for increased ICT utilisation. Politics set the frame for private sector led development, which resulted in an improved supply of ICT and ICT related services in considerable scope. The supply pattern, however, shows that not only the general framework and business performances have influenced the overall development record, it has also been, of course, the demand for ICT that makes a difference. The availability of good products has been vital, like the different growth paths of mobile telephony and Internet services have indicated. Mobile telephony has become a mass product, because of a broad demand for convenient, appropriate and cheap telephone services. Internet services, in contrast, grew steadily, too, but never boomed like mobile telephony. Notwithstanding the relevance of Internet services as facilitator for broad based developments in various sectors, mobile telephony proved that good products are essential for increased ICT utilisation. This also means, that policies in support of ICT have to be not only supply, but also demand oriented. Thus, the next chapter deals with ICT utilisation in different sectors and the question of possible positive feedback mechanisms for better policies derived from the ICT experiences from the ground.

5. The Case of Tanzania: Assessing ICT Utilisation

This chapter aims to identify typical patterns of ICT utilization in Tanzania. The previous chapters have shown that the general environment for ICT has gradually improved which led to better market efficiency characterised by improved ICT service delivery and an overall increased diffusion of ICT in all segments of the market. At the same time, growth patterns, pace and scope of ICT penetration have been distinctive, firstly with regard to urban-rural and income disparities in ICT utilisation and secondly, with regard to service types, where telephone services, particularly mobile telephony, grew faster and computer, Internet or networking related services were less dynamic. Basic services targeting the mass market by providing appropriate low end products have increasingly penetrated the market and usage patterns have indicated a still unmet demand for basic communications. The provision of value added ICT services, in contrast, grew slower, because services were not only more complex and costly, ICT providers and data operators also focused on a rather small customer segment of corporate and institutional users or organizations representing the high ends of the Tanzanian market. Though these have offered higher margins for the individual providers and demanded sophisticated ICT solutions to improve their working conditions and capabilities, their contribution to an overall quantitative ICT record has been rather limited. Basic services have gained momentum due to broad demand, computer and Internet services had started as niche products, and gradually increased their reach, but diffusion depended to a greater extent on typical patterns of sectoral utilisation capacities. Advanced ICT usage has been closely related to the type of individual applications in different sectors. This is why this chapter looks at individual ICT utilisation patterns, focusing on computers and Internet usage, as well as on electronic networking. Areas covered are ICT in governance, health, and education, the role of ICT utilisation for NGOs and SMEs and finally, the prospects of rural ICT utilisation will be elaborated on. Since efforts to increase ICT in governance, health and education have been a relatively recent endeavour and are in a planning or early implementation phase, the analyses approaches the topics by qualitative terms. The remaining three topics can also draw on the results of quantitative research results, collected by means of sectoral surveys conducted within this research project. Before turning to individual segments, however, some general comments can be made: Firstly, with regard to the main user segments as main proponents of ICT developments, secondly with regard to urban Internet cafes or public access facilities that, for the first time, made ICT services available to a broader population.

5.1. Prime Users of ICT

Figures 5.1 and 5.2 give a broad overview of ICT user distribution. Figure 5.1 is based on a vendor survey conducted in 2002[315], and identifies main groups of computer users by asking vendors about their main customer basis. Figure 5.2 is based on a website survey done in 2003[316] and allows a first orientation on Internet user groups in the country. The diagrams reveal distinctive features of computer and Internet usage in the country. Whereas computer distribution is more evenly distributed with 37 percent corporate, 20 percent government and 20 percent NGO users, the moderate prevalence of private sector users shifted towards an excess prevalence in website distribution. Here, the private sector underlined its dominance, accounting for 60 percent of Tanzanian websites with the largest proportion coming from the tourist sector. Though the availability of a website is not an inclusive indicator, the figures mirror the overall ICT utilisation readiness of the different segments. In this regard, the private sector has taken the lead, irrespective of local or international background of the companies. Government agencies and NGOs, with a less distinctive usage pattern for NGOs operating on the local level only, have been secondary users. However, these are only quantitative approximations. Considering the lower number of government institutions, universities, parastatals, and the like, their relative ratio of ICT utilisation might reach a similar usage level to that of their private sector equivalents. Thus, ICT utilisation was boosted by organisational

[315] The survey was conducted in cooperation with UNDP Tanzania and the Dar es Salaam Institute of Technology. The survey was conducted online. The questionnaire was sent to 60 computer vendors, 22 completed the survey forms, which has been a fairly good returning rate of 37 percent. Although the results cannot claim to be representative, they reflect some strong tendencies within the IT-vendor market in Tanzania. Apart from the results with regard to the customer basis, the survey revealed that small and medium local companies overwhelmingly concentrated in Dar es Salaam have dominated the vendor market. Most vendors offered ready-to-use hard- and software packages. Value added services like the provision of network solution were provided by only a small number of companies (nine percent). The great majority of the businesses have valued their own business performances as good or satisfactory (36 percent / 59 percent) and similar figures could be identified with regard to general market prospects (41 percent / 55 percent). Major impetus for the vendors had been the telecommunications sector reform in 1994 and the arrival of the Internet in 1996. Roughly 64 percent of the vendor businesses came into being during the period between 1994 and 1999, thus benefited from the improving political environment (c.f. Nielinger 2002).

[316] The website survey based on 487 Tanzanian websites including *.tz and *.com domains that were selected from in-depth web researches and various national or international web directories or link collections, as well as advanced data mining via common search engines. Therewith, the pre-selection focused on the most perceptible and accessible websites, which allowed a fairly good reflection of Tanzania website distribution.

usage and corporate bodies of the private and public sector became prime users and early adopters of ICT.

Figure 5.1: IT Vendor Customer Basis

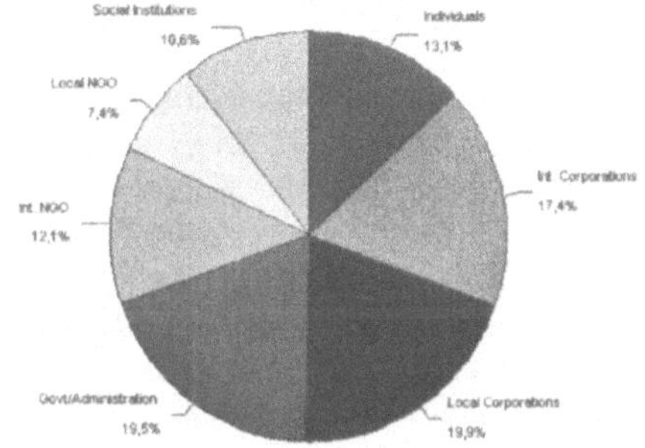

Figure 5.2: Website distribution

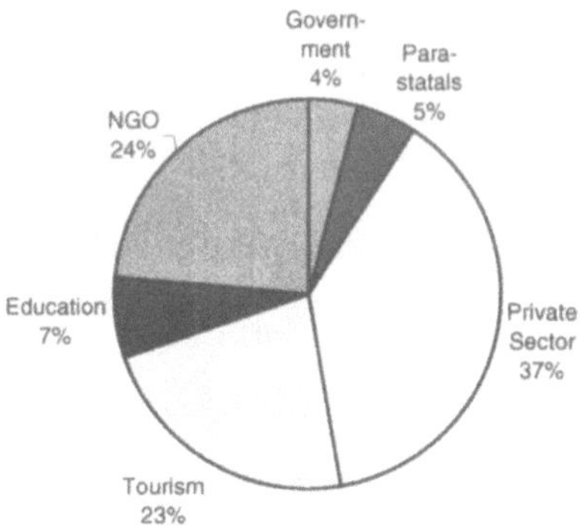

5.2. The Role of Internet Cafes

Internet utilisation started in government agencies, private companies, institutions of higher education and NGOs. Broader ICT diffusion rates could only be observed since 2000, when an increasing number of mainly privately run Internet cafes started to accelerate Internet usage in the country, especially in Dar es Salaam, but successively also in major secondary towns and beyond. Internet cafe operations took advantage of an overall improved availability of

Internet services, special targeting schemes for cafes by the ISPs, particularly Raha.com and Cats-Net,[317] and the removal of import duties and sales taxes for computers and peripherals in 2001 (see chapter 4.3.2.3.). The changing environment for ICT also created conditions for new Internet businesses. Privately run cafes became the dominant form of public access facilities and, therewith, an important place, where the broader population could approach and gather first experiences with the technology.

There are no exact figures on the number of Internet cafes in the country. In 2000, COSTECH estimated a number of roughly 1,000 Internet cafes in the whole country, with about three quarters of them (= 750) in Dar es Salaam (COSTECH 2002). In an ICT survey prepared for Sida in 2001, Miller Esselaar and Associates estimated 300-400 cafes in Dar es Salaam (Miller Esselaar Associates 2001: 17) and, in the same year, the ICT news provider Balancing Act estimated in a special issue on Tanzania about 100 Internet cafes in Dar es Salaam and about 150 for the whole country (BA, News update 125). The largest figure was given by the BMI Techknowledge Group, which estimated a total of 1,542 Internet cafes in Dar es Salaam for 2002.[318] Hence, figures vary between 100 and 1,500 Internet cafes in Dar es Salaam. Though all surveys only attempted to approximate some concrete, and were not designed to provide representative figures, one difficulty in quantifying the number of Internet Cafes might have related to the size and the business orientation of most cafes. The COSTECH survey revealed that most cafes have been fairly small,[319] so that Internet provision was often a supplementary, sometimes even rather hidden business difficult to include in a survey. A realistic figure, also supported by the observations of the author, might coincide with the Miller Esselaar study estimating between 300 and 400 cafes in Dar es Salaam, and with the Dar es Salaam-upcountry distribution, as made by COSTECH, with two third of the cafes in the commercial capital leaving 150 to 200 for the rest of the country.

---

[317] Connectivity for Internet cafes was mainly provided by Raha.com and Cats-Net, who have charged between 25-40 US$ per terminal. Therewith, the size of the cafe was the main variable for billing procedures, largely ignoring connectivity as the main cost factor. Advanced billing models that would charge per bandwidth used are planned, but not yet implemented (Int Sheriff 5.6.2002, Int D'Souza 29.5.2002).

[318] Translating the cafe numbers into user numbers, Tanzania would have, according to the reputable BMI Techknowledge Group, the largest number of Internet users in the SADC region, second only to South Africa (BMI Techknowledge Group 2002: 13).

[319] 63 percent of the total sample of 106 Internet cafes covered in the survey had less than ten computers, only 37 percent had between ten and 20 computers (COSTECH 2002).

The bottom line, however, is the sharp increase of privately run Internet cafes between 2001 and 2003.[320]

The rates for Internet usage have been relatively stable between 500 and 1,000 TSh. per hour in Dar es Salaam, with some hotels charging up to 6,000 TSh. per hour. Upcountry rates have varied between 1,000 and 2,000 TSh. per hour Internet usage. Therewith, the Dar es Salaam prices were remarkably cheap. According to an analysis of cash requirements made in the Miller Esselaar study, these rates were not sufficient to recover the actual costs of Internet cafe operations including equipment, connectivity fee, rent, personnel , electricity and the like (Miller Esselaar and Associates 2001: 46). Similar results were presented by Issmail Nnafie, who states that high competition has driven the costs of Internet access to an unsustainable level, so that most cafes, in fact, have relied on additional services – from beverages to secretarial services – to work profitable (Nnafie 2002: 58). Though unsustainable business operations might have been supported by inexperienced entrepreneurs and the supplementary character of Internet services, the main reason, however, has been the highly competitive environment that hardly left room for price increases. Nevertheless, a market consolidation and an adjustment of prices seems inevitable. Signs of further market consolidation got stronger in recent time, when particularly the larger, better equipped, faster and more reliable Internet cafes gained acceptance. Price adjustments, on the other hand, most likely will not affect the end-consumer, but necessary savings will be made by business reorganisations and, especially, new and cheaper modes of connectivity to be negotiated between Internet cafe operators and ISPs.

Focusing usage patterns in Internet cafes, a prevalent pattern appeared. The main user group is male (roughly 64 percent), young (about 85 percent under 30 with the largest segment between 19 and 25 years), and relatively well-educated (about 83 percent had been to secondary school, roughly 14 percent were degree holders).[321] The main purpose for visiting the cafes is email usage and web-surfing, the former particularly to keep in touch with the family and friends, both in Tanzania and abroad, the latter predominantly relating to various forms

[320] A development that also caught the attention of the international media, e.g. the BBC made a story on a woman entrepreneur setting up a small Internet business (BBC World News 18.11.2002, http://news.bbc.co.uk/2/hi/technology/2487821.stm).

[321] Typical user characteristics of ICT utilisation in Dar es Salaam Internet cafes had been surveyed by COSTECH (2002), Esselar/ Stavrou/ O'Riordan (2004), Nnafie 2002, and Mercer (2002). All of them found the prevalence of a male, young, well educated user group.

of entertainment (Mercer 2002, Nnafie 2002).[322] Some further aspects of Internet utilisation in public access facilities, especially drawing upon differences between rural and urban usage, are discussed in chapter 5.8.2. In this context, it is sufficient to emphasise the role of Internet cafes as engines for increased Internet utilisation. They originated in some new entrepreneurial spirit and benefited from an improving environment for ICT, so that Internet cafe services made the technology available to a broader population. The cafes supplemented previous diffusion patterns by including individuals as beneficiaries to the existing institutional user groups as described above. Thus ICT utilisation was extended, though with a view on user groups and usage patterns in a very focused way. Young, urban, male, fairly well educated took most advantage of new services, mainly to keep in touch with family and friends and for entertainment. Almost like everywhere else in the world, this group are early technology adaptors and became, if services are available, a strong customer basis for Internet cafes. Furthermore, the role of e-mail as major cause for ICT utilisation cannot be overemphasised. For an overwhelming majority, e-mail has been the main reason to visit the sites, other activities were only secondary. Of those that followed, entertainment has been a strong factor for technology usage and ICT based entertainment became a field to test the technology.

5.3. ICT in Governance

The government was the first user of computer technology in the country. The record dates back to the early time of informatics and the first computer installation in the treasury in 1965 (see chapter 4.1.2.). Ever since, the government had a continuous record of computer utilisation, particularly in payroll and pension systems. With increasing availability of ICT in the country, the approach to ICT for government procedures changed and made proactive e-government promotion a central issue on the government's ICT agenda. This was not only because of potential efficiency gains within the administration, but reflected the attempt to make the government a model user for ICT. Since the government focused on the general environment for ICT, the public service, itself, became the only ICT target area other than policy provision or guidance, in which the government could show clear commitment to ICT development. As a result, e-government projects have multiplied and deepened in recent years. Parts of the old administrative systems had consistently been upgraded, new applications had been introduced and a number of new websites improved the

[322] In addition, Internet telephony deserves some attention. Though banned by the regulator, many Internet cafes have offered Internet telephony – with the service often being a factor for sustaining many cafes (Miller Esselaar and Associates 2001: 17).

government's external affairs and communications. For approaching e-government effort in Tanzania it is useful to make a distinction between a mere increase in individual ICT governance applications and more comprehensive and integrated efforts to make e-government subject to a deliberate strategy on the one hand, and applications that target internal working procedures or external government interactions on the other hand (c.f. Heeks 2001, United Nations Department of Economic and Social Affairs 2003).

5.3.1. Approaching E-governance

Forerunner applications of an integrated e-government approach have been the administration of human resources and payroll, as well as an Integrated Financial Management System (IFMS) meant to keep constant surveillance of all financial transactions. With ongoing diffusion of computer technology, the installation of computer workstations equipped with common office applications started at the key government institutions – State House, President's Office (including Planning and Privatisation – the former Planning Commission – and the Civil Service Department), Prime Minister's Office, the Ministry of Finance and the Bank of Tanzania. Subsequently, all major central government agencies increased their ICT utilisation profiles and e-government and potential applications of ICT in government caught the attention of high-level politicians. Several key documents made direct reference to e-governance opportunities and various activities indicated that e-government was perceived as an increasingly important issue. Thereby, the government particularly considered the IFMS as the fundament of a future e-government strategy. In his budget speech 2001, the Minister of Finance, Basil Mremba, emphasised major advancements as accounts of all ministries have been connected to the IFMS and announced the short-term objective to connect all regional sub-treasuries and the long-term objective to incorporate all district councils as soon as possible – with 28 already connected in 2001 (Mremba 2001). In his budget speech 2003, he reiterated the efforts to further strengthen the IFMS as all ministries, central and regional government departments were connected to the system (Mremba 2003). The Tanzania Assistance Strategy action plan referred to improvements in public financial management largely attributed to the introduction of the IFMS (URT 2002a: 4) and restated the objective to further improve the framework for effective data reporting and analysis within government (URT 2002a: 8). The third PRSP progress report highlighted the advancements of the system to assist in the preparation of the budget, to effect payments and to increase efficiency in the allocation of funds needed to procure equipment and services in a more transparent way (URT 2004a: 58), and also the IMF's first review under the Poverty Reduction and Growth Facility acknowledged advancements in the government's data processing capacities that made use of a computer network aiming at better communication between central government and regional or local government (IMF 2004: 36). At the same time as the IFMS was gradually extended, new payroll and human resource systems had been installed at the Civil Service Department in 2000,[323] ICT government applications gained new recognition[324] and, accordingly, e-governance activities proliferated further. Senior government officials participated in the global e-government for development conference held in Palermo in April 2002 to seek input of how to streamline e-Government in the process of economic and social development.[325] Increasingly, various sectoral policies targeted ICT related actions, such as the Small and Medium Enterprise Development Policy that called for government action aiming at the establishment of business centres to provide ICT services for SME or support SMEs in the acquisition and adaptation of technologies (MIT 2002: 19), or the National Trade Policy adopted in 2003 that concluded: "The key to better performance in the services industry lies in the adoption and popularisation of ICT in enhancing performance in services delivery. Tanzania has the potential to build on the rapid expansion in trade in services" (MIT 2003: 12). The national trade policy, approved in 2003, demanded expedited computerisation of the tax system (c.f. MIT 2003: 40), an objective already envisaged by the Tanzanian Revenue Authority (TRA)[326] that considered a computerised and effective administration with strong enforcement capacities and with an integrated approach to the administration of taxes as a prerequisite for a modern tax administration – an objective that, according to TRA's corporate plan 2003, shall be accomplished by 2008 (c.f. TRA 2003: 2).

The various statements reflected an increasing awareness of e-government, and justified individual applications in different areas. On the other hand, they also showed that the various approaches were still rather compartmentalised and fragmented. In order to streamline ICT for government procedures and to offer some strategic leadership, two approaches can be identified. Firstly, a general outline of the overall objectives was provided in the ICT policy of 2003, and,

---

[323] The integrated human resources and payroll systems covered 280,000 public servants and cost around US$ 6.5 million. According to a 2002 Comnet-IT study, "the savings already accrued in improved management, reduced ghost workers, improved control and accuracy, mean that the project has already paid for itself" (COMNET-IT 2002).

[324] On the occasion of the launch of the modernised payroll system in April 2000, President Mkapa praised the interlinked payroll – IFMS as "the first of their kind north of the Limpopo, and south of the Mediterranean Sea" (quoted in: IT Vision [Dar es Salaam] Issue 01).

[325] The Tanzanian delegation was headed by Salim Msoma, permanent Secretary in Ministry of Communication and Transport. For conference details, see http://www.innovazione.gov.it /eng/egov_sviluppo/palermo_2002/index2.shtml

[326] TRA's website can be found at www.tra.go.tz.

secondly the technical capacities to implement ambitious ICT projects were extended. The Civil Service Department, which had already been responsible for most computerisation in public administration, developed an overall approach to e-government integrating activities in different target areas into one overall approach that included network design, as well as hardware and software solutions (Int. Sawe 11.4.2003).

However, it was firstly the ICT policy that clarified the basic approach. It divided the various past approaches into main project fields and, thus, tried to introduce them as six different focal areas of a structured e-government strategy. These were:

1. Integrated Financial Management System,
2. Personnel Control and Information System ,
3. Establishment of national TV Tanzania,
4. Tanzania Global Development Learning Network,
5. Creation of government websites,
6. National Payments System (Moct 2003: 7).

Furthermore, it included a number of policy statements outlining the general roadmap to e-governance. According to the proposed actions, taken from different parts of the ICT policy, the government will:

- be a model user of ICT by deploying ICT systems within the public administration itself to improve efficiency, reduce wastage of resources, enhance planning, raise the quality of services and access global resources.
- support the application of ICT to promote good governance, transparency and accountability, and awareness of the implications of long-term ICT investment and total cost of ownership
- deploy ICT extensively to strengthen law enforcement, security and national defence capability
- deploy ICT to monitor and respond to environmental disasters and to collect and disseminate information on environmental problems
- review its operational processes and institutional structures with a view to making them amenable to ICT application and deployment.[327]

[327] Taken from the ICT Policy 2003 (MoCT 2003), compiled by Sawe (2004).

Thus, e-governance has been considered as a long-term task and an incremental process. The stated objectives are ambitious and mirror the international discussion on the potentials of e-governance focusing on internal workings of government by data processing and supporting external workings of government by processing and communicating data (Heeks 2002: 4, see also Bertelsmann Stiftung 2002, InfoDev / The Centre of Democracy & Technology 2002). Though all policy objectives target a more efficient service provision by the public administration, they are very different in their character either calling generally for good and transparent governance or targeting specific application areas like defence and environmental management.[328] Thus, identifying the main e-government statements of the ICT policy, shows that side-by-side approaches still prevail. The lack of coherence also pointed towards the infancy of the e-government implementation process in the country. Since the general government announcements did not go beyond some common places, it makes sense to look at e-government from a different perspective. This will focus not so much on how the potentials of e-governance have been perceived, but how computer applications have evolved in public administration, because the use of computers in public agencies is, of course, much older than the discussions on e-government, which is a rather new term.

5.3.2. Public Service Reforms as Origins for e-Governance

E-government activities have been under the supervision of the Civil Service Department. Apart from the ministry of finance, the CSD was the first government body dealing with information technologies and they were in charge of the installation of different computer applications within the public service, starting from the payroll and human resources systems concentrated in a few central offices, to broader moves towards office automation among civil servants. Setting the course, the CSD evolved as a recognised protagonist of ICT. The CSD, and therein the Management Information Systems Department headed by David Sawe, developed into a hub for all government related ICT activities. In addition, all important strategy papers that paved the way to the national e-government initiative endorsed by the cabinet in April 2004 (see below), were drafted within that department.

The evolution of the CSD was closely associated with general approaches to public service reform. First attempts to make the public service more efficient already commenced in the late 1980s, coinciding with the general shift in the Tanzanian development approach. Thereby, the reform of the civil service

[328] The notion of defence can be explained by the history of the current e-government approach, which is built on a military communications network (see below).

became a priority, and a Civil Service Reform Programme (CSRP) was launched in 1991. As the project failed to deliver the expected outcomes,[329] the Public Service Reform Programme (PSRP) replaced the CSRP in 1998. First project measures commenced in 2000. The programme aimed at increasing the government's efficiency and overall improving the performance and service delivery by inter alia adjusting the size and affordability of government operations, introducing meritocratic recruitment procedures and to increase the responsiveness of government services (URT 2000b). The CSD was mandated to oversee and implement the reform. To this end, the CSD was divided into several subdivisions, among them the Department for Management Information System, which developed an information system strategy which subsequently became the core of the e-government approach. This identified a lack of accurate, reliable and timely information as one major impediment to an effective civil service, and aimed at "reversing this situation by exploiting opportunities offered by new information technologies (IT) (...). Further, the component aims to revolutionise public service delivery systems by using IT to catalyse the re-engineering of business processes" (CSD 2002). The strategy was linked to the individual measures already conducted under the CSRP,[330] and streamlined them into one approach comprising the following core systems to be implemented in a coordinated and incremental manner, starting with State House and the four central ministries – CSD, the Prime Minister's Office, Treasury, and the Planning and Privatisation Department (CSD 2002):

- Personnel, payroll and pension systems,
- Paper-based records management systems,
- Budgeting and planning systems,
- Expenditure and revenue systems,
- Estates, assets, and inventory management systems, and
- Audit and inspection systems.

[329] The CSRP implementation report of December 1997 identified shortcomings in capacity building, weak coordination of individual project measures and inadequate resource allocation as major reasons for CSRP's failures (c.f. Teskey / Hooper 1999). An assessment of capacity building in the civil service in the late 1990s can be found in: van Arkadie et al. (2000).

[330] Notable initiatives under the CSRP had been (i) personnel control and information systems project, (ii) government accounts development programme, (iii) national website, (iv) establishment of a common IT standards committee, (v) institution-specific management information systems projects in various ministries and (vi) records management project (CSD 2002).

All major IT-components within the PSRP targeted working procedures within the government. Though the individual core systems comprise complex software solutions, the scope remained rather limited and included a handful of key agencies only. Notwithstanding the responsibility of each agency in the implementation process of tailored software solutions, the process remained under guidance of the CSD (PSRP 2004). To this end, the CSD offered IT consultancies, strategy recommendations, creation of ICT awareness or advising and giving specifications on procurement, installation and upgrading of IT equipment (PSRP 2004). Thereby, the CSD pursuit a threefold strategy: Firstly, targeting a focal group for enhanced ICT utilisation,[331] secondly extending the programme to other central government agencies, and, thirdly, linking ICT to the Local Government Reform Programme, particularly with regard to human resource databases and payroll data (PRSP 2004). Therewith, the fundament for ICT in governance was introduced. Though still limited in scope, individual applications already brought some success. This was highlighted by Mary Nagu, Minister of State, on the occasion of a SADC seminar on e-governance, held in Dar es Salaam in September 2004, when she recalled some ICT experiences in public administration.

> "For instance, between the President's Office for Public Service Management and the Ministry of Finance, we have an initiative to jointly manage the deployment and payment of all Public Servants through an Integrated Human Resource and Payroll Management System. While simply starting out to install the new system, we were able to identify and delete close to ten thousand ghost workers. Simultaneously we detected a sharp increase in the number of deletion requests coming in among the routine monthly payroll amendments. After commissioning the new system in June 2002, in addition to efficiency gains and greater accuracy of payments, we also gained notable benefits in the reduced volatility of our annual wage-bills against budgeted amounts, partly due to tighter controls on expenditure and partly to more accurate budgeting data" (Nagu 2004).

Thus, ICT as a tool for governance became increasingly recognised. Running installations were increasingly considered a success and further ICT roll-out has been among major intervention during the 2003/04 PSRP implementation schedule. By 2004, the core system, the so-called Performance Management System that includes undertaking situational analyses, planning assistance tools and monitoring and evaluation applications was installed in ten lead ministries and 24 additional government agencies (URT 2004c: 2-1). ICT procurement,

[331] Thereby, a prioritisation of institutions occurred: State House, Planning and Privatisation and the President's Office – Public Service Management became primary targets. Emphasis was placed on empowering them to establish Management Information System units, building capacity to handle ICT tools and to develop sound ICT strategies and make sure that ICT standards, policies and security are put into practice (c.f. PSRP 2004).

installations and capacity building measures were among the four largest budget items in the approved PSRP budget for 2004/05,[332] indicating that ICT utilisation in public administration will continue to be a priority area in the programme. Despite this progress in PSRP and the steady increase of ICT in governance, the focus was still on central government agencies. Efforts to target a broader approach came only with the proposal to integrate the various applications into a country-wide government communications network, that would make ICT tools available to regional and district administrations. The proposed upgrading of existing network capacities became the fundament of government's e-government initiative launched in 2004.

5.3.3. The e-Governance Initiative

On 20 April 2004, the cabinet approved an approach to develop the infrastructure for e-government covering the whole country with an array of different applications. The approved strategy has initially focused on the following points:

- establish a focal point to coalesce and reinforce existing, but fragmented and independent, e-Government initiatives;
- build a national wide-area network to carry efficient voice and data communications among all Government offices; and
- create awareness of these opportunities across the public service (Nagu 2004).

The proposed government network, that is, the hardware component of the project, was the explicit fundament of the initiative. A joint communication platform promised new opportunities of integrated electronic networking, vertically among government agencies on the central level, horizontally between headquarters and local branches, integrated in a coherent approach customised to the to different needs of government agencies at the central, regional or district level. Thereby, building the hardware infrastructure has been regarded as prerequisite for all follow-up activities. The network would be based on an existing satellite hub, used by the Ministry of Defence, that came into being after many years of cooperation between the government of Belgium and the Tanzanian Ministry of Defence.

[332] For example, the Management Information System and Technology component accounted for 11.1 percent of the overall budget and totalled roughly US$ 2.8 million (c.f. URT 2004b: 2-2).

The Belgian government had assisted the Tanzanian armed forces in improving their communication capacities over the last ten years. The project started with radio-communication and was subsequently upgraded to satellite communication. As it became apparent, however, that roughly 90 percent of the installed communication capacities were not used, the Belgian government insisted on opening the network for civil applications. In 2002, the Belgian side received a proposal for a third phase of the project, worth some US$ 60 million, to considerably upscale the whole project and develop the network as fundament for broad government usage (Int. Brems, Int. de Schelleskens 17.7.2002). The proposal, written by Siemens, submitted by the local technology firm MEECO International as subcontractor and mutually agreed between the Tanzanian government and the constructors,[333] offered a large scale state of the art communication network. It would be built upon the existing systems, particularly in the ministry of finance and comprise a network management centre at the CSD, from where it would link up to the satellite hub of the ministry of defence and national services (MoDaNS) to provide connectivity to 22 regional and between 71 and 120 district headquarters. Additionally installed local area networks would forward connectivity to 22 regional hospitals, other health facilities, the police or local branches of the Tanzania Revenue Authority (TRA) and other agencies to supply them with voice and data services. A mix of fibre optic and microwave radio link connectivity would provide data links to an overall 21 ministries and government agencies (see figure 5.3).

The realisation of this network would be a quantum leap for e-government in Tanzania. Though it must be left open, who initiated the government network in the first instance, the government or the constructor, the submitted proposal based on a thorough needs assessment, was deliberately tied to the ongoing project activities in the ministry of defence, and took into account the Belgium demand for civil use of the military system. In sum, the proposal offered a very attractive package hardly to be disapproved by the Belgium government. However, some critical remarks can be made, too:

1. The government network implies a complete new telephone and data network, that would bypass existing capacities, and would, to a large extent, double the efforts of upcountry infrastructure roll-out. It also excludes emerging local technology companies from project participation, let it be TTCL or other data communications providers.[334] Furthermore,

[333] Both partners had already realised the first two phases of the project, for an overview see Siemens (2003).

[334] Particularly TTCL objected the deal, emphasising the negative effects on local companies. Since the project would transfer a significant amount of telephone and data traffic to the government network, TTCL would loose one of its major customers. However, all lobbying

the project design called for insourcing of a complete national data and voice backbone, so that the government would become one of the largest communication operators in the country, again.

2. MEECO International has been the local subcontractor of the project. The company is also shareholder in Zantel, the second national operator from February 2005 onwards. Thus, it should be asked, if MEECO is allowed to take advantage of the government network and, for example, use the capacities for their own deployment strategy on the mainland – this would be a considerable subsidy to assist Zantel market entrance expected in 2005.

3. Government procurement regulations normally require an open tender for all purchases over US$ 5,000 (c.f. BMI Techknowledge Group 2002: 3); however, there never was a public bidding process with regard to the network or components thereof.

Figure 5.3: Proposed Government Network

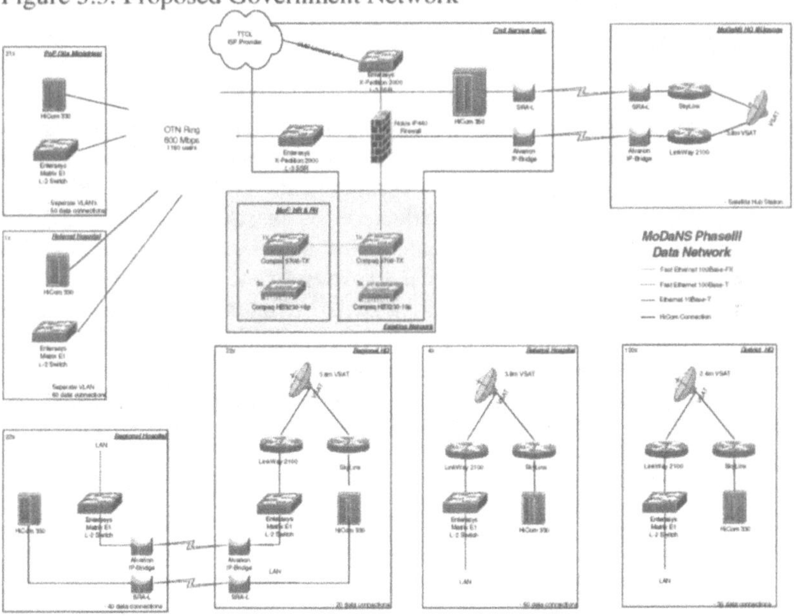

Source: Sawe (2003)

---

against the project, especially brought forward before any settlement in the TTCL privatisation dispute by the then TTCL CEO, Fred van der Voort, did not bring any concessions, so that today's project approach would for the most part burden TTCL.

Thus, there are some pending issues relating to the details of the network: what are the likely consequences of a strategy that would bring the government back in the operative business of communication services, what has been the role of local subcontractors participating in the project and what will be the impact on the general business environment, when the role of government is extended to communication service provision, again, and, thus, contradicting the spirit of an intended private sector led growth strategy? Notwithstanding the pivotal relevance of these questions, the cabinet approval in April 2004 cleared all formal project hurdles, so that project implementation can be expected in the near future, and all possible reservations have been deferred for the time being. The overall attractiveness of the project might explain why the government has opted for an almost complete new network offering reliable voice and data connections within, to and from the central government level, and thus, providing new opportunities for government communications.

The proposed network will only be the hardware fundament. Therein included is a telephone network, which alone will strengthen the communication links between government agencies. Beyond basic voice services, the provision of data communication services will be the second major pillar of the project. This will be a major challenge, because the joint hardware infrastructure must be complemented by a joint software infrastructure, that needs to address security issues and must distinguish between individual applications tailored for specific usage, only, and joint interlinked communication platforms used by various agencies for the same purpose simultaneously (Int. Sawe 11.4.2003). Figure 5.4 gives an overview on the intended e-government architecture. All applications will be based on a joint hardware and software fundament, on which pillars of individual e-government applications will be built. These comprise already existing applications as developed under the PSRP since 2000 – human resources, payroll and financial management – and new applications targeting sectoral activities – education, health and agriculture – to allow the sectoral ministries to make use of individual applications in support of their sectoral mandate and goals.

By now, some advanced e-governance applications have been installed in only a number of key government agencies, traditionally equipped with the better financial and human resources. In contrast, the sectoral approaches to ICT, for example, in health, education or agriculture, are in their infancy, that only started to move horizontally towards the lower ends of government administration. Furthermore, bearing in mind that ICT in itself cannot reshape government functions, this can only be done by those who apply the technology, the work still to come is enormous and the greatest challenge might appear with regard to existing gaps between the e-government design and the public sector realities. In fact, Richard Heeks has already identified these gaps as main causes for e-government failures in Africa. He stated that such gaps arise particularly "because e-government concepts and designs have their origins in the West; origins that are significantly different from African realities. Some best practices are outlined, that may help to close design-reality gaps and, hence, may help to improve project success rates. This will only happen, though, if they too are appropriate to African realities" (Heeks 2002). Only if e-government applications, however useful they might be in theory, are cautiously and deliberately introduced, based on well-defined targets and appropriate sector strategies, if there will be sufficient capacities to implement them and, thereby, recognising the ways the public sector is actually working, there will be some prospects of success.

Figure 5.4: Approach to e-Government

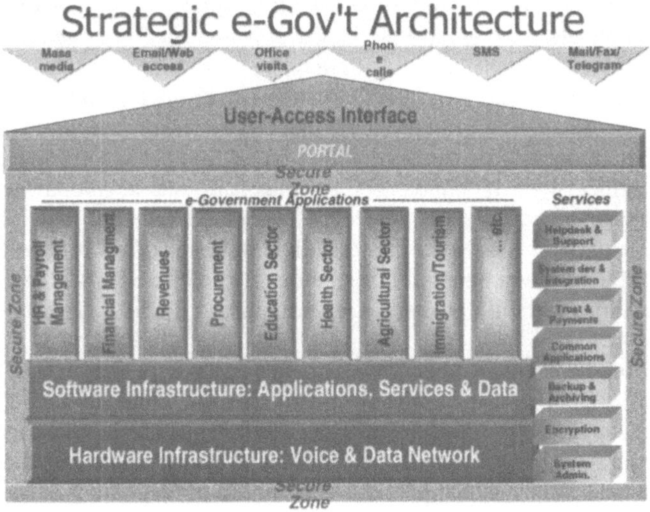

Source: Sawe (2004)

5.3.4. Supplementary e-Government Approaches

The proposed network has been only one, though decisive, step on the Tanzanian e-government agenda. There have been a number of supplementary e-governance projects that, only if viewed together, reveal the actual scope of the endeavour. Projects have included different objectives from administrational procedures to transparent governance. Furthermore, in some instances, ICT had already been introduced to the local government level, indicating that even a poor country like Tanzania can deal with different project levels at the same time.

Firstly, the number of resourceful government websites has increased gradually, today all major government agencies have built an attractive and capable web presence. Among them the official government website (www.tanzania.go.tz) that presents an overview of all major government agencies;[335] various ministries like finance or communication (www.mof.go.tz, www.moct.go.tz) and State House (www.statehouse.go.tz) have developed a resourceful web presence. The parliament offers a document retrieval system in which all acts of parliament, back to the early 1960s, are available online (www.parliament.go.tz). In- depth information about the Public Service Reform Programme can be found at www.estabs.go.tz, the Tanzania Communications Regulatory Authority offers plenty of information at www.tcra.go.tz and the Tanzania Revenue Authority give details of their work at www.tra.go.tz. National school examination results are regularly published at www.tanedu.org and the National Board of Accountants and Auditors publish their exam results at www.nbaa.or.tz. Thus, in the end of 2004, almost all major government agencies publish some detailed information on their work, their corporate plans and provide contact details for further inquiry. In addition, the information portals TanzaniaOnline (www.tzonline.org) and the Tanzania development gateway (www.tanzaniagateway.org) provide online guidance and full text retrieval on various development issues, so that the dissemination of information is easier and knowledge of government policies more spread than only two or three years ago, increasingly with bilingual services available, in English and Swahili.

Secondly, the Tanzania Global Development Learning Centre (TGDLC), was established in 2000 as a government distant learning centre (www.tgdlc.go.tz). The centre, supported by World Bank, has been planned as an integrated part of the PSRP, targeting mainly high- and mid-level decision-makers and endowed them with state of the art technologies ensuring up-to-date and timely sharing of knowledge, skills, and attitudes.. The TGDLC offers computer courses, mainly typical office packages – Word, Excel, PowerPoint – but the core functions relate to executive classes via videoconferencing on various development relevant topics,[336] aiming at contributing to capacity building for social and

[335] The initiative of a government website dates back to 1996 when President Mkapa directed Planning and Privatisation as the lead agency to coordinate the setup of a national website. Officially, the website was launched on 9 December 2001 (IT Vision [Dar es Salaam] Issue 04).

[336] The course schedule for July – December 2002, for example consisted of the following courses: Fighting HIV/AIDS through Information and Strategic Communication, Youth for Good Governance Programme, Community Driven Development, Jungian Psychology,

economic development. In the first two years of operation, the centre had about 600 course participants (Int. Senkondo 30.7.2002). Considering the design – two computer labs, one of them equipped with videoconferencing technology – and the capacity utilisation, the centre is obviously an elite tool with direct benefits to a few course participants only. Beyond, however, it addresses main goals of the PSRP, that is to increase the professionalism of high- and mid-level civil servants level. Thus, the TGDLC functions as a supplementary tool for human capacity building; it strengthens the information retrieval and exchange of decision-makers, it disburdens them from time-consuming conference travels and it recognises that course participants are important multipliers of the content supplied to make it applicable to their local job environment.

A third layer of e-government approaches has related to local e-government projects on the district and municipal level. These are meant to explore the potentials of ICT for local administrations. Projects are spearheaded by the Tanzania Commission for Science and Technology (www.costech.or.tz). Since the early 1990s COSTECH has been a pioneer of ICT in Tanzania, when it started to host the first Fido node for email communication in 1992 (see chapter 4.1.2.3.). After the business of ICT service delivery was taken over by the private sector, COSTECH shifted towards a more project oriented approach, especially exploring the impact of ICT in rural and low end utilisations. For example, the commission has been the lead agency in the set-up of the Sengerema telecentres or the Kasulu Internet Project (see chapter 5.8.2.). With regard to governance, COSTECH advocated the introduction of ICT to local governments following the objective to enable them to create and manage information through electronic means for good governance and to integrate ICT in administrative procedures. Two main projects have been implemented under this scheme, the Kinondoni Municipal Council and the Zanzibar Municipal Council project.

Kinondoni, a Dar es Salaam municipality, was chosen as a pilot project, to prepare a model approach of ICT for good governance at the local level – following the recommendations drawn during a Dar es Salaam roundtable meeting on ICT organised by IICD, which later also provided main support for the project. The project commenced in 1999, when the municipality acquired first computers and conducted related staff training. The first phase until 2001 focused on statutory registration, as well as health and education records; since 2002, the project has been expanded to include more target areas, namely property, tax administration, land administration, business licence, commerce,

---

Climate Change: An African Perspective, Economic and Business Journalism, Introduction to Business Finance or Public-Private Partnership for Urban Environment.

rural development, waste management, economic planning, human resources and financial management (COSTECH/IICD 2002: 9).[337] A parallel approach was chosen for the Zanzibar Municipal Council project, which was implemented between 2001 and 2002 with UNESCO and DANIDA as main donors. The project provided ICT infrastructure and user training to enable five departments of the municipality to create, keep and exchange information electronically, among others, land rates, requirements of obtaining various business licenses and investment opportunities, were made available to the public through the municipal website (http://www.zmc.go.tz).

Following a fairly modest approach to e-governance, both projects have, according to COSTECH, improved the day-to-day operations. Immediate gains related to overall simplified administrative procedures, particularly with regard to record keeping and an improved information flow. Registers, for example, covering deaths, births, agricultural production, health data or school inventories and student numbers became a more reliable data base for municipal planning. New equipment significantly improved routine services, like business registration or license renewals. Henceforth, data sets are entered once and are available for further and continuous processing, avoiding lengthy procedures of handwritten forms that have caused long queues for hours in front of offices (Int. Mlaki 23.4.2002, Int. Nguli 10.4.2003). Thus, a relatively modest input led to good improvements in the overall service delivery.

At the same time, COSTECH tested the scalability of this approach and started a village ICT pilot project covering two villages Lunga-Lugoba in Bagamoyo District (www.lungalugoba.or.tz) and Dakawa in Morogoro District, both villages with between 3,000 and 4,000 population. The pilot attempted to bring to the community affordable ICT to assist the village government functionaries in utilising a village information system for village governance and to support the ward[338] to maintain and consolidate overall statistics for health, education and population distribution.[339] Though first experiences with the installed

---

[337] For details see also the Kinondoni website available at http://www.kinondonimunicipality.go.tz

[338] Tanzania has adopted a three-tier administrative structure comprising of the Central Government, the regions and the Local Government at district level. Each district, there are 106 districts in Tanzania, consists of a number of villages and is grouped into wards. The ICT village pilot serves two basic village units: Lunga Village (in the Lugoba Ward) and Wami village (in the Dakawa Ward).

[339] Typical administrative activities include: education records, health and agriculture data or collection of water dues and other revenues, such as water dues of 100 TSh. per person per month; 100 TSh. per visit to dispensary with treatment or 200 TSh. daily licence for market space (90 per cent of revenue goes to the District and the Village retains 10 per cent), for a

demographic information (houshold register) database must be awaited, the main obstacle has been a lack of funds (Int. Nguli 10.4.2003), COSTECH has been fairly optimistic on project prospects, because of easy-to-use and well adjusted hard- and software solutions and the enthusiasm of the villagers willing to make the pilot a success.

5.3.5. Concluding Remarks on e-Government

Subtracting the visionary element of e-governance that often comes with a notion of fully automated administrative procedures and complete online interaction between government and citizens, the term e-governance describes an evolutionary process that starts with the provision of basic ICT utilities like telephones, computers or printers from where it can be developed further, gradually targeting more sophisticated stages of advanced technology and service levels – wherever they are considered to be appropriate and can be implemented on reasonable terms. The examples have shown that e-government is no luxury concept, well suitable for the developed world, but unfitting for poor countries ostensibly not ready to exploit technology for better governance. It is not the point to withhold advanced technology solutions from the developing world, but to find ways how to make most of new ICT related opportunities. Thereby, suitable technologies, available resources and capacities are important factors for e-government success. However, above all it is the human factor, that is, the attitudes, knowledge and skills of those in government, administrations and the civil service, who actually have to apply the technology, will prove most critical for the prospects and sustainability of e-government initiatives.

Bearing that in mind, the Tanzanian e-government approach has strengths and weaknesses. On the positive side, e-government has been closely linked to broader approaches of public service reform targeting improved public service delivery as an overall goal, where introducing ICT in governance has been only one component of the reform agenda. Secondly, the emergence of an e-governance agenda benefited from a fairly strong strategic leadership, mainly originating from the intellectual input from the CSD, where the concept of e-government, or the translation of a general concept into a Tanzanian approach, has been developed. The persuasiveness of this preparatory work, presenting a coherent approach to ICT in governance, raised awareness on the issue in the first instance, particularly among senior government officials, who started to

---

detailed project description see the text (anonymous and without date) at http://portal.unesco.org/ci/en/file_download.php/191643ab9c5791b69b337f3893c5799bTanzania+Report.doc.

address the issue in the individual ministries, departments and agencies to seek out the potentials of ICT in their work environment. Thirdly, advancements in e-government have followed an evolutionary concept comprising different approaches and methods. Starting with individual applications in some key government agencies, subsequently including the line ministries and the senior government level, the scope of applications has been successively extended and supplemented by other new means like an increasing availability of resourceful government websites, the establishment of TGDLC as a government learning centre and implementation of pilot projects targeting the lower ends of government at the district, and in individual cases, even the village level. Lastly, the proposed government network, approved in April 2004, marks a new quality of e-government, because it significantly increases the potential scope of e-government activities, includes regional and district administrations to government networking and can be utilised according to the needs of each agency connected, from basic telephone services to advanced ICT solutions. Thereby, the government became one of the biggest providers of local content and, thus, an important catalyst for ICT, which has also fostered further developments in the local ICT industry.

However, despite the undisputed progress in the last years, there are some weaknesses and a number of major challenges. Firstly, restrictions must be made with regard to the proposed network. Though there can be no doubt that the government acted deliberately to realise the project, the story behind the project is largely untold. Almost the complete planning process, the consultations between key government players, as well as between the government and the constructor, necessary to develop such a large scale and well-fitting network, had a rather disguising character that presented both the financier and other local market participants with a fait accompli. Therewith, the government not only missed the opportunity to include emerging local IT companies in the project, it also approved the building of a complete new infrastructure, to a large extent doubling the efforts of upcountry infrastructure roll-out. At the same time, when the Minister for Communications and Transport requested from the large parastatals TRC, Trazara and Tanesco to make their underutilised data capacities available for resale agreements with the private sector (c.f. Mwandosya 2003), the government itself has been in the process of creating significant new data capacities without any access opportunities to private data or ISP providers to assist their upcountry business development. Thus, the network project has been largely unresponsive to the needs of the local private sector, and, in fact, might tend to reduce the business opportunities for private sector data providers, because the most promising customer segment on the regional and district level comes from the public sector, and these will make their businesses with the new government operator.

The major issue, however, will be to broaden and deepen the impact of e-governance. The existing ICT infrastructure already contributed to improved and more transparent government procedures, but both, the development of new tools and the gradual geographic expansion in the availability of ICT, will be a challenging task. Today's ICT solutions in governance are hardly more than small project islands that – however successful in the individual case – and will be difficult to replicate taking alone the vast number of public agencies. Apart from continued donor commitment, that will be critical for success, the challenge will be to prioritise and sequence different project components to identify a sustainable cost ratio balance and to find the right relationship between the high ends, such as advanced ICT applications, and the low ends, such as the provision of basic communication services, to increase the positive effects of ICT in governance.

5.4. ICT in Health

According to the government, health is one of the priority sectors. The National Health Policy of 2002 asserts the goal to improve the health and well being of all Tanzanians by ensuring "an effective, efficient, and quality curative, preventive, promotive and rehabilitative health service delivery at all levels of the health system" (c.f. MoH 2002: 4).[340] To this end, the government has approved a number of policy strategies, mainly aiming at improving what is labelled basic essential health care, which, among others, seeks to ensure the accessibility and equitable distribution of health care and health resources. Focus has been put on reproductive and child health, prevention and control of endemic, epidemic and pandemic diseases, fighting malaria, HIV/AIDS and other appropriate health care packages for treatment of common diseases and injuries (MoH 2002: 9). Thus, primary health care has been at the fore of the government's health agenda. The role of information and communication technologies has been only a secondary issue, however, with a growing recognition, particularly with regard to health administration and education. This has been why a growing number of ICT for health projects have been launched in different areas over the past years. Most of them are in early stages, so that it is still too early for conclusive outcomes. Project approaches differ considerably

[340] Under the current administrative set-up, health service provision is divided into three levels – national, regional and district. Under supervision and administration of the Ministry of Health, the national level is made of consultant referral hospitals, training institutions and executive agencies, at the regional level, the provision of health services is vested with the regional administrative secretaries and regional hospitals, and, finally, at the district level, health services are provided by district hospitals, health centres and dispensaries (MoH 2002: 17). There are eight consultancy/specialised hospitals, 17 regional hospitals, 16 health centres, and roughly 663 dispensaries in the country (c.f. Fahamu 2002: 99).

and, at least, three different project types can be identified: Firstly, projects that address the general environment for health. These are mainly government projects dealing with aggregated patient data, administrative management procedures, staff matters or drug storage and supply. The major project has been the MTUHA Health Information System[341] with support from DANIDA. The MTUHA serves as the ministry's national health information system, and has been introduced by all districts. Secondly, medical programmes in the narrower sense, hosted at the major hospitals aiming at improved medical treatment and research cooperation, and thirdly, a range of donor projects attempting to set up pilot projects to investigate the potentials of ICT in health for individual issues like fighting malaria or AIDS/HIV or ICT health applications in remote and disadvantaged regions.

First level projects seek to improve the general environment for health care delivery. The MTUHA Health Information System, broadly aimed at designing, developing and implementing computer-based health information systems for primary health care in the country, is meant to become the health pillar of an overall multi-sectoral e-government approach. As paper-based system, MTUHA was introduced nation-wide between 1994 and 1997 to improve health services through the provision of necessary and sufficient information needed by health managers to monitor, evaluate and plan their activities. Since 2002, first district health facilities were equipped with computers, in 2004 the Ministry of Health has installed computer aided health information programmes in all districts.[342] The main objective has been to improve performances of the health sector through the promotion of appropriate analyses and use of health information for decision-making. Specific measures have included the improvement of data collection, strengthening the capacities of health workers to validate, analyse, and interpret health data and the enhanced dissemination of information to other stakeholders with emphasis on feedback to the lower levels (c.f. Fauteck 2004). However, the practice has diverged considerably from this claim. The official health review of 2003 stated:

> "The Health Management Information System in place (MTUHA) is not performing well. Data are very patchy in coverage, are of unreliable quality, and are not collated, published and disseminated promptly enough to be of much use in planning and management [...]. Data that are not used at the facility level for management purposes are hardly likely to be collected with much enthusiasm. MTUHA training concentrated on the data collection dimension rather than data use for management purposes. It was

[341] The Health Management Information System (HMIS) (Swahili: Mfumo wa Taarifa za Uendeshaji wa Huduma za Afya (MTUHA).

[342] According to Heiner Fauteck, representative of the Tanzanian German Programme to Support Health in a private email dating from 16.12.2004.

widely agreed that MTUHA should be less comprehensive in the scope and volume of data to be collected, and that further training should be oriented towards information needs regarding decentralised health service planning and management" (URT 2003d: 4).

"Less comprehensive" is the key phrase of the statement. If ICT projects are to be deployed in health management, and, thereby, particularly targeting the district and community level, applications should be designed in accordance with the available capacities and well acquainted with the ways, the health system works.

The second major approach on the ICT for health agenda has been networking for health research. Building networking capacities, that is, putting in place means for the retrieval and exchange of health information, has been limited to a number of research centres, so that expanding research capacities by increased ICT utilisation have been recognised as a relatively easy to deploy project type. One of the first projects fell under the Ministry of Health, which has coordinated an academic research network connecting Tanzania to neighbouring Kenya and Uganda as well as to research centres in Europe and the United States.[343] The project commenced in 2000 and has been funded by various medical research centres in Scandinavia and the US to improve the data situation on some of Africa's most serious infectious diseases like malaria, HIV/AIDS, and others, to bring together the research resources of first class medical centres in the north with the knowledge of centres in the south that are closer to the source of the diseases (IT Web [Johannesburg] 10.4.2001).

Research activities that comprised some kind of ICT based networking have also been hosted at the National Institute for Medical Research (www.nimr.or.tz), the Amani Research Centre in Tanga (http://www.amanet-trust.org), the African Malaria Vaccine Testing Network (www.amvtn.org) and the Commonwealth Regional Health Secretariat for East, Central and Southern Africa based in Arusha (www.afronets.org/crhcs.php) (c.f. Fahamu 2002: 99). Explicitly focusing on ICT, the Muhimbili Health Exchange Forum (MUHEF) was created in 2003 as an initiative to facilitate communication and the flow of health information between health-workers in the districts and regions and health experts and postgraduate students of Muhimbili University College of Health Science. The project, funded by the GTZ, makes use of a website that provides

[343] The central data warehouse, from where the collected data can be sent overseas by the Internet, has been located in Dar es Salaam. According to a representative of the main hardware supplier of the project, the decision to chose Dar es Salaam as a main hub for the project was largely due to the relatively advanced IT infrastructure in Tanzania compared with the situation in Kenya and Uganda (c.f. IT Web [Johannesburg] 10.4.2001).

in-depth information ranging from treatment to facility management, discussion forums, online articles and an ask-a-question section.[344] For district health agencies without Internet connectivity, a CD-ROM version of MUHEF is distributed twice a year.

Institutions of higher health education have also extended their ICT capacities. The four medical schools in Tanzania, the Muhimbili University of Health Science of UDSM (www.muchs.ac.tz), the Hubert Kairuki Memorial University Dar es Salaam (http://www.hkmu.ac.tz), the Kilimanjaro Christian Medical College in Moshi (http://www.kcmc.ac.tz), and the Vignan's International Medical and Technological University, have run programmes to setup ICT capacities for research communication and networking (c.f. Fahamu 2002: 99). The most advanced status can be assigned to Muhimbili, which has been part of the ICT capacity building projects at UDSM (see chapter 4.3.6.). With funding from the GTZ, Muhimbili also offers a master programme in public health that includes a considerable distant learning component in collaboration with the Department of Tropical Hygiene and Public Health at the University of Heidelberg (Int. Mey 18.6.2002). Furthermore, Inwent (Capacity Building International) and the Ministry of Health are currently developing a modular diploma course for Council Health Management to be taught at zonal training centres. The six core modules are health sector reforms and district health systems, promoting partnerships in the district, management of health resources, quality management and multisectoral disease control. According to the current plans, a number of modules will have a distance learning element.[345]

Individual programmes run under specific donor portfolios make the third type of ICT based health projects. They range from single health components within a broader approach like IICD and COSTECH's piloting ICT for local government project or a health component within the Finnish RIPS project located in Mtwara and Lindi (Int. Eirola 8.7.2002, Int. Moens 18.7.2002). More focused ICT for health projects have also included attempts to experiment with interactive telemedicine consultations. The Hubert Kairuki Memorial University and Mission Mikocheni Hospital, both in Dar es Salaam, conducted a trial online telemedicine consultation in August 2002 to explore the opportunities for:

- Delivering diagnosis and treatment between hub sites and spoke sites,
- Delivering state of the art nursing training as an addition to the curriculum at Kairuki Memorial Hospital.

[344] c.f. Msamanga (2004), the website can be found at http://www.muhef.or.tz.

[345] According to Helen Prytherch, GTZ Health Coordination Office, in a private email dating from 14.12.2004.

- Delivering ongoing training to health care workers and patients at the village level.
- Facilitating the exchange of state-of-the-art health care information worldwide (Fielding Graduate Institute n.d.).

Within its sectoral focus on health, the GTZ has initiated a number of rural ICT projects, for example, the 'Connect and Learn Initiative' in Tanga, a pilot project seeking to establish the extent to which digital technologies can be used to disseminate health information and knowledge in Tanga region. As part of the pilot project in Tanga, GTZ has given out a large number of computers to various social centres in the region – secondary schools, health centres, hospitals, youth centres, folk development outlets and drop-in-points of the Tanga AIDS working group – and started training particularly women in ICT to find out how computer and Internet technology can boost the access of disadvantaged areas and people to educational materials (c.f. Sebastian 2004).[346]

Other projects take into consideration the increasing district level responsibility for health planning and service delivery. To support the districts, a number of district resource centres have been developed to respond to the increasing need for information on health practice and management. Since 1990 the Centre for Educational Development in Health Arusha (www.cedha.ac.tz), and Healthlink Worldwide – with funds from DFID – have established a network of resource centres in 43 districts to provide health information for district level health workers. Since 2003 access to Internet and video resources has been introduced in a small number of centres after health workers increasingly requested access to new forms of information retrieval (Hammond / Richardson 2003). Finally, the Ministry of Health and Canada's International Development Research Centre (IDRC) have started the Tanzania Essential Health Intervention Project (TEHIP).[347] TEHIP has been one of the oldest health ICT projects. Since 1997, the project has been implemented in two typical, rural districts, Morogoro and Rufiji. By strengthening health research, building capacities for health management and provision of funds for health care delivery, TEHIP has developed and refined several tools for collecting and analysing information, for example, to show health needs at the community level, to map district health budgets and to establish cost information systems to pinpoint, for example, technical inefficiencies (TEHIP n.d.). Therewith, it assisted council health management teams at the district level to develop local plans based on local data

[346] A project website can be found at www.chezasalama.com.

[347] The TEHIP website can be found at www.idrc.ca/tehip.

rather than merely implementing plans or programs delivered from a superior, central planning level.

Thus, there have been considerable project activities in various fields of the ICT for health agenda. Generally, the range of ICT utilisation has increased in recent years, not only as a result of specific projects targeting ICT in health, but also due to an overall increasing availability, accessibility and affordability of ICT. The health sector, like others, benefited from the dynamic developments in the ICT sector. On the other hand, ICT in health has been hampered by still insufficient infrastructure and little awareness. According to a survey conducted by the Muhimbili Health Exchange Forum, only one third of doctors have access to the Internet (Msamanga 2004). This reflects the subordinate role of ICT in health. This is notwithstanding the existence of potential target areas, like health statistics, education and research, and even some room for rudimentary telemedicine – understood as the provision of telephone, Internet and email to consult experts or referral hospital staff to get some advice on diagnosis or treatment. However, a health system, relying to a large extent on dispensaries and health centres, where clinical officers and health-workers have to treat the most common diseases, must find a balance between the fundamentals of health care delivery and the potentials of ICT to achieve goal providing essential health care for the broad population.

Assessing different types of ICT in health projects, a twofold project focus appears, firstly, geographically, covering mainly urban areas with only some remote sites running as pilot projects, secondly, by application, that predominantly focused on administration, education and research. Applications for delivering direct health services for the population have only been marginal and, if applied, projects generally were an integral part of broader approaches tackling learning and access to educational materials on various issues. Apart from infrastructure or the lack thereof, which is an essential prerequisite for effective networking, the prevalence of the urban projects hosted at central health agencies also benefiting from higher education standards which make it significantly easier to run ICT related projects. Here, the technology has proved to be a valuable tool for capacity building and information exchange. Nevertheless, it is not technology alone that makes the difference. For example, collecting national health data – a difficult task in itself, because a significant number of health care facilities did not submit the necessary data in the past (Int. Hingora 9.4.2003) – is without any value, if there are no capacities to analyse the data accordingly. Often, the actual output, meant to assist decision-makers, has been a user unfriendly collection of raw data in lengthy uncommented tables, hardly appropriate to draw any conclusions from. Even if insufficient data flows have been identified as a main problem in health care provision, lacking computerisation or replacing a defunct manual by a defunct computerised reporting system is not at the core of a poor health system. Though computers can ease procedures, they can neither provide essential clinical and public health packages, nor well trained and motivated staff. Thus, computerisation will remain a supplementary issue on the national health agenda, so that balancing efforts of core measures to deliver essential health care and supportive measures to ensure quality and effective health information and deliver effective health education using a mix of different tools will remain a challenging task in the Tanzanian health sector.

5.5. ICT in Education

Education has been a further area of increased ICT utilisation. This includes formal tertiary and secondary education as well as vocational training. Most progress has been made in tertiary education, where the universities were successively connected the Internet and have built capacities to utilise the technology for various purposes – administrative procedures, research networking, provision of teaching materials, and the like. Developments were led at the University of Dar es Salaam from where the reach of ICT in higher education was gradually extended; firstly to the individual faculties and, secondly, to other universities of the countries. Newly installed ICT applications ranged from general email and Internet availability to advanced applications, as already introduced in chapter 4.3.6. In addition, ICT based education programmes have been included in sectoral approaches, like the above mentioned programmes in health education or in the framework of the TGDLC to provide ICT based training for political decision-makers (see chapter 5.3.3.).

The introduction of ICT to the broader education system, however, is in its absolute infancy. Already in 1997, the Ministry of Education and Culture (MoEC) had issued a syllabus for computer studies for secondary schools, which was revised in 2002 (Senzige / Sarukesi 2003). But whereas some private schools managed to introduce computers, the public schools almost failed completely to implement the syllabus. Only a very few students have taken the courses and, according to the Minister of Communication and Transport, Mark Mwandosya, the lack of a programme for training teachers on computers and other multi-media utilisation has been identified as a major reason for slow take up of computer studies in secondary education (Mwandosya 2003: 11). Furthermore, the computerisation of schools faces faces difficulties with regard to scalability and finance. Considering the number of roughly 1,080 secondary schools serving about 206,751 pupils (Senzige / Sarukesi 2003), the challenge of scalability becomes evident. Even if promising pilots have commenced in individual schools, targeting the schools as a whole, would be a complete new task. The situation has been further exaggerated by a deteriorating education system in general. The poor record is reflected in a shortage of qualified teachers, a lack of teaching materials in schools, very modest school facilities often without electricity, and, above all, by declining enrolment rates (c.f. Mendes et al. 2003). Increasing difficulties by the universities to recruit students that meet the most basic entrance qualifications for higher education mirror this alarming state of educational affairs (see chapter 4.3.6.).

Nevertheless, the number of projects aiming at improving computer utilisation in schools and school management have gradually increased in recent years. Most active in this field has been IICD. The Dutch NGO acted as an important facilitator in creating awareness for opportunities of ICT in education[348] and it has also supported a number of individual projects to improve the ICT environment for school education. For example, IICD provided support for Distance Learning Education Services (DILES),[349] an NGO, that had emerged from the IICD-supported round table in 1998 to explore the room for ICT in various fields. DILES has focused on the provision of distance learning tools and services, ranging from paper to the Internet. Also with support from IICD, the above-mentioned Kinondoni governance project comprised an educational component, particularly targeting the education administration with a focus on teaching materials and maintenance of students' records and performances. (Int. Mathew 16.7.2002, see also Intelecon 2002: 10).[350]

Thus, ICT activities in education have come with considerable vertical – from secondary to tertiary – and horizontal variations – from the provision of basic communication services to complex ICT solutions for distant learning. Considering the volume of the projects, both with regard to numbers and related costs, most projects address issues of higher education and education administration. Primary education is hardly covered, secondary education has

[348] For example, the IICD, jointly with the Ministry of Education organised a round table on distant education and education services in February 2002 that brought together practicians and representatives from the ministry and the heads conference to elaborate on the prospects of ICT in education. On that occasion, participants developed eleven project proposals aimed at enhancing the learning and teaching process in primary and secondary education with ICT. Among others, project proposals included improved teacher training, development of model schools in each districts to be started in two pilots, rising ICT awareness among students and development of ICT based training and teaching materials (c.f. IICD/MoEC 2002).

[349] See also the DILES website at http://www.distancelearning-tz.org.

[350] Other IICD projects have included the support for the Tanzania education website (http://www.tanedu.org), the Global Teenager Programme

increasingly gained significance, but is still secondary. Thus, a prevailing bias in favour of the high ends of the education system continues to exist, so that, at least in the medium-term, ICT in education might be less a tool to address social inequities by providing learning opportunities for disadvantaged people, but more suitable to serve the higher ends of the education system. This argument can be made even more pronounced; whereas the creation of ICT capacities is a necessary prerequisite for institutions of higher education to endure in an increasingly international learning environment, secondary schools can benefit from ICT in individual cases, but providing the needed equipment and capacities in each school would overburden the education system. Prioritising ICT in education, a focus should be on the institutions of higher education and post secondary school education. In this area, significant achievements have been made in the last years; universities were not only among the pioneers of ICT in the country in the early years of commencing computerisation, they still are a major pillar of ICT in the country. Particularly the University of Dar es Salaam has developed into a nucleus for ICT in training, education and ICT development. Similar, positive effects can also be expected from an improved education administration with regard to planning, curricula development and supply of teaching materials that base on ICT or ICT related services.

5.6. ICT Utilisation in Small and Medium Enterprises (SMEs)

In contrast to the previous chapters, that covered recent developments and thus approached the issues on more qualitative grounds, the following utilisation assessments had a longer record – or relevant data on ICT usage was more easily accessible – so that some quantitative finding can be presented, too. To start with, this section deals with ICT utilisation of small and medium enterprises (SMEs). SMEs have gained some attention in the ICT for development debate, because they are regarded to account for the majority of firms and a large share of employment in many developing countries. For this reason, and because SMEs were considered to be particularly responsive to technological change, SMEs have materialised as one potential target sector to harness ICT for development. (c.f. Kennedy/Hobohm 1999, Hallberg 2000, Levitsky 1996). The potential impact of ICT on enterprise efficiency and productivity explains why the utilisation of technology presumably has strong linkages to the enterprise's competitiveness. This is applicable to different levels: (i) improvement of internal business procedures, (ii) lowering of transaction costs, and (iii) provision of timely and accurate market information that allow a better understanding of both demand and supply. Furthermore, new business opportunities seem to appear, like the formation of new business-platforms, creation of information-sharing and networking facilities, and prospects of cluster-building in emerging market segments (c.f. Schmitz 1995). However, in most cases such descriptions refer to the potential of SMEs, not their actual performances. They draw round a vision and offer a roadmap to pave the way for the road ahead. But to what extent do they reflect the situation on the ground? Are SMEs ready to take advantage of ICT or is it just wishful thinking of some IT-enthusiasts? This chapter provides some empirical data on the ICT utilisation of SMEs in Tanzania.[351] It tries to identify the size of the gap between reality, i.e. which applications are used for what purpose to what extent, and opportunity, i.e. areas which offer a starting point for further and more elaborated ICT utilisation.[352] The surveyed SMEs came equally from the tourism, food processing and textile sectors.

Figure 5.5: ICT diffusion in Tanzanian SME

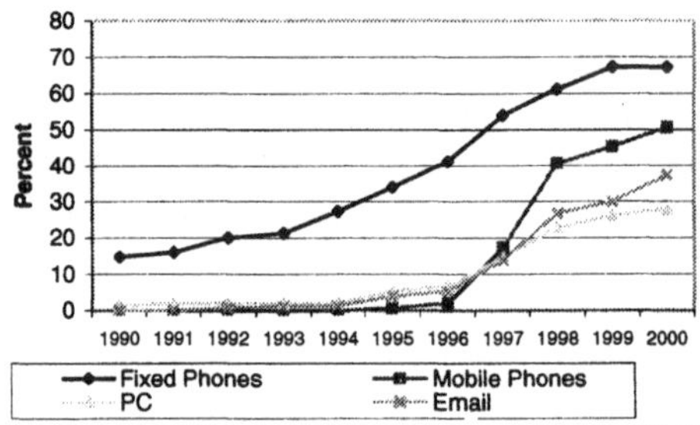

Broader ICT utilisation in SMEs started in 1996. As figure 5.5 shows, fixed line phones have a longer record and remained the most prevalent technology. Their steadily increasing number coincided with the rehabilitation of the public telephone network. Mobile phone, PC and email penetration took off in 1996. Whereas the average ICT diffusion in Tanzania considerably caught up since 2000, SMEs demonstrated significant growth rates between 1996 and 1998.

351 A detailed version of this chapter has already been made available in an online article on the subject, see Nielinger (2003b). I am highly indebted to Francis Matambalya (University of Dar es Salaam), who conducted the SME survey in early 2000 and generously made the raw data available. His excellent work has provided the basis for this new analysis, in which the given data have been re-edited and re-organised and accordingly processed to meet the new objectives. The output of Francis' work, that focused on the impact of ICT on micro-level competitiveness, can be found in: Matambalya / Wolf (2001) and Matambalya (2003).

352 The survey data covers a sample of 150 Tanzanian SMEs that were randomly selected and came equally from the food processing, textile, and tourist sector. Surveys were conducted in Dar es Salaam, Arusha, Mwanza, Zanzibar (Urban West) and Mara regions, covering both urban and rural enterprises.

Thus, they can be described as early adopters of the new technologies. They responded to new opportunities as soon as services were available and, underlined their ability to react flexibly and immediately to a changing environment. Especially mobile phone subscription increased rapidly from almost zero in 1996 to remarkably 40 percent in 1998. This suggests that there was mainly a demand for basic (= telephone) communication, either in addition to fixed line phones or as their substitution that boosted the subscription numbers. PC and email diffusion also grew substantial since 1996, but less significant than telephone growth rates. The similar growth pattern implies that email introduction has been a strong impetus to investment in computer equipment. However, ICT diffusion rates have slowed down since 1999. Possible explanations might be that (i) those, who are able to utilise the technology, have already been sufficiently equipped, (ii) prohibitive costs prevented any further advanced ICT utilisation, or (iii) any further ICT utilisation were not regarded as a means to increase the enterprises' economic opportunities.

Figure 5.6 represents the perceived significance of different ICT for economic performance. Apparently, the perceived significance of technologies coincided with the rate of their diffusion, hence telephone ranked highest and the WWW lowest. 60 percent of respondents recognised the telephone and 38 percent the cell phone as 'very significant' or 'significant'. Computer based applications were led by email (28 percent), computer applications came second (25 percent), and last is the World Wide Web with 12 percent. It is also noteworthy that responses referring to an 'average' or 'less significance' relatively decreased the better the IT equipment, which can be interpreted that users of computer-based communication were quickly convinced of the relevance of the deployed technology.

Figure 5.6: Significance of ICT for Economic Performance

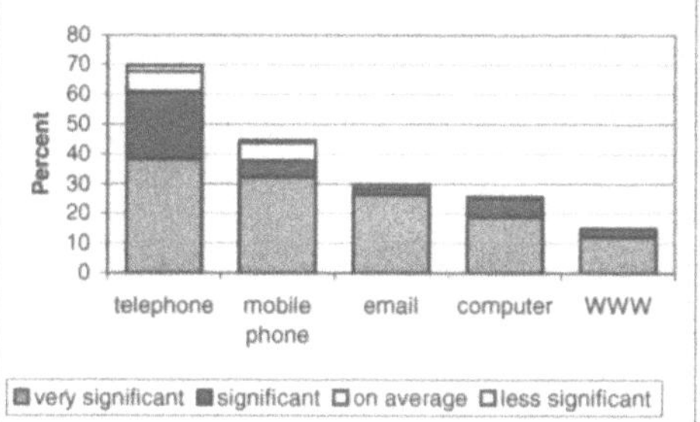

ICT are unevenly distributed with regard to both their sectoral and regional distribution as figures 5.7 and 5.8 illustrate. Mean investment in ICT equalled roughly 1,100 US$ per company with the largest proportion going to computers (890 US$). With regard to sectoral differences, the tourist sector has been the most advanced sector. It accounted for more than two thirds of all investment in ICT among Tanzanian SMEs. The investment rate of the food processing sector was similar to the Tanzanian average, leaving the textile sector far behind. These figures correspond to the significance attained to the sectors in the Tanzanian economy, a dynamic growth pattern in the tourist industry, some renewed efforts to promote food processing and an overall declining textile sector during the last years. Though the extent to which the tourist sector leads the investment scale might come as a surprise at first glance, it reflects two particularities of the sector: (i) the outbound perspective of tourism that comes with an intensive demand for communication, and (ii) the fact that the Tanzanian tourist sector predominantly targeted the premium tourist segment, which, in turn, necessitated a more sophisticated service delivery.

Figure 5.7: Investment in ICT according to sector in '000 TSh.

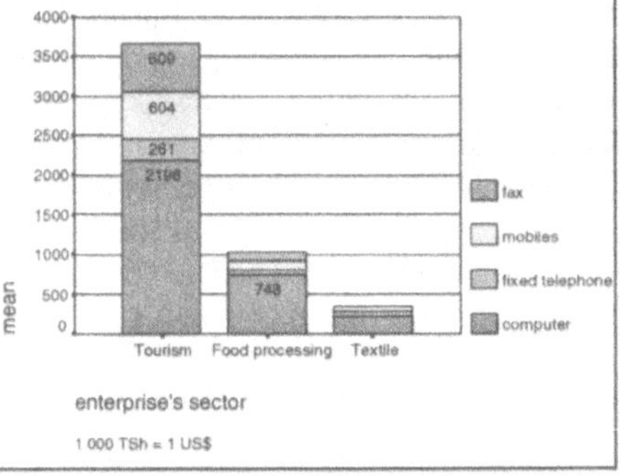

Figure 5.8: Investment in ICT according to region in '000 TSh.

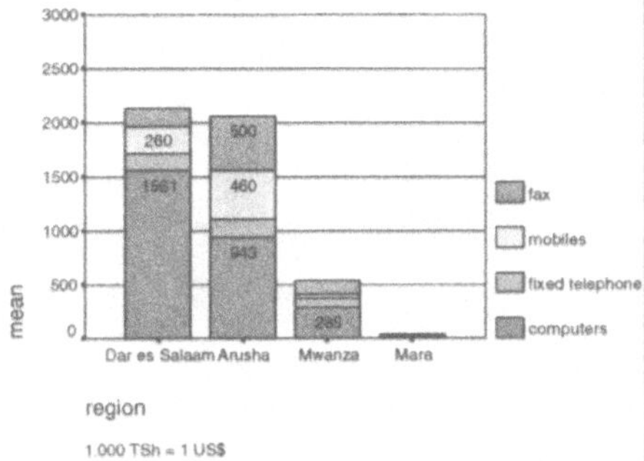

The regional ICT investment pattern demonstrates not only a strong urban bias, but also a bias in favour of just the two main cities: Dar es Salaam, as Tanzania's commercial capital, and Arusha, as the tourist capital of the country. The total amount of investment in these two cities is similar, but the composition is remarkably different. Whereas in Dar es Salaam three quarters of investment go into computers, in Arusha more than one half of investment is spent on telephone and fax. Thus, the availability of different services is reflected in the investment ratio, or, to put it in other words, as soon as computer based ICT were available, the enterprises have started to invest in new technologies in order to take advantage of new business opportunities.

Figure 5.9 provides information on the type of computer applications utilised in SMEs and rank respective applications. With roughly 46 percent word processing is the most widespread application. Internet communication (email, WWW) and other advanced office, management or controlling applications are secondary with only a modest rate of diffusion. Herein, bookkeeping and order processing as basic input and output control have some significance, advanced applications as stock listings or other managerial tools for decision making are only rarely utilised. Thus, the more complex the software, the less it is applied. This might be due to the training needed to use the software efficiently, but also to the costs of more complex software solutions.

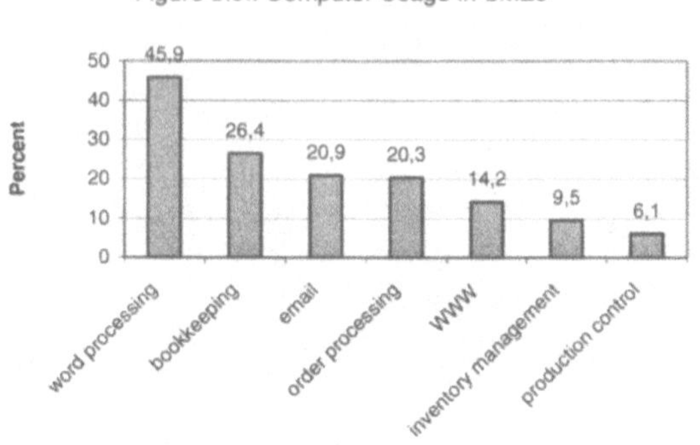

Figure 5.9.: Computer Usage in SMEs

Despite an increasing ICT utilisation, the survey has also revealed a substantial predominance of traditional means of communication. Even the notable growth period in the late 1990s, where some early adaptors have developed ICT capacities, cannot conceal that the overall diffusion has remained low. If SME development is regarded as a suitable instrument to raise economic growth and issues of SME promotion are to be addressed, any approach needs to deal with the challenging mix of business obstacles and the heterogeneous character of different types of enterprises. Focusing on the intersection between SME development and ICT promotion, different starting points can be identified. Generally, the idea of cluster-building and the facilitation of joint action of dispersed and isolated firms might bring new opportunities, such as information and resource sharing, joint marketing or joint lobbying. So attained "collective efficiency" (Schmitz 1995) could bring about better market access, flexible provision of services and products, and scaling up of business activities in emerging and lucrative market segments. However, such approaches are more feasible for more developed and well-managed SMEs. For the majority it would mean to do the second step first. If ICT capacities are sequenced, most SMEs in Tanzania are in a start up phase. First, there must be a necessary understanding of the potentials of ICT before basic application training can make a difference and can eventually expand to advanced applications to assist efficient decision-making procedures.

The individual SME might be overburdened to implement a sustainable and well-tailored ICT strategy. Decentralised SME-service centres could assist SMEs to cope with the challenge. Computer training that meets the different demands from basic to advanced business application, Internet connectivity and email services could be only one part of broad consulting and advisory services covering issues like business registration, financing, regulatory matters or business lobbying. The idea of a one-stop agency to facilitate business, for example, has a fairly good record with respect to the Tanzania Investment Centre (TIC) to attract foreign direct investment. SME service centres could form an equivalent to the domestic market.[353]

5.7. NGO ICT Utilisation

Apart from the private sector, NGOs have been important facilitators of increasing ICT diffusion. In Tanzania, like in Africa in general, NGOs have been pioneers and major facilitators of ICT (c.f. Levey/Young 2002). However, it would be short-sighted to take this observation as general evidence for the scope of their ICT utilisation. Statements, such as "most NGOs have access to the web", or "the use of technologies such as e-mail has become the norm rather than the exception within NGO communities" (Esterhuysen 2002), are still premature. Empirical evidence used in this section[354] suggests the opposite. At this point, it is only a minority of NGOs that have access to the Internet and among these few, effective use of the technology is rather limited – even though, in fact, overall growth and diffusion rates are impressive.

To some extent, such misperceptions might be the result of a maturing debate. The early focus was on individual cases to highlight the potentials of ICT and to create pilots or model approaches. This generated a considerable amount of case studies[355] – in most cases success stories – that, due to their sheer mass, might have conveyed the false impression of a largely connected NGO community in Africa. It is only today that the discussion about NGO's ICT utilisation is raised to a new level of a more comprehensive and systematic approach targeting the broader picture, working comparatively or applying common terms of

[353] Service centres must neither be costly nor complex. Their set up could be structured alongside different service modules, where each centre provides only suitable parts of the whole package of services. Furthermore, such centres could make use of existing infrastructure as provided by organisations like SIDO (Small Industries Development Organisation), SEDA (Small Enterprise Development Agency) or the regional branches of TCCIA (Tanzanian Chamber of Commerce, Industry and Agriculture).

[354] Results of the NGO ICT utilisation survey have already been made available in an online article, see Nielinger (2003c).

[355] This is not to question the relevance of the individual case. Case study collections like NGLS (2000) or TechKnowLogia (2001) offer a good starting point and reflect the various periods of the ICT diffusion process. But as soon as a critical mass of ICT users is reached, analyses aiming to target NGOs and their ICT utilisation in general must go beyond the individual project approach.

references.[356] This chapter provides some empirical data on the actual scope and extent of ICT utilisation among Tanzanian NGOs.[357]

The surveyed (connected) NGOs mirror their heterogeneity. Their sectoral focus goes through all societal issues, with most NGOs active in more than one sector. Furthermore, no particular ICT intensive working area could be detected, though NGOs working in agriculture turned out to be slightly less connected. But overall, the numbers of connected NGOs follow the overall distribution of NGO target sectors in the country.[358] Similar patterns appear with regard to the overall budget and the number of staff. Both variables show a considerable range. Annual budget starts from 500 US$ and goes up to more than one Mio. US$ - the median annual budget has been 50,000 US$. Staff ranged from only two persons working on a voluntary basis to 100 salaried employees working in nationwide NGOs, which results in eleven staff per NGO in the mean. From this general starting point, there is no identifiable pattern of ICT usage. Those who use the technology come from all sectors and do it regardless of their financial and human resources. Furthermore, the overall quantity of connected NGOs must be put into perspective. Only 20 percent of Tanzanian NGOs have a working email account. Though this might be a smaller number than expected, it

---

[356] For a good example, see Thioune (2003). A critical view of NGO ICT utilisation in Tanzania can be found in Mercer (2004).

[357] The data provided are based on two samples. The first contains 957 entries from NGOs that were randomly selected from several national, regional or sectoral NGO directories and targeted all NGOs regardless of their ICT or email profile. This sample covers the overall picture of email usage among Tanzanian NGOs and was used to prepare the data given in figure 5.9. Out of the 957 NGOs, 255 or 25 percent provided an email address. While editing the data set, a further 62 or 6.5 percent dropped out of the sample because of invalid, unknown or ineligible email accounts. The newly adjusted sample left 193 or 20 percent valid email addresses.
The second sample contained 1.228 entries. Because the questionnaires were submitted online, and since the objective was to work with the largest possible sample, considerable research has been conducted to compile an NGO email directory from several sources such as the WWW, various discussion groups or fora as well as offline sources. Naturally, this sample was heavily 'email biased', because only NGOs that provided an email address were added to the list. This sample was used to analyse the ICT utilisation of NGOs as given in figures 5.10-5.12. From the 1228 NGOs in the directory, 526 (42.8 percent) provided an email contact. Again, a considerable number of 124 (23.6 percent) email accounts were invalid or unsuitable. Finally, the remaining sample left 402 (32.7 percent) valid entries, which were taken as the base of the survey. Of these, 77 or 19 percent submitted a completed survey form, which is a satisfactory response rate. Though the results cannot claim to be representative, they indicate some strong tendencies of NGO ICT utilisation in Tanzania.

[358] 'Sector' refers to health, education, children, women, community work, environment, democracy/human rights, agriculture, ICT and umbrella organisations.

should be recognised that, at the same time as some NGOs started to use the technology, the total number of NGOs has also witnessed an exceptional increase.[359]

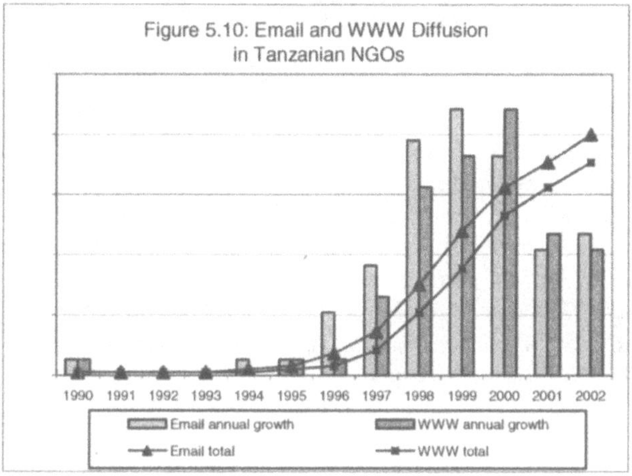

Figure 5.10 shows the chronological email and WWW diffusion among Tanzanian NGOs between 1990 and 2002. The line charts represent the overall number of email and WWW users; the bar charts reflect the annual growth rates of the process. Technological penetration took off in 1996. The number of email users grew sizably for the first time and introduced a phase of accelerated ICT development, also in conjunction with constant improvements in the provision of the telecommunication infrastructure. Exceptional growth was reached between 1998 and 2000 – the boom time for NGO connectivity. Diffusion growth rates amounted to 15 percent or even 22 percent annually. Since 2000 growth has slowed down, but remains on a level of above 10 percent, while email and WWW diffusion are approaching 20 percent and 18 percent, respectively. Email applications have been the engine for growth. Their annual growth rates exceeded the WWW figures until 1999. This pattern of ICT adoption, email first and web later, changed from 2000 onwards. The 2000 peak level of WWW annual growth might reflect a WWW backlog demand of some NGOs at a time when the web has become more and more popular and an increasingly valuable source of information. It also coincides with the introduction of public access facilities, where WWW applications have been required to access one's webmail account.

359 For Tanzania, Mutakyahwa (2002) reports that there were over 1.000 NGOs in 1995 and now there is an estimated number of 4,000 NGOs in the country.

Figure 5.11 quantifies different means of connectivity. Accordingly, more than half of the NGOs had an Internet connection in their office, roughly 30 percent connected privately and 35 percent used public access facilities, like Internet cafes or telecentres. Some 15 percent rely on more than one access point and use a mix of connectivity options. The high quantity of NGOs using their office equipment to connect to the Internet indicates that office solutions were implemented wherever available or affordable. In addition, a significant number of 35 percent used public access facilities to connect to the Internet. Of these, 26 percent solely rely on public facilities, 9 percent use them to complement their office connection. This number highlights the outstanding importance of public access, especially where no alternative solutions are available.

Figure 5.11: Means of Connectivity

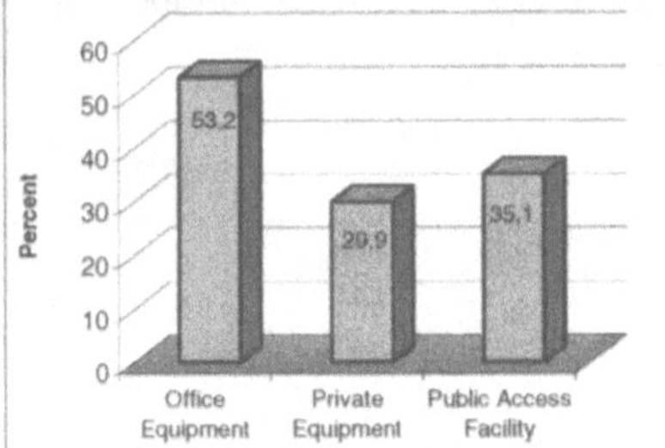

The various sources of financing IT equipment are covered in figure 5.12. The bars stand for the different sources, the staples within the bars quantify the extent of assistance given as 'main', 'secondary' and 'modest' contributions. The international donor community has been the largest finance source for IT equipment. Roughly 50 percent of NGOs rely on the donor community as their main sponsor. Only insignificant 4 percent of the NGOs reported donor funding as a secondary source, and roughly 9 percent receive modest support from the donor community. National donors were rarely the main sponsors of IT equipment (3 percent), but normally a secondary source (16 percent) or a modest contributor (8 percent). Many international donors make national contributions a precondition for their funding, so that national donors usually appeared as secondary sources. Company donations played a subordinate role in the acquisition of IT equipment. Only 8 percent of NGOs quote them as their main source, 1.3 percent as secondary and 8 percent as a modest source. Raising a loan to finance IT equipment is virtually nonexistent. Here, the bar is not so much connected to the formal banking sector, which can hardly be accessed by NGOs anyway, but, for example, to micro-finance schemes or supplementary funds designed to facilitate the work of NGOs or small enterprises. Though the usefulness of such lending facilities depends on their appropriateness and availability, their limited accessibility might mirror NGO's typical business approaches and self-conception; NGOs rather preferred to rely on donor support than to develop alternative business-like approaches. A considerable number of NGOs collected membership fees to improve their financial situation. These fees were a main source for IT financing for 13 percent, a secondary source for 12 percent and a modest source for 17 percent of NGOs. This is a fairly high proportion of internal resources used to acquire IT equipment. Membership fees have been the most prevalent means to raise internal funds, often complementing granted donor support.

Figure 5.12: Finance Source of IT Equipment (in %)

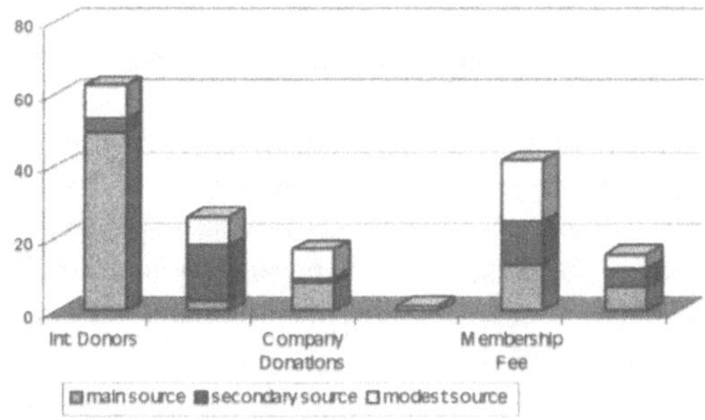

Figure 5.13: Significance of ICT

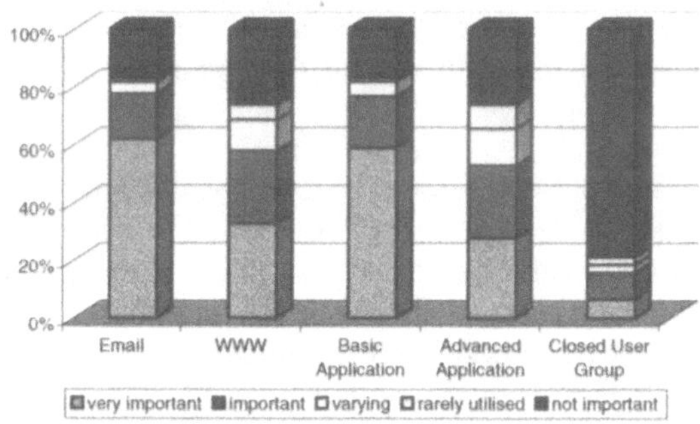

Figure 5.13 shows the significance of ICT for NGO's performance. The bars differentiate various applications corresponding to different grades of complexity; the staples quantify the significance, which the surveyed NGOs attributed to the respective applications ranging from 'very important' to 'not important'. Email and basic applications (word processing, etc.) were the most powerful ways to make use of the computer. Email usage was very important for 61 percent and important for 16 percent of NGOs. Roughly the same holds for basic applications; 58 percent emphasises they are very important, and 18 percent regarded them as important computer tools. Thus, the two represent the

most frequent applications and the main driving force of increasing ICT diffusion. However, noteworthy exceptions appear even in these heavily used segments, that is, 17 percent of NGOs did not assign any importance to email usage or basic applications at all. Bearing in mind that only connected NGOs were covered in the survey, this relatively high share highlights the existence of a considerable number of NGOs for which electronic communication and data processing might not be an appropriate tool to deal with every day working routines.

A second cluster involving similar responses can be observed with regard to the WWW and advanced applications (accounting, payroll, etc.). For almost one third of the surveyed NGOs both applications were very important; roughly one fourth quoted them as being important. Though to a lesser extent than in the email/basic application cluster, both the web – excluding webmail services – and advanced applications enjoyed substantial popularity among Tanzanian NGOs. Still, 16 percent (WWW) and 19 percent (advanced application) of NGOs mentioned that they have used both tools only occasionally or have only little experience dealing with them. On the other hand, 26 percent state that both applications were of no importance for their working procedures. More advanced applications like the provision of closed user groups were used by a minority only. These supposedly skilful user groups, deliberately deploying complex ICT solutions, complete the picture of different user groups as they represent the opposite extreme of the majority of small-scale or modest ICT users.

In addition to the close questions, the questionnaire also provided an opportunity to comment openly on the subject of ICT. The submitted comments have reinforced the findings of the quantitative questions given above and contain a wide range of perceptions, obstacles and future strategies. Two main observations can be made: Firstly, most comments combine an identification of main obstacles, i.e. a lack of funds and a lack of skills to utilise the technology more effectively, with an emphatic reference to the importance of the technology highlighting the benefits already achieved. Secondly, though the comments share some common themes, there is a need to differentiate between different NGOs and their different scope of activity, capacities and objectives. The quintessence of the open comments can be summarised as follows:

- While most NGOs suffer from financial restrictions, this observation has different implications: For large, well-established NGOs operating nationwide, the main challenges have been to update and upgrade their equipment and build the necessary skills to operate sophisticated network solutions. For them, high costs of connectivity have hampered their working routines, but have not prohibited their ICT utilisation. More or less large NGOs can cope with such difficulties and they make use of the technology even under difficult circumstances.

- In contrast, for small NGOs, grassroots organisations, or single issue lobby groups especially in rural areas, high connectivity and equipment costs have been a prohibitive burden to develop minimal IT capacities or to introduce basic ICT equipment. In these cases, high costs have segregated them from the potential benefits of ICT. The lack of qualified personnel has exacerbated this situation.
- Public access facilities have been only partly in a position to counterbalance such tendencies. Limitations of telecentres or Internet Cafes have been identified with regard to time and cost considerations.
- Despite these difficulties, most NGOs have emphatically underlined the significance of ICT. For many NGOs the technology has opened new communication channels that simply did not exist before. Furthermore, gains of networking and the idea of joint social and political action were frequently mentioned as a key to increasing NGOs' ability to play a role in the political process. Or, as one contributor put it, 'The arrival of email and the Internet has made life a lot of easier'.

In conclusion, the image of a largely connected NGO community is a misperception. Only 20 percent of Tanzanian NGOs had a working email account. The heterogeneous nature of NGOs was echoed throughout the survey and reflected a broad range of different user profiles. These can be divided into at least three main user groups: A first group consists of professionally managed, large NGOs highly visible in the political process and well embedded in large-scale networking interactions. This group uses ICT as an accepted and understood tool supporting daily working procedures. A second group, often well acknowledged in the respective working field, has also developed a considerable ICT profile. These NGOs might be less sophisticated than those in the first group, and they might not utilise the technology in all potential working areas, but, within their limits, they value the technology as an important tool, especially with respect to national and international networking, Internet based research and the like. The third group, at the same time the majority of Tanzanian NGOs, consists of small NGOs, often focusing on a single issue, coming from both urban and rural areas, many of them working at the grassroots level. They have developed a rudimentary ICT profile that focuses on email usage and some basic application. Though low profile and low intensity usage, this group has identified their basic ICT needs and applies the applications that are available and affordable. Also, in the low profile third user group, even limited access to the technology has shown a positive impact on daily working procedures.

Apart from the differences, all NGOs share a set of common ICT difficulties: high costs, lack of infrastructure and skills. But whereas for the first two groups these presented – sometimes even serious – business obstacles, for the third group they were often a prohibitive burden to make use of the technology. This has been why, for many NGOs, Internet usage depended on the availability of public access facilities. These centres have been main facilitators of the technology, particularly in those cases, where only little ICT supply was demanded or public access centres were the only access point available. Furthermore, public access facilities also had limitations to meet organisational demands. They were restrictive in the provision of necessary and affordable airtime and largely ignored the fact that data processing procedures have often been difficult to conduct outside the office – taking alone issues of data security and protection. In consequence, public access facilities are well equipped to meet the individual or small organisational demand, but they are not yet ready to offer scalable solutions meeting larger organisational demands.

5.8. Rural ICT Utilisation

Bringing information and communication technologies to rural or remote areas is a key challenge to harnessing the potential of ICT for development. Tanzania represents many of those aspects that make rural Internet connectivity so demanding. That is first, the size of the country. With a total size of 886.037 square kilometres of land area Tanzania is one of the larger territorial states in Africa, covering more land area than its neighbours Kenya and Uganda combined. The sheer size of the country makes it very difficult to deploy viable connectivity solutions for countrywide ICT access. A second exacerbating factor is the high proportion of rural population at 67 percent (UNDP 2003). Translated into possible connectivity solutions, it means establishing access points in many different dispersed locations only serving a very limited number of people. On the other hand, Tanzania also stands for positive developments. The country attempts to promote a deeper ICT diffusion and the government and other stakeholders have started to address the issue of ICT and the development potential of the technology. The establishment of a universal access programme has been in the making, the e-government initiative explicitly includes the district level and a number of sectoral approaches, such as in health or education, seek to provide at least some basic ICT services at district level.

Drawing on experiences of three different rural Internet facilities, this sections further analyse the prospects of rural Internet utilisation in Tanzania.[360]

Figure 5.14: Location of surveyed Internet sites

adapted from CIA Factbook (2002)

5.8.2. Developing Public Internet Access Facilities in Kasulu, Magu, and Sengerema

The public access facilities covered, are located in Kasulu, Magu and Sengerema in the western and north-western parts of Tanzania (see figure 5.14). All three towns are district capitals and all three locations correspond to the main challenge to bring sustainable ICT solutions to remote areas, which are poorly served with basic infrastructure such as transport, electricity or communication. However, beyond such similar starting position, Kasulu, Magu and Sengerema also represent different ICT approaches.

The Sengerema telecentre was established within the ITU, IDRC, UNESCO framework as a multipurpose community telecentre focusing on ICT and ICT

[360] For a detailed version of the paper, see Nielinger (2003a). The data provided was collected during visits to Kasulu, Magu and Sengerema in April 2003 and based on background interviews and user surveys in all three locations. Interviews were conducted with Alphonce Mbonigabha, principal of Kasulu Teacher Training College (2.4.2003), Habby Bugalama, Manager of Sengerema Telecentre (23.4.2003) and Naomi Maselle, Director of Crop Marketing Bureau (24.4.2003). Further information was provided by Geoff Calder from the Global Catalyst Foundation and coordinator of the Kasulu Internet Project, of which one part is the Teacher Training College. With regard to the user surveys, a total of 173 interviews were conducted – 73 in Kasulu, 40 in Magu and 60 in Sengerema. In relation to the estimated user basis in all three locations, the survey covers one third to half of the overall user basis. These were randomly selected, hence the user figures are representative for each facility.

related services, aiming at benefiting the community as a whole.[361] Initial steps to establish the telecentre date back to 1997 and had followed the emergence of the ITU, IDRC, UNESCO approach, implemented by COSTECH to establish a rural multipurpose telecentre in Tanzania. The telecentre started operation in 2000 in a preliminary premise and in September 2002, it moved into a newly built spacious building.

Table 5.1: Type of ICT access site

	Operator	Core Business	Type of Town	Project start	Full Internet access since
Kasulu	**Kasulu Teacher's Training College**	**Teacher Training College**	**District Capital**	**1999**	**2000 through TTCL line Since 2002 VSAT-connectivity**
Magu	**CROMABU – Crop Marketing Bureau Ltd**	**Agro-Marketing**	**District Capital**	**1999**	**2001 through TTCL line**
Sengerema	**Sengerema Telecentre**	**Multipurpose Community Telecentre (MCT)**	**District Capital**	**1997**	**December 2000 in temporary premises through TTCL line. Since September 2002 through VSAT in new premises**

The Kasulu ICT access point was launched at the Teacher Training College (TTC). First ideas to make use of ICT already came up in 1995, when the then Minister of Education had asked all principals to introduce computers in classrooms. First concrete plans materialised in 1998 and the TTC benefited from the existence of the nearby Mtabila refugee camp, which served Burundian refugees forcibly displaced from their homes. The Global Catalyst Foundation, whose mission has been to improve people's lives through the effective application of information technologies, approached the UNHCR and discussed the feasibility of bringing Internet access to residents of a refugee camp to provide education, health, and economic development resources to the refugees. In order to include the local population, too, the project scheme was enlarged to serve the local community as well. TTC made a proposal to the foundation, and

[361] The ITU / UNESCO / IDRC framework is introduced in Ernberg (1998).

the project was finally approved in 1999.[362] Initially only as a computer and not an Internet project, TTC has improved its ICT capacities since then – preliminary Internet connectivity provided by TTCL was set-up in 2000, in April 2002 full satellite connectivity has been installed. The Internet cafe opened in October 2002. Though still focusing on teacher training, TTC has embarked on a strategy to turn the college into a local IT centre. The strategy had three components: firstly, to make use of ICT as a tool for teacher training, secondly, TTC successfully submitted an application to become a Cisco Academy[363] aiming to train future IT specialists, and thirdly TTC serves as a public Internet access and training point. The extension of the core business has not only strengthened the ICT profile of TTC, but has also generated extra income from added ICT services.

The Magu Internet project originated from a price information system for cotton, firstly deployed as a traditional offline information system, later based on ICT. The formation of the Crop Marketing Bureau (CROMABU), the operator of the site, can be traced back to a 1999 ICT roundtable in Mwanza, which was organised by the Tanzania Chamber of Commerce, Industry and Agriculture (TCCIA) and IICD, to identify ICT projects and tailored ICT solutions for the Mwanza region. CROMABU submitted a proposal to the roundtable to develop an agro marketing information system and after approval by IICD,[364] the bureau registered as an NGO, went through a founding process and became operational in 2001; from the very beginning it was meant to be a model approach to be replicated in other parts of the country.

The core of Magu has been the agro information system meant to enable pfarmers to assess market opportunities and farm inputs such as fertilisers,

[362] The Kasulu Internet Project (KIP) comprises three Community Internet Centres: The first is in a secondary school at the Mtabila Refugee Camp, the second at the Kasulu Folk Development College and the third at the Kasulu Teacher Training College. Some background on KIP can also be found on the website of the Global Catalyst Foundation (www.global-catalyst.org).

[363] Some details of the Cisco Networking Academy Programme are given in chapter 4.3.6. In early 2003, the Kasulu Cisco Academy had 12 fully registered students, and a waiting list of 50. Course fees are 200 US$, in contrast to e.g. 500 US$ in Dar es Salaam.

[364] On part of IICD, the support for CROMABU has been embedded in a broader project approach aiming to bring ICT to rural farmers, traders, and small enterprises. As a coordinator and information broker on a national scale, Business Informations Services (BIS), a small Dar es Salaam based consulting company, functions as a focal point of a countrywide agro information system, of which CROMABU serves as the Mwanza link of a national network. For details see the BIS-website (www.bistanzania.com) and Intelecon (2002). For more details, see the recently uploaded CROMABU website www.cromabu.com.

seeds, production techniques or packaging. One main challenge for the operator had been to identify a formula to work with the rural poor. Early attempts to collaborate with local church organisations and the district council as liaison agencies to rural farmers brought only minor success. Mobilisation remained low and the CROMABU management decided to seek direct linkages in the villages. This resulted in the creation of focus groups. Since then, women, farmers and youth have been at the centre of the Magu ICT project. They receive individually tailored training courses and most of the opening hours are assigned to the focus groups. As in Kasulu, the Internet cafe has been an extension of the core business and a tool to generate additional income. However, unlike Kasulu, ICT is also the nucleus of the business in Magu.

5.8.2.1. Starting Basis of the Public Access Sites

Table 5.2 gives an overview of the starting basis of the three sites, which are located in district capitals with a population between 18,000 and 50,000 inhabitants. Before operations commenced, all facilities had to address issues of infrastructure, that is, power and connectivity. In the first instance, the availability of telephone lines and connection to a power grid were a prerequisite for project outset. Even if alternative connectivity solutions like VSAT were proposed, a telephone line had always provided preliminary connectivity. Similar applied to power, but since Kasulu had no connection to the national power grid, project set-up was more complicated. For power supply, all sites of the Kasulu Internet Project (KIP) were forced to develop alternative power solutions – at TTC, a biogas digestion system that powers the IT equipment.[365] This innovative power supply also caught considerable attention from the international media.[366] However, a look at the details also reveals shortcomings. The capacity of the biogas system is fairly limited, providing only roughly 40 percent of the needed energy,[367] and leads to reduced daily operating hours compared to the other two sites.[368]

[365] Mtabila refugee camp and Kasulu Folk Development College are powered by solar panels.

[366] See e.g. the coverage by the BBC, 'Cow pats fuel computers', http://news.bbc.co.uk/2/hi/technology/2957488.stm

[367] For the roughly 60 percent remaining, the generator is fuelled with diesel.

[368] This is notwithstanding the fact, that unreliable power supply is a major obstacle for all three facilities.

Table 5.2: Starting basis of the Internet access facilities				
	Population (town)[369]	**Connection to the national power grid**	**Connection to the National Telephone Operator TTCL**	**Location**
Kasulu	**34,400**	**no**	**yes**	**20 – 30 minutes walking distance from city centre**
Magu	**17,700**	**yes**	**yes**	**city centre**
Sengerema	**49,800**	**yes**	**yes**	**city centre**

Finally, the location of the ICT facility, itself, influenced business operations, because the physical location turned out to be an important issue. Magu and Sengerema took advantage of setting up an entirely new site and opted for a prime location at the centre of the town.[370] The new buildings are well visible and attractive, very appropriate for the comfort of the customers and for hosting the site. In contrast, the Teacher Training College in Kasulu is located 20-30 minutes walking distance from the city centre, which has regularly produced inconveniences for the customers.

5.8.2.2. Connectivity Solutions

After project approval, all three facilities had set up temporary connectivity solutions provided by TTCL. These were voice subscriber lines not designed to carry large data traffic, hence there was a need to migrate to more capable connectivity solutions. Today, each of the sites offers sufficient bandwidth, but performance limits are reached especially during peak hours. Kasulu has deployed a VSAT – purchase and installation for approximately 7,000 US$ - which is capable of 25 kbps up and 125 kbps downlink. The monthly connectivity fee is 550 US$. Magu is connected by TTCL telephone lines. The leased TTCL line

Table 5.3: Internet Connectivity			
	Kasulu	**Magu**	**Sengerema**
Type of connectivity	**VSAT connection with downlink in Germany**	**TTCL leased line**	**VSAT connection operated by a national data operator Afsat with a link to Dar es Salaam to Afsat's international gateway**
Connectivity costs per month	**550 US$**	**432 US$**	**300.000 TSh. (~ 300 US$)**

[369] Data taken from government census data, c.f. URT (2002b).

[370] It is noteworthy that the prime locations in Magu and Sengerema were made available by considerable financial and other supportive commitments of the local community.

costs 432 US$ per month. Sengerema has also deployed a VSAT, but in contrast to Kasulu, Sengerema is linked to COSTECH's national network operated by Afsat. Data traffic is sent to Dar es Salaam and there linked to Afsat's international gateway. Connectivity fees are paid in Tanzanian Shillings and amount to 300,000 TSh. Thus, distinctive solutions were chosen in each location and identifying an appropriate connectivity model was one of the first challenges. The choices have been limited, and for Sengerema and Kasulu, which are not yet connected to the national Internet backbone, VSAT remained the only solution. However, it has not been the costs of the purchase of the equipment or its installation that have troubled the operators. In fact, prices for hardware have gone down considerably in recent years and, with strong backing by the donor community (see below), all could install appropriate solutions. The main problem for the operators has been the monthly connectivity fee. They create a heavy cost burden threatening the sustainability of the business models. Thereby, the Sengerema connectivity option might indicate a way forward. Instead of bargaining connectivity costs on a bilateral basis, joint solutions and shared infrastructure might be a way out of the connectivity cost trap.

5.8.2.3. Services and Equipment

All three facilities were equipped with between 16 and 13 computers. Magu and Sengerema have allotted them to two rooms to allow parallel working sessions. Kasulu only had one computer room, but, since April 2003, it has also received the UNDP sponsored Cisco lab. The latter is exclusively meant for Cisco students and is comprised of six computers and five routers. Opening hours reflect the overall ICT approach. As a multipurpose community telecentre,

Table 5.4: Services and equipment

	Kasulu	Magu	Sengerema
Equipment	**13 computers in classroom (also serves as Internet Café)** **6 computers and 5 routers in the Cisco lab**	**13 computers**	**7 computers in the Internet Café, 9 computers in training room (plus 9 old computers)**
Opening hours	**Monday to Friday 7pm – 10 pm** **Saturday to Sunday 11am to 1pm**	**Monday to Friday 8am – 4pm** **(8am-2pm for focus groups, 2pm to 4 pm for the general public)**	**Monday to Friday 8am – 7pm** **Saturday to Sunday 9am – 7pm**
Internet usage fees	**1,500 TSh. per hour**	**1,000 TSh. per hour for the general public** **500 Tsh. per hour for focus group members**	**1,000 Tsh. per hour**
User Basis /user per 100 pop.	**~130** **0,38**	**70-110 (est.)** **0,51**	**80-130 (est.)** **0,21**

Sengerema has the most extensive opening hours from 8am to 7pm on weekdays and from 9am to 7pm on weekends. At all hours, public Internet services and training sessions can be conducted simultaneously. The Magu facility is open between 8am to 4pm on workdays only. Thereby, the time from 8am to 2pm is for exclusive use of the focus groups. The general public is only allowed between 2pm and 4pm. In Kasulsu, primarily due to limited power supply, operations are constrained. During the day, some constricted training sessions for TCC students and teachers take place. The Internet cafe opens late from 7pm to 10pm on workdays and from 11am to 1pm on weekends.

Usage fees of the sites are between 1,500 TSh. per hour in Kasulu and 1,000 TSh. in Magu and Sengerema. Generally, half-hourly billing is possible. Sengerema offers its usage fee to every customer, Magu charges 1,000 TSh. for the general public and a discount rate of 500 TSh. per hour for focus group members. With 1,500 TSh., the Kasulu charge is highest among the three. The new rate was introduced in April 2003. Previously, they had applied a membership scheme, in which members had to pay 700 TSh. and non-members 1,000 TSh per hour. The price increase was a reaction to the scheduled withdrawal of the GCF, who had agreed to pay connectivity fees for a six-month period. The phasing-out of the cost transfer has left the operators in despair to pay the costs of connectivity by themselves. Since most of the Kasulu users were members, the majority of customers were confronted with an over 100 percent price increase, which in turn led to tensions between customers and the management. The increase might prove to have been a hasty decision, because it discourages new customers and might also overrate the role of user fees as an income basis. This is proven for the Sengerema example. According to Intelecon (2002: 13), it is chiefly computer training that has been the main source of income. For 2001, the Intelecon report stated that the total income of the Sengerema telecentre had amounted to only a moderate 11,000 US$, of which 10,000 US$ were from computer training.[371]

However, the effect of the new prices will only become visible at a later stage. During the old membership scheme, the user basis in Kasulu was roughly 130 users, most of them as regular customers. For Magu 70-110 users are estimated, Sengerema has approximately between 90 and 140. Taking the short period of time and the difficult circumstances of operation, these numbers can be noted as a success. Particularly if compared to the average Tanzanian Internet density of 0.23 (ITU 2004: A-74), the overall record has been good – with Magu (0.51)

[371] Also with regard to Sengerema, the same source adds that the first 2002 quarterly revenues amount to 2,800 US$, which is almost twice the average for 2001. At least these new figures are a clear indicator of further increases, see Intelecon (2002: 13).

and Kasulu (0.38) above the national average and only Sengerema (0.21) slightly below.[372]

5.8.2.4. Partners and Funds

Strong donor backing was a prerequisite to raising the needed resources to carry out the projects. Kasulu and Sengerema represent a joint donor approach. With respect to international donors, Kasulu has been affiliated with the Global Catalyst Foundation, SchoolNet, UNDP, and UNHCR. Sengerema has been part of the original telecentre approach and is linked to ITU, UNESCO, and IDRC, accordingly. Magu has only partnered with IICD. As a national partner, COSTECH has had a stake in Sengerema and Kasulu, and acts as a national supervisor and coordinator for both sites.

Considering the business models and the volume of funds, Sengerema has been the largest and most ambitious project. Three international and eight national partners indicate the exceptional significance that has been attributed to the Sengerema approach. At the same time, the multi-donor approach also caused considerable frictional losses. A National Management Committee comprising representatives of all organisations involved, was established and, especially in the starting up phase between 1997 and 2000, progress has been slow. The identification of an appropriate finance mechanism, assignment of responsibilities and the question of future ownership status had paralysed the process.

In addition, financial flows were erratic. Donors, government, and the local community had pledged to invest 570,000 US$ over a two-year period. Until the end of 2002 only 50 percent of this had been remitted and some partners are reported to be having trouble meeting their commitments.[373] In comparison to the other sites, large-scale and rather costly telecentre approaches, like in Sengerema, tend to carry considerable risks. It has not only been the challenge to seek a balance between high expectations and tangible achievements, it has also been the requirement to turn high investments and considerable running costs into a viable business model.

---

[372] Calculated on the basis of estimated user numbers and the size of the towns.

[373] See Intelecon (2002)

Table 5.5: Funds and Partners

	Kasulu	Magu	Sengerema
Received Funds (as available)	120,000 US$ from GCF for the whole KIP for 2000-2003 plus 6 months of connectivity	122,000 from IICD for a three-year period (2000-2003)	In 2002 pledges exceeding 570,000 US$ over a two-year period; to date 50 percent had been remitted
Interna-tional Partners	Global Catalyst Foundation SchoolNet UNDP UNHCR	IICD (International Institute of Communication and Development)	ITU UNESCO IDRC (International Development Research Centre)
National Partners	COSTECH (Commission for Science and Technology)	-/-	COSTECH (Commission for Science and Technology) TTCL (Tanzania Telecommunication Company) TCC (Tanzania Communication Commission) OUT (Open University of Tanzania) TCCIA (Tanzania Chamber of Commerce, Industry and Agriculture) IAE (Institute of Adult Education) TLSB (Tanzania Library Service Board) Prime Minister's Office

Kasulu also brought various donors under one roof. But here, donor coordination has been easier, because the schedule of responsibilities had been clearly assigned. The Global Catalyst Foundation was committed to establishing Internet connectivity in the region and took the lead in the overall management of the project, School Online was providing hardware and software, as well as the funds needed for professional development and capacity building. UNHCR provided the transport of equipment and logistical support, and UNDP supported the Cisco lab. The Global Catalyst foundation contributed most funds to the project, amounting to 120,000 US$ within a three-year funding commitment for all three implemented Internet facilities in Kasulu. Operations started within one year, a time span that indicates an effective collaboration of all actors. The approach has focused on ICT provisions for the college, and additional services, like the Cisco lab, were attached wherever suitable expendabilities were identified.

Magu has represented a very focused and rather small-scale approach to ICT. IICD has contributed 122,000 US$ over a three year funding commitment. More than the other two projects, Magu has applied a bottom-up approach, that, to a large extent, took up the proposals of a 1999 Mwanza roundtable on ICT. Both the size of the project and the commitment of local stakeholders contributed to a smooth implementation. Thus, the project has not only rooted in the local set up, but also within a wider national network of agro marketing. Nevertheless, the contractual phasing out of donor assistance leaves the operator alone with the unresolved problem of connectivity – a common theme in all three surveyed facilities.

5.8.3. Users and User Profiles of Rural ICT Utilisation

The following section seeks to assess the user and utilisation profiles of rural ICT utilisation in Kasulu, Magu, and Sengerema.[374] Figure 5.15 mirrors the gender ratio of rural ICT usage. Obviously, there has been a significant gender bias regardless of the location of the Internet site. But, whereas in Dar es Salaam every female roughly faces two males, the proportion goes down to 1:3 in rural areas and reflects the difficult and disadvantaged living conditions of rural women.

Figure 5.15: Sex of Rural Internet Users

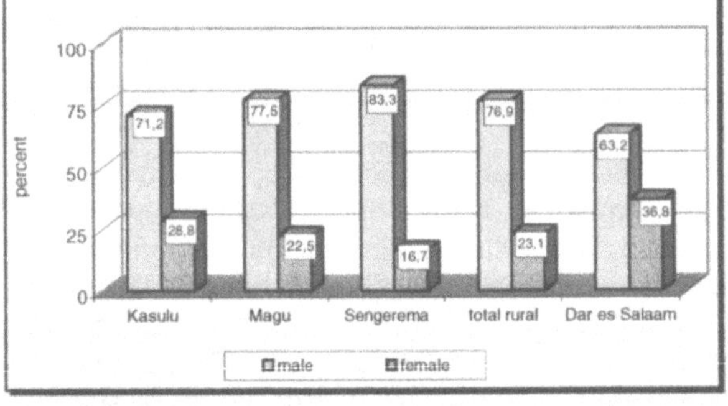

Figure 5.16 addresses the age structure of rural IT users and respective numbers for Dar es Salaam. Apparently, the majority of Internet users in all locations is between 21 and 30 years old, in rural areas, this age group has counted for an average 45 percent, in urban areas, as represented by Dar es Salaam, it has been even 49 percent. Users in their twenties represent the dominant user group.

[374] For a better comparison, analogous data from Dar es Salaam Internet cafe users have been included into the profiles in order to pay extra attention to urban-rural distinctions. The Dar es Salaam data is owed to Issmaïl Nnafie, Eindhoven University of Technology, who generously made his raw data available. Issmaïl surveyed 346 Internet cafes in Dar es Salaam in early 2002. His report can be found in Nnafie (2002).

Focusing on urban-rural variations, urban users have a strong user basis in the youth segment. Taken together, 84 percent of the urban users have been younger than 30 years. Rural users tended to be older than their urban counterparts. The age group between 31-40 has been well represented and even the above 40 years added up to roughly 16 percent of users. Thus, there is a different ICT usage pattern in urban and rural areas. While it has been the youth who define the picture in urban public access facilities, the age structure in rural areas has been more balanced.

Figure 5.16: Age of Rural Internet Users

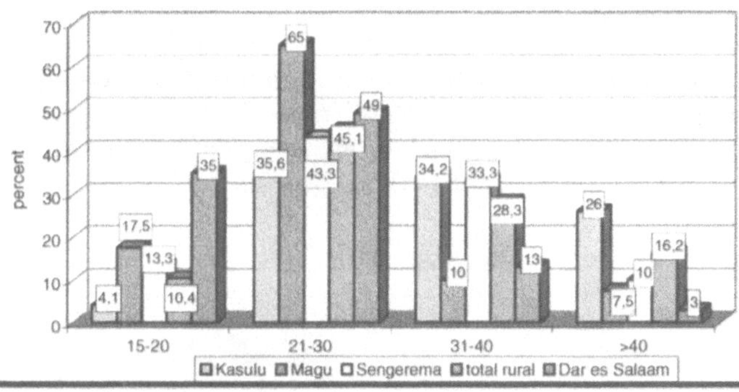

Beyond average rural users figures, interesting variations appear within the three sites. Eye-catching is the significant youth representation in Magu. This might be founded in the special target group scheme pursued in Magu. However, at the same time, users from older age groups are underrepresented. If only limited capacities are available, extra targeting of one user group tends to be at the expense of other groups. Whereas the age structure in Sengerema is close to what could be considered a normal distribution, i.e. a broad majority from the middle age values with some fringes to the under 21 and above 40 margins, again, the Kasulu user structure is different. The middle body of user groups is once again strong, the very young ones are hardly represented, while the group above 40 counts for a considerable fraction of 26 percent of the overall user basis. Once more, the host shapes the structure of the customer basis. Recruiting is mainly from the Teacher Training College, an institution of adult education.

The educational background reveals a very consistent structure of rural Internet users as shown in figure 5.17. Neither urban-rural distinctions nor variations within a rural focus produce any significant effects on the user structure. The great majority comes with advanced secondary education, at certificate or diploma level. They are concentrated at the 50 percent level, slightly higher in rural, a little less distinctive in urban areas.

The user groups rating second are either connected to secondary education or are holders of a university degree. The average rural numbers show 'secondary education' and 'university degree' in equal strength at roughly 21 percent. Focusing on rural sites only, some variations appear with a higher proportion of university degree holders in Kasulu, reflecting the composition of personnel at a Teacher Training College. In contrast, the ranking of Dar es Salaam based users is more distinct, putting 'secondary education' with 30 percent in second position, followed by 16 percent university degree holders in third position. The latter also corresponds with the predominance of young age groups as main Internet cafe users in Dar es Salaam. Only a very small percentage of users, ranging from 2.5 percent to 7 percent, have a primary education background. These modest results might reflect both, from an operator's perspective, the difficulty to reach this group, and, from a user's perspective, a lack of appropriate services that would promise any benefits from ICT usage.[375]

Figure 5.17: Educational Background

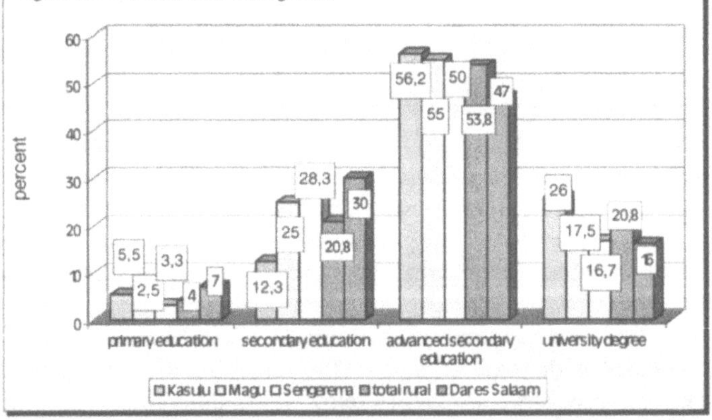

Figure 5.18 quantifies the average monthly income in Tanzania Shilling of rural (and Dar es Salaam) public access facility users. Though the figure does not say anything about the average income situation, it reaffirms the prevalence of widespread poverty in the country. A national average of 19,9 percent of the population lives below a poverty line of 1 US$ a day; 59,7 percent are below 2

[375] The questionnaire also provided an opportunity to tick 'no formal education'. But only 4 users in all 173 submitted questionnaires have marked the category. Consequently, the category has been omitted from the chart – but it remains a strong indicator for an existing educational bias in ICT utilisation.

US$ a day.[376] The latter totals an income of approximately 60.000 TSh. per month, an amount that corresponds to a middle income in the given income scale. The chart reveals the interesting fact that a low income is not a significant impediment to ICT utilisation. Each income group is fairly represented, slightly in favour of an upper middle-income segment (based on a general low level). But this rather mirrors the normal income distribution than any particular access barrier.

Figure 5.18: Income of Rural Internet Users (in TSh.)

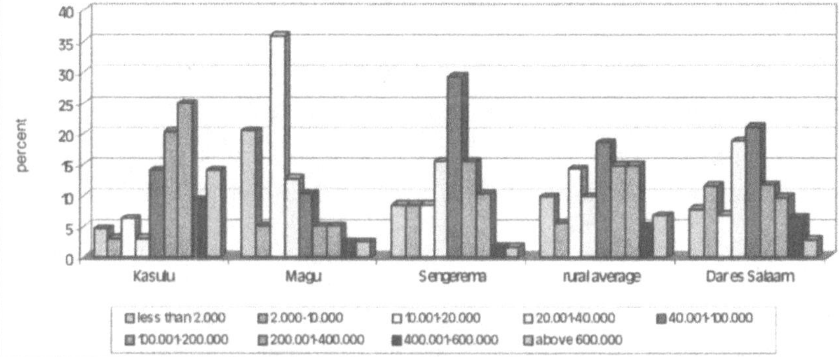

Furthermore, the different characteristics of the three rural sites are striking. Sengerema very much equals a normal distribution similar to the Dar es Salaam figures. Considerable user numbers come from lower income groups, a strong segment from middle and upper middle income users and a few from the 'upper' income groups. The figures from Kasulu and Magu are different. Kasulu users are concentrated in the upper middle and upper income groups with only little usage in the lower income segment, Magu represents the opposite, with a strong user basis in the lower income groups and only little representation of the middle and upper income groups. Taking together the three public access facilities, they provide a strong argument for the assertion that income is not the decisive factor of ICT utilisation in rural areas. As indicated earlier, the distinctive approach of each respective facility to ICT significantly shapes the composition of its user basis, in each case further predisposed by individual local conditions. The prevalence of higher income groups in Kasulu mirrors the composition of staff and teacher trainees in an adult education institution, as well as the high NGO density in the region. Magu users are to a large extent predetermined by the explicit pro-poor approach with the special target group

[376] Figures according to UNDP (2003).

design while Sengerema's broad multi-purpose approach might equal an Internet diffusion pattern not influenced by specific targeting.

Figure 5.19: Applications utilised on a regular basis

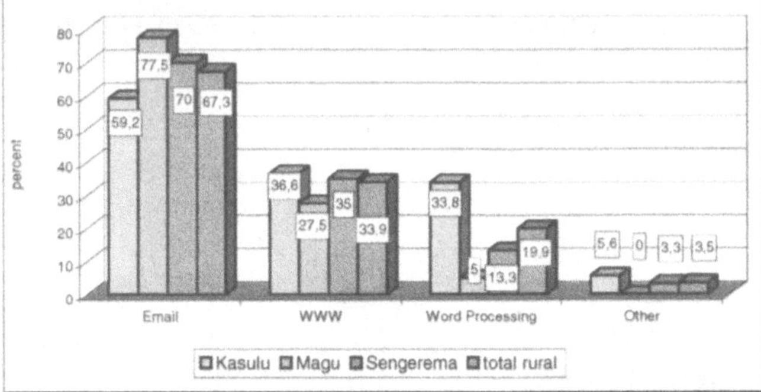

The figures above have identified some key characteristics of those who make use of public access facilities. The figures below seek to quantify ICT utilisation and the purposes of ICT utilisation. Figure 5.19 shows the applications used on a regular basis – permitting multiple responses. Obviously, email is by far the most prevalent application. For 78 percent of the visitors in Magu, 70 percent in Sengerema and 60 percent in Kasulu, email is the most frequently utilised tool. Therefore, email is the main engine for increasing ICT diffusion in 'rural' Africa. WWW usage, excluding webmail products, ranges from 28 percent in Magu to 37 percent in Kasulu, figures that place the WWW second in a general application ranking. The WWW serves many different purposes, but, although there some variations, a basic usage pattern reappears in all three surveyed locations, i.e. the main reason for surfing the web is to access news and education-related information – just in contrast to Dar es Salaam usage patterns, where entertainment has been the main reason to surfing the net. News is mainly international, to a lesser extent national news and very rarely regional news. Educational matters range from general inquiries, e.g. study opportunities at home and abroad, sponsorships and application procedures, to background research on a given topic. Thus, Email and the WWW are the applications most frequently demanded in public access facilities.[377] Other than that, only word processing has some significance. This applies especially for Kasulu, where 37 percent of the visitors make use of it regularly. In Sengerema, only 13 percent of

[377] The usage pattern is even more pronounced in Dar es Salaam, where email is used by 94 percent of the visitors on a regular basis and the WWW by 63 percent of the visitors (Nnafie 2002).

users apply word processing on a regular basis, in Magu it is only 5 percent. Basic applications are only a by-product, although for some users, word processing might be an important tool. Other applications, e.g. spreadsheet and desktop publishing, are only used sporadically.

The questionnaire also provided an opportunity to comment openly on the subject of ICT. In most cases, the comments were related to problems and difficulties faced by the users. Thereby, the following issues, shortcomings and positive receptions, have been identified:

- *Services:* Roughly 33 percent of the given responses were related to services. Dividing the issue into main subcategories, responses covered the demand for more training courses and a corresponding availability of qualified teachers, the need for more computers and extended opening hours to overcome time limitations to using the facility. Further noteworthy are the considerable numbers of users who appreciated the existence of the telecentres, very often highlighting the professionalism and amiability of the personnel. Customer satisfaction has been particularly high in Magu, some reservations relating to sub-optimal customer care have appeared in Sengerema.

- *Infrastructure:* Infrastructural problems have been identified mainly with regard to power supply and power interruptions, network stability, and reliable and fast connectivity solutions. Power issues have been particularly pressing in Kasulu. Periodic system crashes have been reported from all sites, always resulting in considerable inconveniences for the users. Though all sites are fairly well-equipped, recurrent customer inconveniences highlight the importance of efficient and available network administration and maintenance.

- *Content / Guidance:* 23 percent of the users have addressed issues of content and guidance. Of these, most users have demanded the provision of specific information covering a wide spectrum ranging from education and medical treatment to news, sports, and so on. Responses from Magu highlighted the impact of what has already been achieved, especially with regard to the crop marketing project, information on farming and cultivation methods and appreciation of tailored on-line training courses. Indeed, many users in Magu have insistently underlined the positive impact of the Internet on their living conditions.

- *Costs:* 14 percent of the users have complained about high costs of using the facility. Of these, the majority addressed the issue of user fees, with only a few users referring to the costs of training courses. The issue was particularly relevant for Kasulu users, where usage fees increased, more than doubling the costs for most users. In contrast, responses relating to costs were moderate in the other two towns. Here, users have mainly emphasized the financial burden that comes with Internet usage, but as a bottom line, accepted the price level, only proposing discount rates for frequent users or the introduction of a 15 minute usage fee instead of the current 30 minute minimum usage fee.

5.8.4. Concluding Rural ICT Utilisation

Rural ICT utilisation is an ambitious concept. It goes much deeper than only providing Internet and computers to remote areas. The main objective aims to bring equal opportunities to marginalized and disadvantaged communities and to provide ICT as an enabler for rural development. Inevitably, the innovative concept has produced high expectations and a gap between the technical and non-technical starting basis and intended outcomes. In rural areas, basic requirements often taken for granted in other environments, aggravate the difficulties of introducing ICT. Poor transport is a logistical challenge, unreliable power supply hampers smooth business procedures and a lack of sufficient bandwidth forces the operators to opt for costly connectivity solutions. Furthermore, the communities themselves are confronted with a completely new technology. Fear of contact literally needs to be overcome and transformed into a positive attitude towards the technology.

In addition, rural ICT facilities differ from their urban counterparts in another aspect. All operations are concentrated under one roof, so that business activities cover a broad spectrum of diverse responsibilities. In urban areas, Internet cafes usually just have to manage their business as a public access facility, in other words, provide for installation and maintenance. Equipment is delivered, generally including an after sale service package, connectivity is usually given by an ISP in a competitive environment, training is offered by many different institutions, allowing one the choice from a broad range of training courses and, finally, special target schemes are pursued by specialised agencies. In short, the challenges and risks of ICT utilisation are shared between many different players. In contrast, rural facilities have to cope with all the challenges by themselves and they have to transfer the sum of individual problems into a coherent and comprehensive strategy.

Rural Internet utilisation is in its infancy. An average user density of 0.37, or, in other words, a user-population ratio of 1:270 in all three surveyed locations, clearly demonstrates the scope of the assigned task. Only a minority makes use of the technology, and the low level user basis not only corresponds to the novelty of the services, but also to little awareness, a lack of appropriate and relevant services and to a perceived weak cost-benefit ratio for ICT utilisation. On the other hand, turning the perspective upside-down to focus on those who use the facilities, the Internet access sites present themselves as functioning and demanded service points, already well-seated within their communities. A further elaboration reveals a typical ICT diffusion pattern, thereby indicating possible roads ahead for future advancements. In all three locations, specific user profiles prevail. A first general classification typifies the average rural ICT user as a 20 to 30 year-old male, who has an educational background of at least advanced secondary education. He is proficient in English and uses the Internet facility mainly for private email communication, but is also familiar with the WWW which he mainly uses for education. He has taken a training course in office applications and visits the Internet site 2-3 times a week. Consequently, the survey data to a large extent reaffirms a user profile that is well-known from other surveys dealing with ICT user profiles throughout the world. A notable exception appears with regard to the income situation. Users come from a wide range of income groups, highlighting the fact that a low income must not necessarily be an obstacle for ICT usage.

Beyond these average and aggregated figures, user profiles also show considerable variations, largely influenced by the nature and orientation of the Internet facility's host and the given local environment. The Kasulu Internet site is located at the TTC and the technology has been introduced for purposes of education. At the same time, the user profiles of the public Internet cafe attached to the college show that a considerable number of customers are recruited from the general environment of the TCC. The same applies for the Magu public access facility, where user profiles have been shaped by the overall business approach, particularly the special target schemes for women, youths, and farmers and the explicit pro-poor strategy. In contrast, Sengerema follows a broad multipurpose approach and features the same basic pattern of the user profiles, but is significantly less distinctive with regard to particular target groups. Beyond such individual characteristics, all three facilities show a large proportion of student users, emphasizing the role of young users and their immediate affinity with the technology. Also sizeable users are NGOs, public administration and the business community. Not subject to any special targeting, these form a small but stable customer group quickly responsive to the newly-offered services.

Each site shows individual user group patterns, mainly caused by the type of access facility and the existence of special target schemes. These observations carry a twofold implication: First, the establishment of a rural public access facility can be built on a fair demand of existing user groups in each community, and, secondly, usage patterns can be directed intentionally depending on the business approach chosen. The strategy towards the promotion of ICT determines the outcomes. Thus, there is no predetermined road to success, each facility must identify its needs and develop its strategy accordingly. Regardless of the tools applied and the strategies chosen, one question must be answered. What is the benchmark of rural ICT utilisation? If it is customer satisfaction, emphatic appreciation of the newly available technology or a full plant utilisation during peak business hours, all three Internet facilities would have already achieved their goals. Moving further on and tackling broader aspects of community development is a more complex challenge. But despite all difficulties, the examples from Kasulu, Magu and Sengerema show that the groundwork has been done and lessons learned included to increase the likelihood of viable and sustainable project designs.

5.9. Prospects of ICT Utilisation in Tanzania

ICT utilisation in Tanzania has increased significantly in recent years and a broad mix of different applications have been installed in many areas serving a variety of purposes. Services became gradually more available, brought new opportunities and resulted in improved working procedures in many different fields. At the end of 2004, ICT has developed into an important tool for an increasing number of people. Increasing ICT diffusion originated in the reforms of the telecommunication sector that commenced in 1994 and set the course for these developments. Macroeconomic adjustments, sectoral policy reforms and new regulative provisions created an environment conducive to private sector led growth in the ICT sector and paved the way for new electronic voice and data services that became gradually more available and affordable. Agents of change have been corporate bodies, foremost private sector companies, public institutions and NGOs. Their commencing ICT utilisation shifted ICT services from a niche product to a level that became a first critical user basis from where advanced diffusion patterns could develop. Internet cafes, stimulated by new entrepreneurial spirit and benefiting from an improved business environment, took advantage of the growing acceptance and started to offer Internet services quickly demanded by an increasing number of the population. Particularly the supply of email services boosted the business of Internet cafes, notwithstanding the visible signs of a temporary oversupply and an inevitable market consolidation. Thus, the demand for email services was the main engine for increased Internet utilisation. The overall prospects are still promising. Urban areas have been increasingly supplied and progress to include the districts has been fairly good. The core customer segments developed into a stable user basis and new customer segments have been targeted successively. Furthermore, due to scaling effects and new low cost technology solutions, costs of infrastructure deployment and service provision will continue to decrease, so that the size of communication networks will expand further. This general trend will also produce network externalities that in the past contributed to an ever growing ICT user basis.

Thereby, access to ICT has been unevenly distributed alongside centre-periphery and high-end and low-end divisions. Corporate bodies started to deploy ICT tools at central management level, from where they have trickled down ever since. Similar applies for individual usage patterns, where urbanity and a good educational background have been the most benefiting factors for commencing ICT utilisation. These overall trends resulted in a prevailing national digital divide, but, at the same time, only created the necessary user basis for ICT advancements in the country. Thus, there has been a group of early adopters that have smoothed the progress of ICT utilisation. However, new user segments have been successively brought in – by growing ICT availability and decreasing costs, as well as better service delivery by the ICT providers. Overall entrance costs to access communication networks declined and attracted new customer segments. The viability of this strategy has been shown with regard to NGOs and SMEs. Their ICT utilisation had often started as soon as some basic ICT services became available. Since both include small scale ICT users often working with only limited resources, their starting utilisation proved the growing demand for ICT. Both groups valued the technology as an important tool for improving their organisational performances, and, particularly the availability of email services, brought new opportunities for both groups. Thus, the existence of a national digital divide still prevails, but the permeability to ICT access opportunities has increased ever since the arrival of the technology.

Thus, increasing market efficiency has been one way to promote ICT and resulted in tangible effects. The second pillar based on special ICT targeting schemes and special programmes to introduce technology tools to individual beneficiaries. Thereby, sectoral ICT strategies became a strategic element of technology promotion. Within this shift towards strategic targeting, the government itself became a focal addressee of ICT endorsement, and, to this end, the government streamlined ICT into a coherent approach and developed sectoral approaches aiming at promoting a goal oriented ICT deployment in the country. One major output has been that the government successfully developed into a model user of ICT. Particularly the recent e-government initiative has introduced a broad integrated e-government approach built upon existing structures developed under the PSRP, successively included sectoral approaches, such as in health or education and finally, will make use of a proposed joint hard- and software structure connecting all government agencies from the central to the district level to one communication network. The initiative has been the latest example of a considerable e-government record, that, like in the general picture, has been implemented as a top down approach starting in a handful of key government institutions and gradually including the line ministries and other public agencies. Focus was on internal working procedures to smooth the administrative process and policy planning, where ICT tools have been increasingly deployed. Furthermore, all major government agencies have developed resourceful websites, making procedures, plans and regulations accessible and thus have contributed to more transparent government. Thereby, strong political guidance, particularly provided by the CSD, has been a key factor for a fairly good performance record. The department was at the centre of all e-government planning and steered the ongoing reform of the public service. Currently, Tanzania is at a threshold of more advanced e-government applications facing the challenge to deepen ICT outreach to individual sectors and to the lower ends of government administrations. Thereby, future challenge will be less the availability of the necessary network infrastructure, but the design of deployed applications, the creation of sufficient user capacities and the strength to actually implement the reform programme. Individual experiences look promising, but could not conceal shortcomings. Sectoral approaches, like in health or education, are still infant and only performed reasonably well at the higher ends of tertiary education or research networking. Tackling the broader segments still lacks orientation. Setting the goal to provide computers to all schools or local health centres rather tend to ignore the realities of the public education and the health system, which are both in a poor state. Here, priority segments need to be identified and possible trade offs must be addressed in order to develop appropriate ICT roll-out strategies that do not focus on ICT as an end in itself, but on improving public service delivery, thus seeking a broader development impact. Nevertheless, sectoral activities have gone beyond pilot phases and, now there is a good amount of best practice experiences allowing to assess what will work and what will not.

In health, a focus should be on the advancement of the health information system, MTUHA, enabling to deal with health data and use them for better health administration, drug storage and supply and for evaluating training needs. Strengthening capacities for research networking and health education are complementary issues. Individual projects dealing with online consultations for diagnosis or treatment will remain subordinate and solitary projects, the provision of basic communication for health-centres, however, would be a realistic and desirable target for the medium term future. In education, focus should be on post-secondary education and education administration, where ICT deployment resulted in sizable effects on overall sectoral performances. Particularly tertiary education benefited from increased technology diffusion. In contrast, targeting secondary schools has been a disappointing endeavour, with some computer installations in private schools, but bypassing the public secondary education system almost completely. Thus, there are only humble prospects for a mandatory syllabus for computer training in secondary schools. This does not mean to withdraw from secondary school targeting, but instead of looking at the schools individually, secondary school access to ICT should be integrated into community or shared usage approaches making use of neighbouring facilities, so that the focus should be on teacher training, developing training modules or content delivery of teaching materials to be supplied to, for example, district access points, rather than deliver hardware to each individual school.

As it is applicable for other sectoral approaches, let it be health, education or service provision for specific target groups, the projection of future ICT utilisation will depend on the force of rural telecommunication development. Since 2003, first elements of a framework for rural ICT development became visible, targeting payphone services at the village and Internet services at the district level. Considering the state of telecommunications in the country, these targets are not overambitious. Though the details of a proposed universal access fund need to be awaited, the basic approach will focus on creating efficient rural telecommunication markets supplemented by subsidised service provision. The existing examples of rural ICT have exposed one of the strongest arguments for rural ICT promotion; there is a considerable demand for ICT and people are willing to pay for communication services. Similar to ICT diffusion patterns on the national level, rural developments developed from a rather narrow user basis only gradually trickling down to broader customer segments. ICT provision without additional measures in rising awareness and training is likely to result in modest effects on the user basis only, but the examples from Magu and Kasulu have shown that rural users have been responsive to targeting schemes, where entrance barriers to make use of the technology have been successfully lowered. However, the user base in rural areas will remain limited, so that sizeable scaling effects will fail to appear, so that focus will continue to be on public access solutions. Regardless whether facilities will be privately run businesses or publicly financed telecentres, both will face the challenge to provide appropriate services to a broad range of potential customers. The rural Internet sites covered in this work have shown that distinctive ways to approach ICT led to different outcomes implying that deliberate planning and good facility management can extend the reach of ICT. On the other hand, the examples have also indicated that focused project approaches are more likely to sustain in the long run. This contradicts the demand for multi-purpose facilities attempting to host broad customer segments. These rather tend to have overburdened the operators, so that a balanced approach providing broad and in-depth guidance at the same time will remain a key challenge for rural ICT utilisation.

Beyond issues of utilisation, content and services, it is paradoxically the government network that could develop into a major obstacle for upcountry ICT deployment, because the government network will bind significant amount of voice and data traffic without any benefits to private sector providers who performed reasonably well in a liberal market environment. A new government operator will almost inevitably lead to sizable market distortions and would turn upside-down the telecommunications development strategy based on private sector participation. Poor service delivery by public operators has been one major reason for the poor state of telecommunication before the sector reform commenced; and, bringing back in the state as large scale service provider without any resale arrangements for unused capacities, could result in an unintentional resurgence of defunct service provision in rural telecommunications.

6. Conclusion

This study has shown the viability of an ICT development strategy based on good policies and effective regulation aiming at creating a private-sector led development dynamic. The study has also shown the gradual increase of ICT diffusion and the positive effects of ICT utilisation in different sectors. Both developments have gathered momentum and formed a process, in which an effective framework is gradually maturing – including individual setbacks – and in which ICT diffusion is steadily increasing. The cause of the momentum was, firstly, due to increased sectoral differentiation and improved interaction, in which politics has gained new functionality with regard to reactive capabilities – to adjust to new market conditions – and proactive capabilities – to provide orientation and guidance to all actors involved; secondly to the continual growth of networks that steadily add new value to the networks as more users were connected through it. These forces have not only resulted in an emerging telecommunication sector – a modern sector employing an increasing number of regular tax payers with model character for other sectoral development policies – but, even more importantly, have created a basis for ever new advancements in ICT application, benefiting from and induced by a conducive policy environment.

At the same time, this is only a first step. Recalling the individual components of a comprehensive ICT for development strategy, a number of shortcomings can be identified. With regard to African developments in general, reform mechanisms of privatisation, competition and regulation meant to improve accessibility and affordability of reliable telecommunication services, have been implemented at different pace and depth, and, in practice, prioritised efforts of privatisation to break up defunct monopoly structures. Introducing competition in the different service segments, however, was often sidelined, having the effect, that monopolies, state-owned or privatised, still are a prevailing feature in African telecommunications. This applies particularly for the core businesses of the incumbents, such as fixed line telephony, international gateway operations and the provision of Internet backbones. Prevailing monopoly structures have been at the core of limitations of sectoral reform projects in many African countries.

Secondly, a market structure dominated by a former public utility is a significant challenge for regulatory agencies that seek affordable access through the creation of a fair and competitive environment. Although the number of regulatory bodies in Africa has grown considerably in the last decade, their capacities often remained weak, so that regulatory capacities requiring a sufficient number of experienced and well-trained regulatory staff able to act independently from government control, significantly deviates from the regulatory needs in a maturing market.

Thirdly, African countries have failed to take up the challenge of rural telecommunication development, sufficiently. This is only partly owed to the complex nature of universal access and an inexperienced environment dealing with such issues. Quick gains of sector reforms in the urban centres have led to an overestimation of market forces, resulting in a political and regulative retreat from pursuing universal access goals, as it could be observed in the country case Tanzania. A rethinking, putting new emphasis on the regulative powers based on a sophisticated mix of incentives and obligations for service delivery in disadvantaged regions, is only slowly reappearing. The instruments and mechanisms have been developed and set the course for new approaches in rural telecommunication development. However, African policy-makers and regulators have yet to provide evidence for their commitment and their ability to implement policies in support of universal access goals.

Thus, only parts of the original reform projects have been realised and faster advancements have been impeded by suboptimal decisions, weak implementation and enforcement capacities, by only unenthusiastic efforts to implement the reforms and by the unwillingness to solve conflicting aims, particularly with regard to the role of the former telephone monopoly. If the reform process is considered in its entirety, not with regard to individual – though even successfully implemented – components, the process can be characterised as a deferred reform. This is relevant in that sense, that pace is becoming a critical factor, not only in the realisation of roll-out goals, but particularly in creating opportunities by building a competitive advantage in developing products and services that have started to target a growing East African and African IT market.

Apart from challenges on the supply side of telecommunication services, there are also challenges on the demand side. Evidently, there are different kinds of demand. The modern urban sectors demand similar infrastructure like their counterparts in the developed world. There are islands of high intensity usage, whose market behaviour is similar to usage patterns in the developed world. There are early technology adopters – though rooted in a difficult environment – who are well integrated in global exchange processes.

However, this does not apply to the majority of users. For them, demand-driven service delivery is a vital component for increased and effective ICT utilisation. Thereby, the argument for demand driven solutions has two aspects. Firstly, empirical evidence has shown that broader ICT diffusion in Africa has followed a typical diffusion pattern, in which basic telephony and email services have been the engines for growth and will continue to be so in the near future. Thus, one must take into consideration the reasons why people value ICT. This does not necessarily follow economic thinking – putting more emphasis on the value-added service segments – but simply the logic to improve people's life by having the freedom to make a telephone call or write an email. This refers to a thought of Amartya Sen who regards the extension of freedom as the primary end and principal means of development (c.f. Sen 1999: xii). Sen highlights the perspective of human capability and the ability of people "to lead the lives they have reason to value and to enhance the real choices they have" (Sen 1999: 293). In this regard, ICT have already enriched people's lives, providing the freedom to make use of those telecommunication services they value most.

The other side of promoting ICT usage, is the provision of support systems to identify the demand and to articulate the needs with regard to ICT solutions. There is still much to be done to understand how ICT can affect the lives and livelihoods of the broader population and how ICT can enhance access to markets, institutions or the political process. A demand-driven support system acknowledges that providing knowledge, skills and supportive organisational and social structures is essential to making effective use of ICT, as well as that access to information is more than just physical, but is "furnishing people with the cognitive capacities and intellectual frameworks than enable humans to interpret, select and utilise information in ways that augment their capabilities to control and enhance the material circumstances and qualities of their existence" (David /Foray (2003: 10). Enabling support systems, addressing user skills, content issues and possible designs of linkages to the offline world, for example, mediating and translating online acquired knowledge to those who do not access telephones or computers, are in their infancy and touch upon questions beyond ICT, because successful systems require not just appropriate technology, but also people, processes, and institutions.

Many of the shortcomings of today's ICT approaches relate to the complex nature of ICT and to limited knowledge of the impact of ICT on economic and social developments. Considering the individual mechanisms of ICT promotion, however, the underlying cause of difficulties are in their majority very specific and can be detected by thorough assessments. These relate to weak project designs, lack of implementation capacities, insufficient technical capacities, poor institutions or a lack of commitment. Therewith, problems in ICT promotion, to some extent, simply mirror the problems well known from other sites of the development agenda. Though this observation might be commonplace, it implies that ICT, of course, are no panacea out of the vicious circle of poverty and marginalisation, but only one component of a much bigger project, that is, improving the performance record of development efforts.

Appendix 1: Selected African ICT Indicators and ICT related Rankings

	Teledensity per 100 pop.
Mauritius	**27.03**
Seychelles	**26.91**
Cap Verde	**15.99**
South Africa	**10.66**
Botswana	**8.72**
Namibia	**6.48**
São Tomé und Príncipe	**4.13**
Swaziland	**3.40**
Gambia	**2.80**
Gabon	**2.47**
Zimbabwe	**2.47**
Senegal	**2.23**
Sudan	**2.06**
Côte d'Ivoire	**2.04**
Equ. Guinea	**1.74**
Djibouti	**1.54**
Comores	**1.35**
Lesotho	**1.32**
Ghana	**1.27**
Mauritania	**1.18**
Togo	**1.05**
Kenya	**1.03**
Benin	**0.92**
Eritrea	**0.90**
Guinea Bissau	**0.89**
Zambia	**0.82**
Cameroon	**0.70**
Malawi	**0.70**
Congo	**0.67**
Angola	**0.61**
Nigeria	**0.58**
Burkina Faso	**0.54**
Mali	**0.53**
Ethiopia	**0.53**
Sierra Leone	**0.48**
Tanzania	**0.47**
Mozambique	**0.46**
Madagascar	**0.37**
Guinea	**0.34**
Burundi	**0.32**
Rwanda	**0.28**
CAR	**0.23**
Uganda	**0.22**
Niger	**0.19**
Chad	**0.15**
DR Congo	**0.02**
Liberia	...
Somalia	...

	Mobile density per 100 pop.
Seychelles	**55.35**
South Africa	**30.39**
Mauritius	**28.91**
Botswana	**24.13**
Gabon	**21.50**
Namibia	**8.0**
Cap Verde	**9.78**
Mauritania	**9.22**
Gambia	**7.29**
Congo	**6.72**
Equ. Guinea	**6.34**
Côte d'Ivoire	**6.23**
Swaziland	**6.10**
Senegal	**5.49**
Cameroon	**4.27**
Lesotho	**4.25**
Kenya	**4.15**
Togo	**3.49**
Benin	**3.22**
Zimbabwe	**3.03**
Djibouti	**2.29**
Ghana	**2.07**
Tanzania	**1.95**
Uganda	**1.59**
Mozambique	**1.40**
Rwanda	**1.36**
Nigeria	**1.34**
Sierra Leone	**1.34**
São Tomé und Príncipe	**1.31**
Zambia	**1.30**
Guinea	**1.18**
DR Congo	**1.06**
Madagascar	**1.02**
Angola	**0.93**
Malawi	**0.82**
Burkina Faso	**0.75**
Burundi	**0.74**
Sudan	**0.59**
Mali	**0.50**
Chad	**0.43**
CAR	**0.32**
Niger	**0.14**
Ethiopia	**0.07**
Guinea Bissau	...
Liberia	...
Eritrea	...
Comores	...
Somalia	...

	Internet density per 100 pop.
Seychelles	**14.52**
Mauritius	**9.91**
São Tomé und Príncipe	7.28
South Africa	**6.82**
Zimbabwe	**4.30**
Togo	**4.10**
Cap Verde	**3.64**
Botswana	**2.97**
Swaziland	**1.94**
Gabon	**1.92**
Gambia	**1.82**
Kenya	**1.25**
Senegal	**1.04**
Lesotho	**0.97**
Ghana	**0.78**
Benin	**0.74**
Djibouti	**0.69**
Côte d'Ivoire	**0.55**
Zambia	**0.49**
Guinea	**0.46**
Comores	**0.42**
Guinea Bissau	**0.40**
Uganda	**0.40**
Cameroon	**0.38**
Mauritania	**0.37**
Equ. Guinea	**0.36**
Nigeria	**0.35**
Madagascar	**0.35**
Rwanda	**0.31**
Angola	**0.29**
Sudan	**0.26**
Malawi	**0.26**
Mali	**0.24**
Eritrea	**0.23**
Tanzania	**0.23**
Burkina Faso	**0.21**
Namibia	**0.2**
Chad	**0.19**
Mozambique	**0.17**
Sierra Leone	**0.16**
Congo	**0.15**
Niger	**0.13**
CAR	**0.13**
Burundi	**0.12**
DR Congo	**0.09**
Ethiopia	**0.07**
Liberia	...
Somalia	...

	Investment in telecom. Infra-structure 1996-2002 (with private sector particip.)
South Africa	**1.0654.8**
Nigeria	**982.7**
Côte d'Ivoire	**827.4**
Ghana	**436.1**
Senegal	**406.8**
DR Congo	**369.7**
Mauritius	**365.6**
Tanzania	**321.0**
Cameroon	**266.1**
Uganda	**204.1**
Congo	**111.9**
Kenya	**107.0**
Mauritania	**99.6**
Benin	**90.4**
Botswana	**80.0**
Guinea	**75.3**
Angola	**75.3**
Zambia	**56.9**
Niger	**52.7**
Zimbabwe	**46.0**
Mozambique	**44.0**
Mali	**42.7**
Eritrea	**40.0**
Burkina Faso	**36.6**
Gabon	**35.0**
Swaziland	**33.6**
Lesotho	**33.5**
Malawi	**25.5**
Sierra Leone	**23.5**
Burundi	**15.6**
Rwanda	**15.6**
Chad	**13.0**
Madagascar	**10.1**
Gambia	**6.6**
Sudan	**6.0**
Togo	**5.0**
Namibia	**8.0**
Somalia	**2.0**
Guinea Bissau	...
Cap Verde	...
Liberia	...
Equ.-Guinea	...
São Tomé und Príncipe	...
CAR	...
Ethiopia	...
Djibouti	...
Comores	...
Seychelles	...

	Telecommunication revenue in Mio. US$
South Africa	**5.338.8**
Kenya	**483.4**
Côte d'Ivoire	**398.7**
Nigeria	**355.4**
Uganda	**275.0**
Tanzania	**218.5**
Zimbabwe	**207.2**
Senegal	**202.0**
Botswana	**176.5**
Sudan	**164.8**
Mauritius	**163.3**
Mozambique	**128.4**
Ghana	**126.1**
Angola	**113.8**
Gabon	**108.0**
Ethiopia	**104.4**
Madagascar	**96.0**
Namibia	**85.0**
Zambia	**69.2**
Cameroon	**68.6**
Burkina Faso	**63.3**
Benin	**61.9**
Mali	**59.2**
Togo	**42.1**
Cap Verde	**41.6**
Seychelles	**40.1**
Malawi	**33.8**
Guinea	**28.5**
Gambia	**26.6**
Swaziland	**26.4**
Mauritania	**25.2**
Djibouti	**22.5**
Chad	**21.3**
Rwanda	**20.1**
Equ. Guinea	**18.8**
Niger	**18.4**
Eritrea	**16.4**
Burundi	**14.7**
Lesotho	**11.9**
CAR	**10.8**
São Tomé und Príncipe	**6.9**
Guinea Bissau	...
Liberia	...
Sierra Leone	...
DR Congo	...
Congo	...
Comores	...
Somalia	...

	Area in 1000 km[2]
Sudan	**2.506**
DR Congo	**2.345**
Chad	**1.284**
Niger	**1.267**
Angola	**1.247**
Mali	**1.240**
South Africa	**1.221**
Ethiopia	**1.104**
Mauritania	**1.026**
Tanzania	**945**
Nigeria	**924**
Namibia	**824**
Mozambique	**802**
Zambia	**753**
Somalia	**638**
CAR	**623**
Madagascar	**587**
Botswana	**582**
Kenya	**580**
Cameroon	**475**
Zimbabwe	**391**
Congo	**342**
Côte d'Ivoire	**322**
Burkina Faso	**274**
Gabon	**268**
Guinea	**246**
Uganda	**241**
Ghana	**239**
Senegal	**197**
Eritrea	**118**
Malawi	**118**
Benin	**113**
Liberia	**111**
Sierra Leone	**72**
Togo	**57**
Guinea Bissau	**36**
Lesotho	**30**
Burundi	**28**
Rwanda	**26**
Swaziland	**17**
Gambia	**11**
Mauritius	**2**
São Tomé und Príncipe	**1**
Seychelles	**0,4**
Cap Verde	
Equ. Guinea	...
Djibouti	
Comores	...

	Population per km²
Mauritius	**597**
Rwanda	**331**
Burundi	**275**
Seychelles	**182**
Nigeria	**146**
Gambia	**139**
Uganda	**125**
Malawi	**114**
Ghana	**89**
Togo	**88**
Sierra Leone	**73**
Ethiopia	**67**
Swaziland	**63**
Benin	**59**
Lesotho	**59**
Kenya	**55**
Côte d'Ivoire	**52**
Senegal	**52**
Guinea Bissau	**51**
Burkina Faso	**43**
Eritrea	**43**
Tanzania	**40**
South Africa	**37**
Liberia	**34**
Cameroon	**34**
Zimbabwe	**34**
Guinea	**32**
Madagascar	**28**
Mozambique	**24**
DR Congo	**23**
Somalia	**15**
Sudan	**14**
Zambia	**14**
Congo	**11**
Angola	**11**
Mali	**9**
Niger	**9**
Chad	**7**
Central African Republic	**6**
Gabon	**5**
Mauritania	**3**
Botswana	**3**
Namibia	**2**
Cap Verde	
Equ. Guinea	...
São Tomé und Príncipe	...
Djibouti	
Comores	...

	GNI per capita in US$
Mauritius	**3.860**
Gabon	**3.060**
Botswana	**3.010**
South Africa	**2.500**
Namibia	**1.790**
Swaziland	**1.240**
Angola	**710**
Côte d'Ivoire	**620**
Congo	**610**
Cameroon	**550**
Lesotho	**550**
Senegal	**470**
Guinea	**410**
Benin	**380**
Sudan	**370**
Kenya	**360**
Zambia	**340**
Nigeria	**300**
Tanzania	**290**
Mauritania	**280**
Gambia	**270**
Ghana	**270**
Togo	**270**
Burkina Faso	**250**
CAR	**250**
Mali	**240**
Uganda	**240**
Madagascar	**230**
Rwanda	**230**
Chad	**210**
Mozambique	**200**
Eritrea	**190**
Niger	**180**
Malawi	**160**
Liberia	**140**
Sierra Leone	**140**
Guinea Bissau	**130**
DR Congo	**100**
Ethiopia	**100**
Burundi	**100**
Cap Verde	...
Equ. Guinea	...
São Tomé und Príncipe	...
Djibouti	...
Comores	...
Seychelles	...
Somalia	...
Zimbabwe	...

Data for 2002; Source: World Bank (2004c), ITU (2004), ITU (2003)

Appendix 2: Map of Africa

Appendix 3: Map of Tanzania

Appendix 4: List of Interviews

Name	Organisation	Date of interview
Abeynayake, Anushka G.	ILO Gender, MSE, Informal Sector	17.6.2002
Akisainen, Riku	UNDP IT Project Officer Development Managment Unit	24.4.2002
Alphonce, Galphonce G.	CTC Mwanza Director	22.4.2003
Amanyiswe, Gofry	REPOA – Research on Poverty Alleviation IT Officer	30.7.2002
Baxter, Jonny	DFID – Department for International Development Education Adviser	26.7.2002
Bayo, Peter	Angonet Executive Secretary	14.4.2003
Bhatt, Harish R.	SoftTech Managing Director	29.5.2002
Brems, Dirk	Belgium Embassy Attaché for International Co-operation	17.7.2002
Bugalama, Habby	Sengerema Telecentre Manager	23.4.2003
Calder, Geoff	Global Catalyst Foundation	1.4.2003
Chagani, Gulam H.	eBiz Solutions Managing Director	29.5.2002
D'Souza, Volven	CatsNet General Manager	29.5.2002
De Lauwer, Guy	Belgium Embassy Ambassador	17.7.2002
Egawa, Akio	Embassy of Japan Minister / Deputy Chief of Mission	17.7.2002
Eirola, Martti	Embassy of Finland First Secretary	8.7.2002
Fournadjiev, Vasco	UDSM Department of Computer Science	23.7.2002
Gewe, Lulinda Andrew	Tanzania Post Corporation Manager Computer&IT	5.6.2002
Goyayi, Frank	Afsat Communications Marketing & Sales Engineer	17.5.2002
Habicht, Frank	AfricaOnline Technical Manager	24.5.2002
Held, George	Mobitel Chief Executive Officer	21.5.2002

Hingora, Ahmed	Ministry of Health Head of Health Sector Reform Secreteriat	9.4.2003
Horneman, Johan	SimbaNet Sales & Marketing Project Manager	21.5.2002
Ickstadt, Christoph	TTCL/Detecon Acting Director Marketing	13.5.2002
Jensen, Nils	SIDA ICT Project Manager	27.3.2003
Kapange, Barnabas W.	Ministry of Agriculture and Food Security Department of Research & Development Scientific Information Officer	19.7.2002
Karhammar, Ralph	Embassy of Sweden / SIDA Regional Infrastructure Adviser	23.7.2002 27.3.2003
Kowero, August	Ministry of Communication and Transport Ass. Director, Posts & Telecommunication Services TCC Comissioner	2.5.2002
Kuwite, Siegfried	Tanzania Investment Centre Research & Information Systems Manager	28.4.2003
Laudridsen, Rasmus H.	Royal Danish Embassy Programme Coordinator Business Sector Support Programme – BSSP	27.7.2002
Lyimo, Bartholomew M.	Tanzania Greenhouse Gas Action Trust-TAGGAT & Environment Content Editor	5.6.2002
Magoha, John	UNDP National UNV Specialist Cisco Networking Academy	11.6.2002
Makwaia, Eduard A.	Bank of Tanzania Director Management Information Systems	5.7.2002
Maltzan, Oskar von	KfW Office Dar es Salaam – German Financial Cooperation Director	2.5.2002
Maselle, Naomi,	Crop Marketing Bureau – CROMABU Director	24.4.2003
Masika, Richard J.	Dar es Salaam Institute of Technology Vice Principal	24.7,2002
Matambalya, Francis	UDSM Faculty of Commerce and Management	17.5.2002
Mathew, Thomas	Education Consultant	16.7.2002
Mbita, Iddi H.	Datel Tanzania Commercial Representative	4.6.2002

Mbonigabha, Alphonce J.	Kasulu Teacher Training College Principal	2.4.2003
Medeye, Goodluck J.S. Ole	Tanzania Communications Commission – TCC Director Finance & Administration	10.6.2002
Messer, Adam	CyberTwiga CEO Chairman COAT	20.5.2002
Mey, Detelef	GTZ Office Dar es Salaam Director	18.6.2002
Miller, Jonathan	Miller Esselaar & Associates Consultant	12.6.2002
Mlaki, Theophilius E.	COSTECH Director of Information and Documentation	23.4.2002
Mloka, Cesear	Celtel Marketing Manager	23.5.2002
Mmuya, Max	UDSM Department of Political Science and Public Administration	15.5.2002
Moens, Nicolaas P.	IICD – Institute for Communication and Development Programme Manager	18.7.2002
Moore, Daniel C.	USAID Team Leader / Environment Programme	18.6.2002
Mruma, Isaac	Tanzania Communications Commission – TCC Public Relation Manager	10.6.2002
Msaki, Jimmy R.	USAID Aid Programme Assistant	18.6.2002
Mufuruki, Ali	Infotech CEO	22.4.2002
Mwakalinga, Hebron	Business Information Service	9.4.2003
Nfuka, Edophence Ngemera	UCC – University Computer Centre	10.6.2002
Nguli, Hamisi	COSTECH Scientific Officer	10.4.2003
Ntiro, Simbo	Consultant	22.7.2002
Nzuki, Margareth	ESRF Project Co-ordinator Tanzania Online Gateway	20.5.2002
Ogembo, Lameck	Mediapost General Manager	24.5.2002
Pastory, Pascal	Newafrica ISP Manager	5.6.2002
Philipp, Udo	ESRF Senior Consultant	6.5.202002

Riad, Hady	German Embassy First Secretary	15.5.2002
Rowberg, Erik	AFAM Managing Director	14.4.2004
Rutihinda, Cranmer	Tanzania Investment Centre Director of Research & Information Systems	28.4.2003
Rweyemamu, Pold	IT Vision DigIT Technical Editor	16.5.2002
Sachdev, Aatish K.	Cybernet IT Executive	14.4.2003
Sangiwa, Bill	CyberTwiga / Mobitel	22.7.2002
Sauvik, Leif B.	Embassy of Norway Minister Counsellor Deputy Head of Mission	17.7.2002
Sawe, David	The President's Office, Civil Service Department Director Management Information Systems	23.4.2002 11.4.2003
Senkondo, Charles Y.	Tanzania Global Development Learning Center Director	30.7.2002
Sheriff, Suhail	Raha Business Development Manager	5.6.2002
Sherwood, Bruce	Vodacom Tanzania Managing Director	7.5.2002
Yonah, Zaipunah	Simunet General Manager	15.4.2002
Yonazi, Enock	East African Community Engineer Planner (Communication)	11.4.2004
Zwan, Leo van der	Royal Dutch Embassy First Secretary Education	26.7.2002

7. Bibliography

AAAS – American Association for the Advancement of Science (1996): Electronic Networking in Tanzania, http://www.aaas.org/:/international/africa-guide/tanzania.htm

AAU – Association of African Universities (2000): Technical Experts Meeting on the Use and Application of Information and Communication Technologies in Higher Education Institutions in Africa, 17. – 19. May, 2000, University of Dar es Salaam, September 2000, http://www.aau.org/english/documents/aau-ictreport-toc.htm

Adam, Lishan (1999): Information and Communication Technologies in Ethiopia: Past, Present and Future Potential for Social and Economic Development, http://www.uneca.org/aisi/nici/ethiopia/Ethiopap7.htm

ADF – African Development Forum IV (2004): Governance for a Progressive Africa, http://www.uneca.org/adf/

ÅF-International (2002): Capacity Development Assistance to the Tanzanian Communications Commission. Final Inception Report, Stockholm

Africa Internet Forum / UNECA / infoDev (1999): Economic Internet Toolkit for African Policy Makers, http://www.infodev.org/projects/internet/010toolkit/afprelim.pdf

Africa Online (2001), What is e-touch?, http://www.africaonline.com/site/Articles/1,3,41144.jsp

African Lakes (2001): Africa Online Tanzania, http://www.africanlakes.com/subsidiaries/tanzania.html

Africanconnection.org (1999): Fact Sheet Tanzania, www.africanconnection.org/docs/factsheets/tanzania.html

AISI – African Information Society Initiative (1996): African Information Society Initiative (AISI). An Action Framework to Build Africa's Information and Communication Infrastructure, http://www.bellanet.org/partners/aisi.htm

AISI – African Information Society Initiative (1999): AISI-Connect National ICT Profile Tanzania, http://www2.sn.apc.org/africa/countdet.CFM?countries__ISO_Code=TZ

AISI – African Information Society Initiative (AISI) (1998): AISI Connect Database. National ICT Profile Tanzania, http://paradigm.sn.apc.org/africa/countdet.CFM?countries__ISO_Code=TZ

AISI (2003): Africa's digital agenda, www.uneca.org/aisi/

Analysis (2000): The Network Revolution and the Developing World. Final Report for World Bank and infoDev, Washington

Analysys (2004): Press Release 1 July 2003. Analysys awarded contract with the Tanzanian Communications Commission, www.analysys.com/default_acl.asp?Mode=article&iLeftArticle=1264&m=&n=

Apoyo (2003): Advice on the Provision of Telecommunications Services by Private Providers for Rural and Low-Income Users – Tanzania. Recommended Strategy and Operational Parameters Report. Consensus Building Workshop, November 2003, Dar es Salaam, http://www.psrctz.com/Press%20Releases/241103-Strategy%20Report%20v[1].final.pps

Apoyo (2003a): Advice on the Provision of Telecommunications Services by Private Providers for Rural and Low-Income Users – Tanzania. Demand Study for Rural Telecommunications in Tanzania, November 2003, Dar es Salaam, http://www.psrctz.com/Press%20Releases/241103-Tanzania%20Demand%20Study%20v.Final2311.pps

Apoyo (2003b): Consensus Building Workshop on the Draft Strategy Report. Draft Design of the Pilot Project, November 2003, Dar es Salaam, http://www.psrctz.com/Press%20Releases/241103-Pilot%20Project%20-%20Tanzania.pps

Augustin, Ulrike (1998): Privatisierung in Tansania, Hamburg

AVU – African Virtual University (2004): Education for the Knowledge Age. Degree Programs, http://www.avu.org/degree.cfm

Bangemann Report (1994): Recommendations to the European Council. Europe and the Global Information Society, http://europa.eu.int/ISPO/infosoc/backg/bangemann.html

Bell, Daniel (1999): The Coming of Post-Industrial Society, New York [first published 1973]

Bertelsmann Stiftung (2002): E-Government – Connecting Efficient Administration and Responsive Democracy, Gütersloh

Best, Michael (2003): The Wireless Revolution and Universal Access, http://cyber.law.harvard.edu/digitaldemocracy/best-wirelessrevolution-sept03.pdf

Bezanson Keith / Segasti Francisco (1995): The Elusive Search: Development and Progress in the Transition to a New Economy, IDRC Ottawa

BIPE (2000): What telecom regulation for low-income African countries? Study undertaken for the European Commission, http://www.africanconnection.org/documents/BIPEStudEnglish.doc

Bittner, Andreas (2001): Africa Online – Mission Possible!, in: Engels, Benno / Nielinger, Olaf (Hg.) (2001): Elektronischer Handel in Afrika, Asien, Lateinamerika und Nahost, Hamburg. S. 85-105

BMI – Techknowledge Group (2002): IT in Africa Handbook 2002. Tanzania, Johannesburg

BoT / TIC / NBS – Bank of Tanzania / Tanzania Investment Centre / National Bureau of Statistics (2001): Tanzania Investment Report 2001, Dar es Salaam

Bratton, Michael / van de Walle, Nicolas (1994): Neopatrimonial Regimes and Political Transition in Africa, in: World Politics, No.4/94, pp. 453-489

Bridges.org (2001): Spanning the Digital Divide. Understanding and Tackling the Issues, Washington

Bridges.org (2003): World Economic Forum-NEPAD E-Africa Commission E-Readiness Policy Programme. Building capacity to narrow the digital divide in Africa from within, Durbanville

Canadian High Commission Trade Office (2000): This Month in South, Southern and Eastern Africa. July 2000, Johannesburg

Caspary, Georg / O'Connor David (2003): Providing Low Cost Information Technology Access to Rural Communities in Developing Countries: What Works? What Pays? OECD Development Centre, Paris

Castells, Manuel (2000): The Rise of the Network Society. The Information Age: Economy, Society and Culture. Volume I, Oxford [first published 1996]

Castells, Manuel (2000a): End of Millenium. The Information Age: Economy, Society and Culture. Volume III, Oxford [first published 1998]

Cave, Martin (2002): Review of Radio Spectrum Management, an independent review for the DTI and HM treasury, http://www.ofcom.org.uk/static/archive/ra/spectrum-review/2002review/1_whole_job.pdf

Chapmann, Robert / Slaymaker, Tom (2002): ICTs and Rural Development. Review of the Literature, Current Interventions and Opportunities for Action, Overseas Development Institute, Working Paper No. 192

Chetty, Matthew (2003): Information and Communication Technologies (ICTs) for Africa's Development, paper prepared for the NEPAD Ministerial Conference on Science and Technology, http://www.nepadst.org/publications/docs/doc08_112003e.pdf

CIA – Central Intelligence Agency (2003): CIA Factbook 2003. Field Listing: Telephone Systems, www.cia.gov./cia/publications/factbook/fields/2124.html

CIA (2002): The World Factbook 2002,
http://www.cia.gov/cia/publications/factbook

Cisco Systems (2002): Cisco Networking Academy Program,
http://cisco.netacad.net/public/academy/cnap_brochure031403.pdf

Cisco Systems (2003): Cisco Networking Academy Program,
http://www.cisco.com/warp/public/779/edu/media/docs/DS_CNAP.pdf

Cline Cole, Reginald / Powell, Mike (2004): ICTs, 'Virtual Colonisation' & Political Economy, in: Review of African Political Economy, March 2004, No 99, Vol 31, pp 5-9

Cling, Jean-Pierre / Razafindrakoto, Mireille / Roubaud, Francois (eds.) (2003): New international poverty reduction strategies, London

COMNET-IT – Commonwealth Network of Information Technology for Development Foundation (2002): Study of E-Governance: Development of Country Profiles, Paris

COSTECH – Commission for Science and Technology (1999): The Development of Information Technology at the Tanzania Commission for Science and Technology, Dar es Salaam

COSTECH – Commission for Science and Technology (2002) E-Readiness Assessment. Summary of Preliminary Results, Dar es Salaam (draft)

COSTECH/IICD (2002): National ICT Programmes and Projects Networking Workshop, Golden Tulip Hotel, Dar es Salaam, 11th December 2002

CSD – Civil Service Department (2002): Management Information Systems, Dar es Salaam

Datel (n.d.): Datel Company Profile, Dar es Salaam

David , Paul A. / Foray, Dominique (2003): Economic Fundamentals of the Knowledge Society, in: Policy Futures In Education. An e-Journal, 1(1), Special Issue: .Education and the Knowledge Economy, January 2003

Deutscher Bundestag (Hrsg.) (2002): Schlussbericht der Enquete Kommission Globalisierung der Weltwirtschaft, Opladen

Diamond, Larry (1994): „Towards democratic consolidation", in: Journal of Democracy, Vol. 5, No. 3, pp. 4-18

DOI – Digital Opportunity Initiative (2001): Creating a Development Dynamic, http://www.opt-init.org/framework.html

Digitaldividends (2000): Is the Digital Divide a problem or an opportunity? http://www.digitaldividends.org

DOT Force (2001): Digital Opportunities for All: Meeting the Challenge. Report of the Digital Opportunity Task Force (DOT Force) including a proposal for a Genoa Plan of Action, http://www.dotforce.org/reports/DOT_Force_Report_V_5.0h.pdf

Dot-Com Alliance (2002): An Innovative Approach to ICT and Telecom Regulation and Policy in Africa, http://www.dot-com-alliance.org/newsletter/nettel4.html

Dravis, Paul (2003): Open Source Software. Prospects for Development?, Washington

Drew, Eileen P. / Foster, Gordon F. (1994) (eds.): Information technology in selected countries. Reports from Ireland, Ethiopia, Nigeria, and Tanzania, Tokyo, http://www.unu.edu/unupress/unupbooks/uu19ie/uu19ie00.htm#Contents

Dymond, Andrew (2002): Telecom and Information Services for the Poor: Towards a Strategy for Universal Access, Paper held at the TRASA, ITU and CTO Regional Workshop on Universal Access and Rural Connectivity, Dar es Salaam, 8.7.2002, http://www.itu.int/ITU-D/treg/Events/Seminars/2002/Tanzania/Documents/Andy-Two%20Gaps.pdf

e-Africa Commission (n.d.): About e-Africa, http://www.eafricacommission.org/about-eafrica.htm

EASSY – Eastern Africa Submarine Cable System (2003): Closing the final link around Africa, http://www.apc.org/apps/img_upload/6972616672696361646f63756d656e74/telkom.ppt

Easterly, William (2001): The Elusive Quest for Growth. Economists' Adventures and Misadventures in the Tropics, London

EIU – Economic Intelligence Unit (2003): Country Profile 2003. Tanzania, London

EIU – Economic Intelligence Unit (2004): Country Report Tanzania, May 2004, London

EMP – Emerging Market Partnerships (n.d.): AIG African Infrastructure Fund, http://www.empwdc.com/EMP_Africa.htm

Equant (2004): About Equant, http://www.equant.com/content/xml/v2_prehome_aboutequant.xml

Ernberg, Johan, (1998): Universal Access for Rural Development from Action to Strategies, paper presented at the First International Conference on Rural Telecommunications, Washington, 30 November - 2 December, 1998, http://www.itu.int/ITU-D/univ_access/telecentres/papers/NTCA_johan.html

Esselaar, Steve / Stavrou, Aki / O'Riordan, Jennifer (2004): VSAT Case Studies (Nigeria, Algeria, Tanzania), Research Report prepared for IDRC, CATIA and GVF, Johannesburg

Esterhuysen, Anriette (2002): „Networking For a Purpose: African NGOs Using ICT", in: Levey/Young (2002)

Ethinktank (2004): Messages, http://groups.yahoo.com/group/eThinkTankTz/messages/

Etisalat (2004): Annual Report 2003, http://www.etisalat.co.ae/AnnualReport/AR2003En.pdf

Fahamu (2002): Healthcare training and Internet connectivity in Sub-Saharan Africa, http://www.fahamu.org.uk/downloads/Nuffieldwebreport.pdf

Fauteck, Heiner (2004): MTUHA in Lindi. Working Plan for the ded/TGPSH work place „Advisor on Health Management Information System in Lindi Region", 03/2004 – 02/2006 (draft)

Fielding Graduate Institute (n.d.): Project Tanzania. Global Telehealth and Telemedicine for the 21st Century, http://www.fielding.edu/research/tanzania/summary.htm

Fillip, Barbara (2001): Digital Divide. Report prepared for JICA-USA, www.jica.go.jp/usa/topics/pdf/digital_divide_eng.pdf

Fink, Carsten / Kenny, Charles J. (2003): W(h)ither the Digital Divide?, http://www.developmentgateway.org/download/181562/W_h_ither_DD__Jan_.pdf

Fink, Carsten / Mattoo, Aaditya / Rathindran, Randeep (2002): An Assessment of Telecommunications Reform in Developing Countries, World Bank Policy Research Working Paper 2909, October 2002

G-8 Okinawa Charter (2000): G-8 Okinawa Charter on Global Information Society, http://www.dotforce.org/reports/it1.html

Grace, Jeremy et al. (2001): Information and Communication Technologies and broad-based development, draft World Bank paper, February 2001

GreenNet (n.d.): In the beginning there was FIDO ..., http://www.gn.apc.org/anniversary/fido.html

GTZ / ZEF (2002): Information and Communication Technologies for Development. Present Situation, Perspectives and Potential Areas for German Technical Cooperation in Peru, Lao P.D.R., Vietnam, Tanzania and Uganda, Eschborn and Bonn, http://www1.oecd.org/dac/ictcd/docs/matrixdocs/DEU_GTZ_paper1.pdf

Guédon, Jean-Claude (2004): The (Dis-)information Society?, in UNRISD News, Spring/Summer 2004, No. 26, pp. 4-5

Gurstein, Michael (2003): Effective use: A community informatics strategy beyond the Digital Divide, http://www.firstmonday.dk/issues/issue8_12/gurstein/index.html

Haddad, Wadi D. / Draxler, Alexandra (2002): The Dynamics of Technologies for Education, in: Haddad, Wadi D. / Draxler, Alexandra (eds.) (2002): Technologies for Education. Potentials, Parameters, and Prospects, Paris

Hafkin, Nancy / Wild, Kate (2003): ICT in Africa: The Challenge to donors in a Global Information Society, in: Levey, Lisbeth A., Young, Stacy (eds.) (2002): Rowing Upstream, Snapshots of Pioneers of the Information Age in Africa, Johannesburg (available online http://www.piac.org/rowing_upstream/tableofcontents.html)

Hallberg, Kristin (2000): A Market-Oriented Strategy for Small and Medium-Scale Enterprises, IFC Discussion Paper No. 40, Washington

Hammond, Sarah / Richardson, Victoria (2003): Keeping information alive: resource centres for district level health workers in Tanzania, in: African Research and Documentation, No. 93 (2003), pp. 19-21

Happy Computers (1986): Mit so wenig so viel erreichen, http://www.stcarchiv.de/hc1986/09_shiraz.php3

Hassan, Ali (1997): Internet Service Providers in Tanzania, http://www.glcom/hassan/isptz.html

Heeks, Richard (1999): Information and Communication Technologies, Poverty and Development, Development Informatics Working Paper Series, Working Paper No. 5, http://www.man.ac.uk/idpm/idpm_dp.htm#devinf_wp

Heeks, Richard (2001): Building eGovernance for Development, iGovernment paper no.12, IDPM, University of Manchester

Heeks, Richard (2002): eGovernment in Africa: Promise and Practice, iGovernment paper no.13, IDPM, University of Manchester

Hirschler, Kurt (2004): Tanzania in Transition – Violent Conflicts as a Result of Political and Economic Reforms, in: Gerhardt, Ludwig, Möhle, Heiko et al. (eds.): Umbrüche in afrikanischen Gesellschaften und ihre Bewältigung. Beiträge aus dem Sonderforschungsbereich 520 der Universität Hamburg, Münster 2004 (forthcoming)

Hoogen, Sascha (1997): Commodore 64, http://www.zock.com/8-Bit/D_C64.HTML

Hudson, Heather E. (2001): Converging Technologies and Changing Realities: Toward Universal Access to Telecom in the Developing World, in: Melody, William H. (ed.) (2001): Telecom Reform. Principles, Policies and Regulatory Practices, Lyngby, pp 387-398

IDRC / GVF – International Development Research Centre / Global VSAT Forum (2003): Satellite Access in Africa, www.ictdevagenda.org/frame.php?dir=07&sd=10&sid=1&id=533

IICD / MoEC (2002): Roundtable on Education in Tanzania. Summary Proceedings Round Table Workshop, Dar es Salaam

Ikeda, Nobuo (2003): Spectrum Buyouts: A Mechanism to Open Spectrum, http://www.rieti.go.jp/jp/publications/dp/02e002.pdf

ILO – International Labour Office (2001): Life at Work in the Information Society, Geneva

ILO – International Labour Office (2002): Learning and training for work in the knowledge society, Geneva

IMF – International Monetary Fund (2003): Aligning the Poverty Reduction and Growth Facility and the Poverty Reduction Strategy Paper Approach. Issues and Options, Washington

IMF – International Monetary Fund (2004): Tanzania. First Review Under the Three-Year Arrangement Under the Poverty Reduction and Growth Facility and Request for Waiver of Performance Criterion, Washington

Imfundo (2003): Improving Access to Education via Satellites in Africa: A Primer, http://imfundo.digitalbrain.com/imfundo/web/tech/documents/satprimer/Improving%20access%20via%20satellite.pdf?verb=view

InfoDev (2000): The Network Revolution. Opportunities and Challenges for Developing Countries, Washington

InfoDev (2000a): Rural Access to Information and Communication Technologies: The Challenge for Africa. Report prepared for the African Connection Secretariat, Washington

InfoDev (2003): ICT for Development. Contributing to the Millenium Development Goals. Lessons Learned from Seventeen infoDev projects, Washington

InfoDev / The Centre of Democracy & Technology (2002): The E-Government Handbook for Developing Countries, Washington

Intelecon (2002): Tanzania. Rural ICT Market Opportunity Report. Report for African Connection, Centre for Strategic Planning, http://www.infodev.org/projects/telecommunications/351africa/TANZANIA percent20Rural percent20Market percent20Assessment percent20-percent20FINAL.pdf

Intelecon (2003): The Next Step. A rural ICT Program for Africa, Vancouver

Intelecon (2003a): A Rural ICT Toolkit For Africa. Report prepared for the African Connection Secretariat, Vancouver

Intelecon (2003b): Guide to Rural ICT Market Opportunities Reports for ten African Countries, Vancouver

International Support Group / TeleCommons Development Group (2000): Internet Use and Diagnostic Study – Tanzania, http://www.telecommons.com/uploaddocuments/Tanzania_final.doc

Intven, Hank / McCarthy Tétrault (eds.) (2000): Telecommunications Regulation Handbook, www.infodev.org/projects/314regulationhandbook

Ishemo, Shubi (2004): Culture & Historical Knowledge in Africa: a Cabralian Approach, in: Review of African Political Economy, March 2004, No 99, Vol 31, pp 65-82

ITU (1984): The Missing Link: Report of the Independent Commission for World-wide Telecommunications Development, Geneva

ITU (1998): African Telecommunication Indicators 1998, Geneva

ITU (1998a): Integrated Rural Development and Universal Access. Brief description of ITU's Buenos Aires Action Plan Programme Nos. 9 & 12 (Valetta Action Plan Programme 3), http://www.itu.int/ITU-D/univ_access/reports/PPstatus981016.html

ITU (1999): Challenges to the Network. Internet for Development, Geneva

ITU (1999a): World Telecommunication Development Report 1999. Mobile Cellular, Geneva

ITU (2001): African Telecommunication Indicators 2001, Geneva

ITU (2001a): Yearbook of Statistics. Telecommunication Services 1990-1999, Geneva

ITU (2001b): World Summit on the Information Society. Geneva 2003 – Tunis 2005, Geneva

ITU (2001c): ITU Internet Reports. IP Telephony, Geneva

ITU (2002): World Telecommunication Development Report 2002. Reinventing Telecoms, Geneva

ITU (2002a): ITU Internet Reports. Internet for a Mobile Generation, Geneva

ITU (2002b): Trends in Telecommunication Reform 2002. Effective Regulation, Geneva

ITU (2002c): Istanbul Action plan, http://www.itu.int/ITU-D/isap/IstanbulActionPlan.html

ITU (2002d): Final Report WTCC-02, Istanbul, Turkey, www.itu.in/ITU-D/WTDC-02FinalReport/index.html

ITU (2003): World Telecommunication Development Report. Access Indicators for the Information Society, Geneva

ITU (2003a): Trends in Telecommunication Reform 2003. Promoting Universal Access to ICTs. Practical Tools for Regulators, Geneva

ITU (2004): African Telecommunication Indicators 2004, Geneva

ITU (2004a): Regulatory database. Regional Profile Africa, www.itu.int/ITU-D/treg/profiles/SepRegAuth.asp

ITU (2004c): Status of the main fixed-line operators – Africa, www.itu.int/ITU-D/treg/profiles/MainFixedOps.asp

ITU (2004d): Helping the World Communicate. World Summit on the Information Society. Geneva 2003 – Tunis 2005, Geneva

ITU-BDT (2000): Review of Spectrum Management and Costing for the Tanzania Communications Commission, 2000, Geneva

ITU-BDT (2000a): Interim Report on Special Programme for the LDCs, http://www.itu.int/ITU-D/tdag/PDFs/3rdMeeting/3rd-13e.pdf

ITU-BDT (2001): Summary Record of the Partnership Roundtable for least Developed Countries. Telecommunication Development Bureau (BDT) Fifth Meeting of the Telecommunication Development Advisory Group (TDAG), 22-23 February 2001, Geneva, http://www.itu.int/ITU-D/tdag/PDFs/5thMeeting/14-e.pdf

ITU-R (2003): World Radiocommunication Conference 2003, www.itu.int/ITU-R/wrc/wrc-3/index.asp

ITU-R (2004): Radiocommunication Sector, http://www.itu.int/ITU-R/information/brochure/brochure-BR.pdf

Jensen, Mike (1999): African Internet Reports: Tanzania, www3.sn.apc.org/africa/Tanzania.htm

Jensen, Mike (2001): The African Internet – A Status Report, www3.sn.apc.org/africa/afstat.htm

Jensen, Mike (2003): The Current Status of Information and Communications Technologies in Africa, in: Okpaku (2003), pp. 55-78

Jensen, Mike / Esterhuysen Anriette (2001): The Community Telecentre Cookbook for Africa. Recipes for Self-Sustainability. How to Establish a Multi-purpose Community Telecentre in Africa, UNESCO, Paris

Kahn, Robert E. (1999): Evolution of the Internet, in: UNESCO (1999), pp. 157-164

Kearns, Peter (2002): Towards the Connected Learning Society. An International Overview of Trends in Policy for Information and Communication Technology in Education, Canberra

Kennedy, Richard / Hobohm, Sarwar (1999): Capacity Building for Private Sector Development in Africa, Unido PSD Technical Working Paper No. 3, Vienna

Kenny, Charles (2002): The Internet and Economic Growth in Least Developed Countries. A Case of Managing Expectations?, UNU / Wider Discussion Paper No. 2002/75

Kenny, Charles et al. (2000): ICTs and Poverty, Draft for InfoDev Working Paper, August 2000

Klugman, Jeni (ed.) (2002): A sourcebook for Poverty Reduction Strategies, 2 Volumes, World Bank, Washington

Kondoro, J.W.A. (2002): Message form the Principal, in: DIT Newsletter, No 4/2002, p. 2-3

Kyarie, James K. (1995): Telematics: Which Way for Kenya?, http://www.sas.upenn.edu/African_Studies/Padis/telmatics_Kyarri.html

Lamtrac (2001): Survey of the Need for a Vocational Training Programme for ICT Professionals in Tanzania, Stockholm

Leiner, Barry M et al. (2003): A Brief History of the Internet, Version 3.32, http://www.isoc.org/internet/history/brief.shtml

Levey, Lisbeth A., Young, Stacy (eds.) (2002): Rowing Upstream, Snapshots of Pioneers of the Information Age in Africa, Johannesburg (available online http://www.piac.org/rowing_upstream/tableofcontents.html)

Levine, Linda (1996): Report to the IDRC on the Use of Information and Communication Technologies in Sub-Saharan Africa in the Area of Governance, www.idrc.ca/acacia/studies/ir-lev.htm

Levitsky, Jacob (1996): Support Systems for SMEs in Developing Countries. A Review, Vienna

LHRC – Legal and Human Rights Centre (2002): Tanzania Human Rights Report 2002, Dar es Salaam

Lie, Eric (2004): Background Paper: Radio Spectrum Management for a Converging World, www.itu.int/osg/spu/ni/spectrum/RSM-BG.pdf

Luhanga, Matthew (1994): Telecommunications in Tanzania, www.vii.org/papers/tanzania.htm

Madete, Lorah / Aiko, Rose (2003): Investment for Development Project. Report for Tanzania. Second Draft, Economic and Social Research Foundation (ESRF), Dar es Salaam

Mallango, Edward (2000): The Role of the Human Resources in the Privatisation Process of Tanzania Telecommunication Company Ltd, Paper presented on the Occasion of the '16th HRM/HRD Regional Meeting for the English Speaking Countries' organised by the ITU and TTCL, 16.-19. October, Arusha

Mallya, Ernest T. (n.d.): A Critical Look at Tanzania's Development Vision 2025, http://unpan1.un.org/intradoc/groups/public/documents/idep/unpan002404.pdf

Manuel, Trevor A. (2003): Finding the Right Path. Africa and the Washington Consensus, in: Finance and Development, September 2003, pp. 18-20

Mansell, Robin / Wehn, Uta (eds.), 1998, Knowledge Societies: Information Technology for Sustainable Development, London

Marson, Ingrid (2004): OpenOffice.org goes Swahili, ZDNet UK, 6.12.2004, news.zdnet.co.uk/software/linuxunix/0,39020390,391179058,00.htm

Matambalya, Francis (2003): "Information and Communication Technologies (ICTs) and Africa's Economic Development" in: Nnadozie, Emmanuel (2003) (Ed.): African Economic Development, Academic Press, Amsterdam et al.

Matambalya, Francis / Wolf, Susanna (2001): The Role of ICT for the Performance of SMEs in East Africa. Empirical Evidence from Kenya and Tanzania, ZEF Discussion Paper on Development Policy, No. 42, Bonn, December 2001

Mayanja, Meddie (n.d.): The African Community Telecentres: In Search of Sustainability, World Bank Institute, http://www.comminit.com/st2003/sld-7725.html

McConnell, Scott (2001): Telecentres Around the World: Issues to be Considered and Lessons Learned, ICT Development Group, Richmond

McLuhan, Marshall (1995): Understanding Media: The Extension of Man, New York.

Melody, William H. (2001): Policy Objectives and Models of Regulation, in: Melody, William H. (ed.) (2001): Telecom Reform. Principles, Policies and Regulatory Practices, Lyngby

Mendes, Shawn / Tuijnman, Albert / Young, Nina (2003): Education and training for ICT development in Sri Lanka and Tanzania. Training for ICT development in Sri Lanka and Tanzania, http://www.sida.se/content/1/c6/02/11/88/TrainingForICTSriLankaTanzania.pdf

Menzel, Ulrich (1995): Geschichte der Entwicklungstheorie. Einführung und systematische Bibliographie, Hamburg

Mercer, Claire (2002) Developing the Networking Society? Information and communication technologies in Tanzania (draft)

Mercer, Claire (2004): Engineering Civil Society: ICT in Tanzania, in: Review of African Political Economy, Vol 31, No. 99, pp. 49-64

MIC – Millicom International (2004): Annual Report 2003, http://www.millicom.com/investor/downloads/mic2003_ar.pdf

Miller Esselaar and Associates (2001): A Country ICT Survey for Tanzania, Report prepared for Sida, Stockholm

Miller, Jonathan (2003): Negotiating the Net in Tanzania, http://www.cidcm.umd.edu/ntn/tanzania/case_study_tanzania.pdf

Minges, Michael (2000): Counting the Net: Internet Access Indicators, http://www.isoc.org/inet2000/cdproceedings/8e/8e_1.htm

MIT – Ministry of Industry and Trade (2002): Small and Medium Enterprise Development Policy, Dar es Salaam

MIT – Ministry of Industry and Trade (2003): National Trade Policy, Dar es Salaam

MoCT – Ministry of Communication and Transport (2001): Status of the Communications and Transport Sector as at February 2001, Dar es Salaam

MoCT (2001a): Information and Communication Technologies (ICTs) Development Initiatives. Status Report on Implementation of ICTs Activities, Dar es Salaam

MoCT (2001b): Tanzania Communications Act 1993. Subsidiary Legislation (Supp. No. 40 of 2nd November 2001), Regulations 2001, Dar es Salaam

MoCT (2003): National Information and Communications Technologies Policy, Dar es Salaam

MoCT (2003a): Status on the Preparation for Building an Information Society in Tanzania, Dar es Salaam

MoH – Ministry of Health (2004): National Health Policy, Dar es Salaam

Moroney, Sean / Hamilton, Paul (2004): Satellite and VSAT: Innovative uses for Rural Telephony and Internet Development, http://www.itu.int/osg/spu/ni/ipdc/workshop/presentations/Moroney.doc

Mramba, Basil (2001): Speech by the Minister for Finance Hon. Basil P. Mramba (MP), Introducing to the National Assembly the Estimates of Government Revenue and Expenditure for the Financiel Year 2001/2002 on 14th June 2001, Dar es Salaam

Mramba, Basil (2002): Speech by the Minister for Finance Hon. Basil P. Mramba (MP), Introducing to the National Assembly the Estimates of Government Revenue and Expenditure for the Financiel Year 2002/2003 on 13th June 2002, Dar es Salaam

Mramba, Basil (2003): Speech by the Minister for Finance Hon. Basil P. Mramba (MP), Introducing to the National Assembly the Estimates of Government Revenue and Expenditure for the Financiel Year 2003/2003 on 12th June 2001, Dar es Salaam

Msamanga, G.I. (2004): Muhimbili Health Exchange Forum, lecture presented at the IFM, Global Development Learning Centre, 26th August 2004

Mutagahywa, Beda (2003): Application and Institutional Integration of Distance Education and ICTs at the University of Dar Es Salaam, Tanzania, Paper given on the occasion of the Consultative meeting on the role of modern distance education and ICTs in enhancing access to quality tertiary Education in Sub Sahara Africa, 26th – 27th September 2003, Accra, Ghana, http://www.worldbank.org/afr/de_ict/ghana_0903/dar_ICT_pres.pdf

Mutakyahwa, R.G. (2002): „The Impact of Non Governmental Organisations in Development", in: Friedrich Ebert Stiftung (2002): Political Handbook & NGO Calendar 2002. Tanzania, Dar es Salaam, pp. 78-82

Mwandosya, Mark (2003): Harnessing ICT to Enhance Business and Public Sector Effectiveness, Speech held by Prof. M.J. Mwandosya, Minister for Communications and Transport at the Commonwealth-Tanzania Investment Conference on 29th May 2003

Myrdal, Gunnar (1968): Asian Drama, New York

Nagu, Mary (2004): Opening Speech by honourable Dr. Mrs. Mary Nagu (MP), Minister of State, President's Office for Public Service Management, to the 5th SADC Executive Programme Seminar on e-Government, held at the Golden Tulip Hotel, Dar es Salaam, 14 September 2004

Nagy K. Hanna (2003): Why National Strategies are needed for ICT-enabled Development, ISG Staff Working Papers, http://www.apdip.net/documents/policy/misc/policy_strategy1.pdf

Nalingigwa, Asupya (2002): Legal and Regulatory framework for ICTs Development, paper presented on the occasion of First Stakeholders' Workshop on Information and Communication Systems Technologies (ICT's) Policy of Tanzania, Dar es Salaam 25.5.2002

Naschold, Felix / Fozzard, Adrian (2002): How, When and Why does Poverty get Budget Priority. Poverty Reduction Strategy and Public Expenditure in Tanzania, Overseas Development Instiute, Working Paper 165

Navas-Sabater, Juan, Dymond, Andrew, Juntunen, Niina (2002): Telecommunications and Information Services for the Poor. Toward a Strategy for Universal Access, World Bank Discussion Paper No. 432, Washington

Nepad (2004): Press Release 23.9.2004, Agreement on broadband ICT networks for Eastern and Southern Africa, Midrand

Neto, Maria Isabell (2004): Wireless Networks for the Developing World: The Regulation and Use of Licence-Exempt Radio Bands in Africa, http://itc.mit.edu/itel/students/papers/neto_thesis.pdfNetwork Startup Resource Centre (NSRC) (1999): NSRC Tanzania, http://www.nsrc.org/db/lookup/ISO=TZ

Nfuka Edophence Ngemera (2002): Introduction to UCC, paper given on the occasion of a Roundtable Conference on Education, Bagamoyo, 20. – 23. February 2002

NGO Policy Forum (2002): NGO Statements. Consultative Group Meeting, Dar es Salaam, 2-5 December 2002

Nielinger, Olaf (1998): Demokratie und Good Governance in Afrika. Internationale Demokratisierungshilfe als neues entwicklungspolitisches Paradigma?, Hamburg

Nielinger, Olaf (2001): Perspektiven der Internet-Ökonomie in Afrika, in: Engels, Benno / Nielinger, Olaf (Hrsg.) (2001): Elektronischer Handel in Afrika, Asien, Lateinamerika und Nahost, Hamburg, S. 47-84

Nielinger, Olaf (2001b): Partnerschaften für Entwicklung? Die schwierige Annäherung zwischen Wirtschaft und Nichtregierungsorganisationen, in : Betz, Joachim, Brüne, Stefan (2001): Neues Jahrbuch Dritte Welt. Entwicklungsfinanzierung, Opladen, S. 127-143

Nielinger, Olaf (2002): Fact Sheet Tanzania IT Vendor Survey, http://www.duei.de/iak/de/content/forschung/pdf/projektnieltext4.pdf

Nielinger, Olaf (2003): Afrika und der UN Gipfel zur Informationsgesellschaft, http://www.duei.de/iak/de/content/aktuelles/pdf/AiB4-03.pdf

Nielinger, Olaf (2003a): Rural ICT Utilisation in Tanzania. Empirical Findings from Kasulu, Magu, and Sengerema, http://www.duei.de/iak/de/content/forschung/pdf/projektnieltext6.pdf

Nielinger, Olaf (2003b): Fact Sheet. ICT-utilisation of Small and Medium Enterprises (SME) in Tanzania, http://www.duei.de/iak/de/content/forschung/pdf/projektnieltext2.pdf

Nielinger, Olaf (2003c): Fact Sheet. ICT utilisation by Non-Governmental Organisations (NGOs) in Tanzania, http://www.duei.de/iak/de/content/forschung/pdf/projektnieltext1.pdf

Nnafie, Issmail (2002): Internet cafes in Dar es Salaam: Problems and Opportunities. Recommendations for eThinktank Tanzania, Master of Science thesis, Eindhoven University of Technology, Faculty of Technology Management, http://www.ourtanzania.com/cafes.pdf

Noll, Roger (2000): Telecommunications Reform in Developing Countries, in: Krueger, Anne O. (ed.): EconomicPolicy Reform: The Second Stage, Chicago

Nunno, Richard (2002): Review of Spectrum management practices, http://hraunfoss.fcc.gov/edocs_public/attachmatch/DOC-229047A1.pdf

Nwachuku, Michael A. (1994): Development of information technology in Nigeria, in: Drew / Foster (1994)

Odedra-Straub, Mayuri (1995): Women and information technology in Sub-Saharan Africa, in: Swasti Mitter and Sheila Rowbotham (1995)(eds.): Women encounter technology: changing patterns of employment in the third world, http://www.unu.edu/unupress/unupbooks/uu37we/uu37we00.htm#Contents

OECD (2003): Donors ICT Strategies Matrix, http://www.developmentgateway.org/download/220519/Fullmatrix.pdf

Oestmann, Sonja / Dymond, Andrew C. (2001): Telecentres – Experiences, Lessons and Trends, in: The Commonwealth of Learning (2001), pp. 1-15

Okpaku, Joseph (ed.) (2003): Information and Communication Technologies for African Development. An Assessment of Progress and Challenges Ahead, United Nations ICT Task Force, New York

OleKambainei, Emmanuel / Sintim-Misa, Mavis A. (2003): Info-communication for Development in Africa. The African Connection Initiative, in: Okpaku (2003), pp. 151-173

Orbicom (2003): Measuring the Digital Divide and Beyond, Montreal

Paconet (2003): Company profile, www.paconet.net

Planning Commission (1998): Tanzania Development Vision 2025, Dar es Salaam, http://www.tanzania.go.tz/vision.htm

Pohjola, Matti (2002): New Economy in Growth and Development, UNU / WIDER Discussion Paper No. 2002/67

Powell, Michael (2004): Information Societies and Social Change, in: UNRISD News, Spring/Summer 2004, No. 26, pp. 1-3

PSRC – Presidential Parastatal Sector Reform Commission (1994): Parastatal Privatisation and Reform. 1993 Review and 1994 and 1995 Action Plan, Dar es Salaam

PSRC – Presidential Parastatal Sector Reform Commission (2000): TTCL Strategic Investor Picked, PSRC Press Release 27.06.2002

PSRC – Presidential Parastatal Sector Reform Commission (2000a):TTCL Financial Bids Opened, http://www.psrctz.com/Press%20Releases/ttcl_financialbids.htm

PSRC – Presidential Sector Reform Commission (2001): TTCL Divestiture Process, www.psrctz.com/Press%20Releases/TTCL%20divestiture-150101.htm

PSRC – Presidential Parastatal Sector Reform Commission (2002): Welcome to the divestiture of Tanzania Telecommunication Company Ltd, www.psrctz.com/Utilities%20&%20Major%Transactions/ttcl.htm

PSRP – Public Service Reform Programme (2004): Information Technology and Systems Component, http://www.estabs.go.tz/info_tech_systems_challenges.html

PWC – PriceWaterhouseCoopers (2003): Tanzania. Doing Business Guide 2003, Dar es Salaam

PWC – PriceWaterhouseCoppers (2001): Largest telecoms privatisation mandate this year signed for $1.3bn, http://www.pwcglobal.com/uk/eng/about/svcs//pfp/nitel.html

Quaynor, Nii Narku (2003): Africa's Digital Rights, in: Okpaku (2003), pp.175-202

Regional Ministerial Conference (2003): Mauritius Conference on Access to ICT for All. Declaration and Action Plan, http://www.icta.mu/conf/decl.doc

RINAF – Regional Informatics Network for Africa (1998): An External Evaluation for UNESCO, (Prepared by Mike Jensen), http://www.unesco.org/webworld/build_info/rinaf.html

Rodriguez, Francisco / Wilson, Ernest (2000), Are Poor Countries Losing the Information Revolution, InfoDev Working Paper, May 2000

Rostow, Walt W. (1960): The Stages of Economic Growth. A Non-Communist Manifesto, Cambridge

Rylander, Sten (2002): Statement by H.E. Mr Sten Rylander, Ambassador of Sweden, on the Occasion of the National Stakeholders workshop on the National Information and Communications Technologies (ICT) Policy, Dar es Salaam, 25.5.2002

SADC – South African Development Community (2000): SADC Regional Frequency Allocation Plan for 20 – 3100 MHz, Garborone

SATQ – Southern Africa Telecom Quarterly (1997): The Southern Africa Regional Telecommunications Restructuring (RTR) Programme. Tanzania: In Profile, SATQ Vol 2, No.2, http://rtr.worldweb.net/tansect.htm

Sawe, David (2004): Presentation on e-Government, Presentation delivered at the Tanzania Development Learning Centre, Dar es Salaam, 27th February 2004

Scharffenberger, George (1999): Telecentre Evaluation Methods and Instruments: What works and why?, http://www.idrc.ca/telecentre/evaluation/html/main.html

Schmitz, Hubert (1995): Collective Efficiency: Growth Path for Small Scale Industry, in: Journal of Development Studies, Vol. 31, No. 4, pp. 529-566

Schuurman, Frans (ed.) (1993): Beyond the Impasse. New Directions in Development Theory, London

Sebastian, Costantine (2004): GTZ initiates "Connect and Learn Initiative" in Tanga, http://www.developmentgateway.org/download/247696/GTZ.doc

Sen, Amartya (1999): Development as Freedom, New York

Senunas, Eric (1997): The 1997 GATS Agreement on Basic Telecommunications: A Triumph For Multilateralism, or the Market?, http://www.bc.edu/bc_org/avp/enacted/st_org/iptf/articles/content/1997111401.html

Senzige J.P. / Sarukesi K. (2003): An Approach to ICT school education in Tanzania, http://www.ssn.flinders.edu.au/global/afsaap/conferences/2003proceedings/senzige.PDF

Sheriff, Suhail (2003): Tanzania Internet Exchange (TIX), Paper given on the occasion of regular Swopnet lectures, 25.6.2003, Tanzania Development Learning Center (TDLC), Dar es Salaam

Sida (2004): Open Source in Developing Countries, Stockholm

Siemens (2003): MODANS has opted for Siemens technology for more than a decade, http://www.siemens.com/page/1,3771,1089081-1-999_0_0-17,00.html

Simbanet (2002): Times @ Simba, Vol. 102, Feb. – April 2002

Simunet (n.d.): Simunet Company Ltd. Fact Sheet, Dar es Salaam

SITA (2004): SITA – Corporate profile, http://www.sita.aero/index.asp?activeDir=/Home/News_Centre/Corporate_profile/&activeFile=index.html

Soete, Luc (1985): International Diffusion of Technology, Industrial Development and Technological Leapfrogging, in: World Development, Vol. 13(3), pp. 409-422

Soltane, Ben (2003): Regional Information and Communications Technologies Developments, in: Okpaku (2003), pp 125-150

Steinmüller, Wilhelm (1993): Informationstechnologie und Gesellschaft. Einführung in die angewandte Informatik, Darmstadt

Talero, Eduardo / Gaudette, Philip (1996): Harnessing Information for Development. A Proposal for a World Bank Group Strategy, http://www.worldbank.org/html/fpd/harnessing/

TCC – Tanzania Communication Commission (2000): Tanzania Communications Commission: Corporate Plan 2000-2005, Dar es Salaam

TCC – Tanzania Communication Commission (2001): The Regulator, Quarterly newsletter of the Tanzania Communication Commission, No. 1, April-June 2001

TCC – Tanzania Communication Commission (2001a): The Regulator. Quarterly newsletter of the Tanzania Communication Commission, No. 2, July-September 2001

TCC – Tanzania Communication Commission (2001b): The Regulator. Quarterly newsletter of the Tanzania Communication Commission, No. 3, October-December 2001

TCC – Tanzania Communication Commission (2001c): Spectrum Charges. Annual RadioCommunication Licence Fees, http://www.tcc.go.tz/Spectrum_charges.htm

TCC – Tanzania Communication Commission (2001d): Revised Guidelines and Procedures for Licensing Telecommunications Service Providers and Operators, Dar es Salaam, http://www.tcc.go.tz/Guidelines_telecom.htm

TCC – Tanzania Communication Commission (2002): The Regulator, Quarterly newsletter of the Tanzania Communication Commission, No. 1, January – April 2002

TCC – Tanzania Communication Commission (2002a): Commission's Determination on Interconnection Rates among the Fixed and Mobile

Telecommunications Networks, Determination No. 3 of 2002, www.tcc.go.tz/Determination%20No.3.htm

TCC – Tanzania Communication Commission (2003): Telecom Contractors 2003-2004, http://www.tcc.go.tz/telecom_contractors.htm

TCC – Tanzania Communication Commission (2003a): List of Telecommunication Operators, Dar es Salaam (unpublished)

TCRA – Tanzania Communication Regulatory Authority (2004): TCRA News Release. TCRA Board Members Appointed, http://www.tcra.go.tz/Press%20release%20Board%20members.pdf

TCRA – Tanzania Communication Regulatory Authority (2004a): Press Release. Decision on Interconnection Rates in Tanzania, http://www.tcra.go.tz/Determination%201%20of%202004.pdf

TCRA – Tanzania Communication Regulatory Authority (2004b): Determination on Interconnection Rates among the Fixed and Mobile Telecommunications Networks, Dar es Salaam

Tech Target Nework (no date): IT-encyclopedia, http://whatis.techtarget.com

TechKnowLogia (2003): Technologies and Learning, Vol 5, Issue 1, January-March 2003

TechKnowLogia. International Journal of Technologies for the Advancement of Knowledge and Learning (2001): Thematic Focus: Technology for Social Action, Vol. 3, Issue 4, July/August 2001, www.TechKnowLogia.org

TEHIP –Tanzania Essential Health Interventions Project (n.d.): The Tanzania Essential Health Interventions Project, Dar es Salaam

Teskey, Graham / Hooper, Richard (1999): Tanzania Civil Service Reform Programme. Case Study, http://unpan1.un.org/intradoc/groups/public/documents/CAFRAD/UNPAN010583.pdf

The Commonwealth of Learning (2001): Perspectives on Education. Telecentres: Case Studies and Key Issues, Vancouver

Thioune, Ramata Molo (ed) (2003): Information and Communication Technologies for Development in Africa. Volume 1. Opportunities and Challenges for Community Development, Ottawa et al.

Thomas, Julian et al. (1999): Governing information and communication technologies, in: UNESCO (1999): World Communication and Information Report 1999-2000, Paris, pp.116-130

Thompson, Mark (2004): Discourse, 'Development' and the 'Digital Divide': ICT & the World Bank, in: Review of African Political Economy, March 2004, No. 99, Vol. 31, pp. 103-123

TIC – Tanzania Investment Centre (2002): Tanzania Investors Guide, Dar es Salaam

Tipson, Frederick S. / Frittelli, Claudia (2003): Global Digital Opportunities National Strategies of "ICT for Development", http://www.markle.org/gdoi.final.report.12.23.03.pdf

Toffler, Alvin (1980): The Third Wave, New York

TRASA – Telecommunication Regulators Association of Southern Africa (1999): Country Profile Tanzania, www.trasa.org/memberstates/tanzania/profile.html

TRASA – Telecommunications Regulators Association of Southern Africa (2003): Country Report Tanzania, http://www.trasa.org.bw/files/attachments/country%20report-%20tanzania.doc

TTCL (2004): About TTCL – Overview, www.ttcl.co.tz/About_TTCL/index.php

UCC – Uganda Communications Commission (2003): RCDF Operations, www.uccc.go.ug/rcdf/operations.html

UCC – Uganda Communications Commission (2003): Rural Communications Development Fund (RCDF). Successful Bidders, http://www.ucc.co.ug/rcdf/s_bidders.html

UCC – University Computer Centre (UCC) (2001): Company Profile, Dar es Salaam

UCC - University Computing Centre (2001): UCC Company Profile, Dar es Salaam

UDSM – University of Dar es Salaam (n.d.): TEIL project activities, http://www.infoteil.udsm.ac.tz/text/TeilActivies.html

UN (2002): Report of the International Conference on Financing for Development, Monterrey, Mexico, 18-22 March 2002, http://www.un.org/esa/ffd/aconf198-11.pdf

UN Department of Economic and Social Affairs (1998): Poverty reduction strategies: a review, New York

UNCSTD (1997): Report of the Working Group on Information and Communication Technologies for Development, Geneva

UNCTAD (2000): Building Confidence. Electronic Commerce and Development, Geneva

UNCTAD (2003): Information and Communication Technology Development Indices, New York and Geneva

UNDP (1999): Human Development Report 1999, Globalization with a Human Face, Oxford

UNDP (2001): Human Development Report 2001. Making new technologies work for human Development, Geneva

UNDP (2003): Human Development Report 2003. Millenium Development Goals: A compact among nations to end human poverty, New York

UNDP (2004): Human Development Report 2004. Cultural Liberty in Today's Diverse World, Geneva

UNDP-Tanzania (2001): Least Developed Countries Initiative. Cisco Networking Academy, Dar es Salaam

UNECA (1996): Private Sector Participation in Telecommunication in the United Republic of Tanzania, http://www.uneca.org/eca_resources/Publications/RCID/old/96_TanzaniaTelecom.htm

UNECA (2003): E-Strategies. National, Sectoral and Regional ICT Policies, Plans and Strategies, Addis Ababa

UNECA-NICI (1998): Country Report Tanzania, www.uneca.org/aisi/nici/tanzania/tanzab.htm

UNESCO (1980): Many Voices, One World. Report by the International Commission for the Study of Communication Problems, Paris

UNESCO (1999): World Communication and Information Report 1999-2000, Paris

United Nations Department of Economic and Social Affairs (2003): World Public Sector Report 2003: E-Government at the Crossroads, New York

United Nations Non-Governmental Liasion Service (NGLS) (2000): Voices from Africa. Information and Communication Technologies. Issue No. 9, Geneva

URT – United Republic of Tanzania (1967): The Arusha Declaration and Tanu's policy on socialism and self-reliance, Dar es Salaam

URT – United Republic of Tanzania (1993): The Tanzania Communication Act, Act No. 18 of 1993, Dar es Salaam

URT – United Republic of Tanzania (1993a): The Tanzania Telecommunications Company Incorporation Act, 1993, Dar es Salaam

URT – United Republic of Tanzania (1997): National Telecommunications Policy, Ministry of Communication and Transport, Dar es Salaam

URT – United Republic of Tanzania (2000): Tanzania Assistance Strategy. A Medium Term Framework for Promoting Local Ownership and Development Partnerships, Dar es Salaam

URT – United Republic of Tanzania (URT) (2000a): Poverty Reduction Strategy Paper, Dar es Salaam

URT – United Republic of Tanzania (2000b): Civil Service Department. Medium Term Strategic Plan 2000 – 2004, http://www.estabs.go.tz/Documents/CSD'StrPln'99 percent20Version percent202.doc

URT – United Republic of Tanzania (2001): Tanzania Communication Act 2001, Ministry of Telecommunication and Transport, Dar es Salaam

URT – United Republic of Tanzania (2001a): The Tanzania Commission for Science and Technology Act, No. 7 of 1986 (revised), Dar es Salaam

URT – United Republic of Tanzania (2001b): Importation and Distribution of Telecommunication and Radio- communication Equipment Regulations, 2001, Dar es Salaam

URT – United Republic of Tanzania (2002): Recent Macroeconomic Developments, Prepared for the Consultative Group Meeting for Tanzania, Dar es Salaam, December 2 – 5, 2002

URT – United Republic of Tanzania (2002a): Tanzania Assistance Strategy. Action plan. Zero Draft, Dar es Salaam

URT – United Republic of Tanzania (2002b): 2002 Population and Housing Census, www.tanzania.go.tz/census

URT – United Republic of Tanzania (2003): Poverty Reduction Strategy. The Second Progress Report 2001/02, Dar es Salaam

URT – United Republic of Tanzania (2003a): Tanzania Country Paper as Input to the 2nd World Summit on the Information Society. Preparatory Committee (WSIS Prepcom II), Geneva, 17-28 February 2003, http://www.wsis2005.org/wsis/images/ pays/documents/Tanzania PC2.doc

URT – United Republic of Tanzania (2003b): Tanzania Communications Regulatory Authority Act No. 12, http://www.tcc.go.tz/Tanzania%20Communications%20Regulatory%20Act-2003.pdf

URT – United Republic of Tanzania (2003c): The Economic Survey 2002, Dar es Salaam

URT – United Republic of Tanzania (2003d): Tanzania joint Health Review. Main Report, http://www.sti.ch/pdfs/swap321.pdf

URT – United Republic of Tanzania (2004): Electronic and Postal Communications Bill. An Act to Provide for and Regulate activities in the electronic and Postal Communications Sectors and for Related Matters, http://www.tcra.go.tz/DRAFT%20ELECTRONIC%20AND%20POSTAL%20COMMUNICATIONS%20BILL%202005.pdf

URT – United Republic of Tanzania (2004a): Poverty Reduction Strategy. The Third Progress Report 2002/2003, Dar es Salaam

URT – United Republic of Tanzania (2004b): Public Service Reform Programme. Revised PSRP Action. Plan and Budget (2004/05 to 2006/07), http://www.estabs.go.tz/download.html

URT – United Republic of Tanzania (2004c): Public Service Reform Programme. Annual Progress Report 2003/04, http://www.estabs.go.tz/download.html

USA – Universal Service Agency (2003): Under-Serviced Area Licencees, www.usa.org.za/usals.html

USAID (n.d.): Leland Initiative. Africa Global Information Infrastructure Project, http://www.usaid.gov/regions/afr/leland/project.htm

Van Arkadie, Brian, et al. (2000): Building Capacitiy in the Government Economic Service. A study prepared as part of the IDF-funded ESRF Government Capacity Building Program, ESRF Discussion Paper No. 25, Dar es Salaam

Vodacom (2004): Annual Report 2003, Midrand

Wallsten , Scott (2002): Regulation and Internet Use in Developing Countries, http://econ.worldbank.org/files/24461_wps2979.pdf

Wallsten, Scott J. (2001): An Econometric Analysis of Telecom Competition, Privatization and Regulation in Africa and Latin America", Journal of Industrial Economics, 49(1), pp. 1-19.

Wallsten, Scott (1999): An Empirical Analysis of Competition, Privatization, and Regulation in Africa and Latin America, http://www.worldbank.org/html/dec/Publications/Workpapers/wps2000series/wps2136/wps2136.pdf

Walther, Miriam / Hentschel, Christine (2002): Armutsstrategiepapiere (PRSP) - Neuanfang in der Strukturanpassungspolitik von IWF und Weltbank? Wirtschaftspolitik und Armutsbekämpfung in den PRSPs von Bolivien, Burkina Faso, Mauretanien, Tansania und Uganda. WEED Arbeitspapier, Berlin, http://www.weed-online.org/publikationen/arbeitspapiere/17706.html

Wangwe, Samuel / Semboja, Haji H. / Tibandebage, Paula (eds.) (1998): Transitional Economic Policy Options in Tanzania, Dar es Salaam

Wenzel, Hans Dieter / Wiedemann, Ralf (1989): Tanzania's economic performance in the eighties. Studies in Applied Economics and Rural Development No. 20, Research Centre for International Agrarian Development, Heidelberg

Williamson, John (2003): From Reform Agenda to Damaged Brand Name. A short history of the Washington Consensus and suggestions for what to do next, in: Finance and Development, September 2003, pp. 10-13

Whyte, Anne (2000): Assessing community telecentres: guidelines for researchers, International Development Research Centre (IDRC), Ottawa

Wilken Afsat (n.d.): Company profile, www.wilken-dsm.com/page4.htm

Wilson, Ernest (2004): Negotiating Internet Diffusion in Africa: Role of International Cooperation, http://www.cidcm.umd.edu/ntn/continent/continent_5-31-04.doc (draft)

Wireless Internet Institute (2003): The Wireless Internet Opportunity for Developing Countries, Boston

Wobst, Peter (2001): Structural Adjustment and Intersectoral Shifts in Tanzania. A Computable Equilibrium Analysis. International Food Policy Research Institute (IFPRI), Research Report 117, Washington

World Bank (1988): World Development Report 1988. Public Finance in Development, Washington

World Bank (1989): Sub-Saharan Africa. From Crisis to Sustainable Growth, Washington

World Bank (1998/99): World Development Report 1998/99. Knowledge for Development, Washington

World Bank (2000): Rural Access to Information and Communication Technologies: The Challenge for Africa, http://www.telecommons.com/uploaddocuments/Final%20June%20Report%2Edoc

World Bank (2000a): Tanzania Third Telecommunication Project. Aide Mémoire, Washington (unpublished)

World Bank (2001): Implementation Completion Report on a Credit to the United Republic of Tanzania for a Third Telecommunication Project, 5 June 2001, Washington (unpublished)

World Bank (2002): Tanzania at the Turn of the Century. Background Papers and Statistics, Washington

World Bank (2002a): Information and Communication Technologies. A World Bank Group Strategy, Washington

World Bank (2003): ICT and MDGs. A World Bank Perspective, Washington

World Bank (2004): Reforming Infrastructure. Privatization, Regulation, and Competition, Washington

World Bank (2004a): World Development Report 2004. Making Services Work for Poor People, Washington

World Bank (2004c): World Development Indicators 2004, Washington

World Bank / IMF (2002): Financing for Development. Implementing the Monterrey Consensus. Paper Prepared by the Staff of the World Bank and the IMF for the Spring 2002 Development Committee Meeting, April 11, 2002, http://www.imf.org/external/np/pdr/FfD/2002/imp.htm

World Bank/IMF (2004): The United Republic of Tanzania. Evaluation of the Poverty Reduction Strategy Paper (PRSP). Process and Arrangements made the Poverty Reduction and Growth Facility (PRGF), Washington

World Economic Forum – WEF (2003): Global Information Technology Report 2003-2004, Oxford

World Economic Forum – WEF (2003a): Southern African Development Community – World Economic Forum Consultation Report on e-Readiness, Cologne/Geneva

World Trade Organisation – WTO (2004): Telecommunication Services, http://www.wto.org/english/tratop_e/serv_e/telecom_e/telecom_results_e.htm

World Trade Organisation – WTO (n.d.): The General Agreement on Trade in Services (GATS): objectives, coverage and disciplines, http://www.wto.org/english/tratop_e/serv_e/gatsqa_e.htm

WSIS (2003): Declaration of Principles. Building the Information Society: a global challenge in the new Millennium, http://www.itu.int/dms_pub/itu-s/md/03/wsis/doc/S03-WSIS-DOC-0004!!MSW-E.doc

WSIS (2003a): Plan of Action, http://www.itu.int/dms_pub/itu-s/md/03/wsis/doc/S03-WSIS-DOC-0005!!MSW-E.doc

WSIS Africa Regional Conference (2002): Bamako 2002 Conference, http://www.geneva2003.org/bamako2002/doc_html/declaration_final_en.htm

Ya'u, Y.Z. (2004): The New Imperialism & Africa in the Global Electronic Village, in: Review of African Political Economy, March 2004, No 99, Vol 31, pp 11-29

Yona, Daniel (1999): Speech by the Minister for Finance Hon. Daniel N. Yona (MP), Introducing to the National Assembly the Estimates of Government

Revenue and Expenditure for the Financiel Year 1999/2000 on 10th June 1999, Dar es Salaam

Yonah, Zaipuna (2001): Aggressive Internet Access Strategies for Empowering Economic and Service Sectors in Tanzania: The Status of and Future Plans for ICT Facilities and Services in Tanzania, Paper presented at the ICS Telecommunications Technologies Expert Group Meeting on "Data Communication for Service Applications in Developing Countries", Trieste, 28. – 31.3.2001

Yonah, Zaipuna (2002): Building Infrastructure for ICT Development in Tanzania, Paper held on the Occasion of the National Stakeholders workshop on the National Information and Communications Technologies (ICT) Policy, Dar es Salaam, 25.5.2002

Zhen-Wei Qiang, Christine /Pitt, Alexander / Ayers, Seth (2003) Contribution of Information Communication Technologies to Growth, http://www.ugabytes.org/ICTs%20and%20growth.html.pdf

Zongo, Gaston (2001): Information and Communication Technologies for Development in Africa: Trends and Overview, http://www.unssc.org/web1/programmes/glnp/Knowledge_Sharing/case_studies/IT_African_Development.PDF

Newspapers, Newsletters, Bulletins, etc. accessed on a regular basis

Africa Analysis (abbreviated as AA)

African Research Bulletin (abbreviated as ARB)

Balancing Act's News Update (abbreviated as BA)(www.balancingact-africa.com)

Business Times (Dar es Salaam)

Cellular News (abbreviated as CN) (www.cellular.co.za)

Daily News (Dar es Salaam)

East African (Nairobi)

Financial Times (Dar es Salaam)

ICDT Newsletter (www.ictd.org)

Indian Ocean Newsletter (abbreviated as ION)
IT Vision (Dar es Salaam)
IT Web (South Africa) (www.itweb.co.za)
Marchés Tropicaux (abbreviated as MT)
Sunday Observer (Dar es Salaam)
The Express (Dar es Salaam)
The Guardian (Dar es Salaam)

FSC
www.fsc.org
MIX
Papier | Fördert gute Waldnutzung
FSC® C083411

Kolibri 360 GmbH
Ferdinand-Jühlke-Straße 7
99095 Erfurt, Deutschland
produktsicherheit@kolibri360.de

Druck:
CPI Druckdienstleistungen GmbH
im Auftrag der
Kolibri 360 GmbH
Ferdinand-Jühlke-Str. 7
99095 Erfurt